Quantum Mechanics of Fundamental Systems 2

Edited by

Claudio Teitelboim

Centro de Estudios Científicos de Santiago
Santiago, Chile
and University of Texas at Austin
Austin, Texas

and

Jorge Zanelli

Centro de Estudios Científicos de Santiago
Santiago, Chile
and Universidad de Chile
Santiago, Chile

PLENUM PRESS • NEW YORK AND LONDON

Library of Congress Cataloging in Publication Data

Quantum mechanics of fundamental systems 2.

(Series of the Centro de Estudios Científicos de Santiago)

"This volume is based on a meeting on Quantum Mechanics of Fundamental Systems, held at the Centro de Estudios Científicos de Santiago (CECS) from December 17 to 20 of 1987"—Pref.

Bibliography: p.

Includes index.

1. Mathematical physics. 2. Quantum theory. I. Teitelboim, Claudio. II. Zanelli, Jorge. III. Meeting on Quantum Mechanics of Fundamental Systems (1987: Centro de Estudios Científicos de Santiago) IV. Series.

QC20.Q36 1989 530.1′2 89-8687

ISBN-13: 978-1-4612-8087-3 e-ISBN-13: 978-1-4613-0797-6
DOI: 10.1007/978-1-4613-0797-6

Softcover reprint of the hardcover 1st edition 1989

A Division of Plenum Publishing Corporation
233 Spring Street, New York, N.Y. 10013

Quantum Mechanics of Fundamental Systems 2

Series of the Centro de Estudios Científicos de Santiago

Series Editor: Claudio Teitelboim
Centro de Estudios Científicos de Santiago
Santiago, Chile
and University of Texas at Austin
Austin, Texas, USA

IONIC CHANNELS IN CELLS AND MODEL SYSTEMS
Edited by Ramon Latorre

PHYSICAL PROPERTIES OF BIOLOGICAL MEMBRANES AND THEIR FUNCTIONAL IMPLICATIONS
Edited by Cecilia Hidalgo

PRINCIPLES OF STRING THEORY
Lars Brink and Marc Henneaux

QUANTUM MECHANICS OF FUNDAMENTAL SYSTEMS 1
Edited by Claudio Teitelboim

QUANTUM MECHANICS OF FUNDAMENTAL SYSTEMS 2
Edited by Claudio Teitelboim and Jorge Zanelli

Contributors

Elcio Abdalla, Instituto de Física Universidade de São Paulo, 20516 São Paulo, Brazil

J. Alfaro, Pontificia Universidad Católica de Chile, Facultad de Física, Santiago 22, Chile

Thomas Banks, Santa Cruz Institute for Particle Physics, University of California, Santa Cruz, California 95064

Laurent Baulieu, Université Pierre et Marie Curie, 75252 Paris, Cedex 05, France

L. Bonora, S.I.S.S.A., Trieste and I.N.F.N., Sezione di Trieste, Italy

H. J. de Vega, Laboratoire de Physique Théorique et Hautes Energies (Laboratoire Associé au CNRS), Université Pierre et Marie Curie, 75252 Paris, Cedex 05, France

M. Dine, Institute for Advanced Study, Princeton, New Jersey 08540 and City College of CUNY, New York, New York 10031

Tohru Eguchi, Department of Physics, University of Tokyo, Tokyo, Japan 113

W. Fischler, Theory Group, Physics Department, University of Texas, Austin, Texas 78712

Marc Henneaux, Faculté des Sciences, Université Libre de Bruxelles, B-1050 Brussels, Belgium and Centro de Estudios Científicos de Santiago, Santiago 9, Chile

Marek Karliner, Stanford Linear Accelerator Center, Stanford University, Stanford, California 94305

Igor Klebanov, Stanford Linear Accelerator Center, Stanford University, Stanford, California 94305

Taichiro Kugo, Department of Physics, Kyoto University, Kyoto 606, Japan

Hirosi Ooguri, Department of Physics, University of Tokyo, Tokyo, Japan 113

Burt A. Ovrut, Department of Physics, University of Pennsylvania, Philadelphia, Pennsylvania 19104-6369

M. Rinaldi, S.I.S.S.A., Trieste and I.N.F.N., Sezione di Trieste, Italy

J. Russo, S.I.S.S.A., Trieste, Italy

Norma Sánchez, Observatoire de Paris, Section de Meudon, 92195 Meudon Principal Cedex, France

Fidel A. Schaposnik, Departamento de Física, Universidad Nacional de La Plata, 1900 La Plata, Argentina

N. Seiberg, Institute for Advanced Study, Princeton, New Jersey 08540; on leave of absence from the Department of Physics, Weizmann Institute of Science, Rehovot 76100, Israel

R. Stora, LAPP, F-74019 Annecy-le-Vieux, France and Theory Division, CERN, CH-1211 Geneva 23, Switzerland

Leonard Susskind, Physics Department, Stanford University, Stanford, California 94305

Claudio Teitelboim, Centro de Estudios Científicos de Santiago, Santiago 9, Chile and Center for Relativity, The University of Texas at Austin, Austin, Texas 78712

L. F. Urrutia, Centro de Estudios Científicos de Santiago, Santiago 9, Chile. *Permanent address*: Centro de Estudios Nucleares, Universidad Nacional Autónoma de Mexico, Circuito Exterior, C.U., 04510 Mexico, D.F.; Departamento de Fisíca, Universidad Autónoma Metropolitana, Iztapalapa, 09340 Mexico, D.F.

P. van Nieuwenhuizen, Institute for Theoretical Physics, State University of New York at Stony Brook, Stony Brook, New York 11794

J. D. Vergara, Centro de Estudios Nucleares, Universidad Nacional Autónoma de Mexico, Circuito Exterior, C.U., 04510 Mexico, D.F.

Preface

This volume is based on a meeting on *Quantum Mechanics of Fundamental Systems*, held December 17–20 of 1987, at the Centro de Estudios Científicos de Santiago (CECS). The meeting was intended to review new developments in the field defined by its purposely vague title.

We were especially interested in communicating important advances in a broad perspective and in a self-contained manner to a wide colleagueship, and in particular to Latin American physicists.

Whatever success we may have achieved in that direction would certainly not have been possible without the help and generosity of many people. For their kind support we thank the Tinker Foundation, the International Centre for Theoretical Physics, the Ministère des Affaires Etrangères et Service Culturel et de Coopération Scientifique et Technique de France and the Centro Latinoamericano de Fisica. We are especially grateful to the Santiago College and to its principal, Rebeca Donoso, for allowing us to hold the meeting at that school when, at the last moment, we were deprived of the previously arranged meeting site.

Last but not least we thank the participants for an extraordinary week and the CECS staff for performing, as usual, beyond any reasonable expectation.

Claudio Teitelboim
Jorge Zanelli

Santiago, Chile

Contents

Chapter 1

Anomalous Jacobians and the Vector Anomaly

J. Alfaro, L. F. Urrutia, and J. D. Vergara

1. INTRODUCTION

Ever since the discovery of current anomalies there have been important applications of this idea to various problems of physical interest: It led to a precise calculation of the neutral pion decay (global axial anomaly) and to an understanding of the $U(1)$ problem: the nonexistence of the ninth Goldstone boson associated with a broken symmetry of quantum chromodynamics [1].

Furthermore, anomalies in gauge currents have been considered for a long time to be unacceptable because the standard proof of unitarity and renormalizability of the theory does not hold in this case. It follows that the only way to make sense of these theories is to cancel the gauge anomalies by selecting the particles contained in the model. Forcing the cancellation of gauge anomalies has led to the prediction that quarks and leptons come in families and to properties of the spectrum of massless fermions in

---

J. Alfaro • Pontificia Universidad Católica de Chile, Facultad de Física, Santiago 22, Chile. L. F. Urrutia • Centro de Estudios Científicos de Santiago, Santiago 9, Chile. *Permanent addresses:* Centro de Estudios Nucleares, Universidad Nacional Autónoma de Mexico, Circuito Exterior, C.U., 04510 Mexico, D.F.; Departamento de Fisica, Universidad Autónoma Metropolitana, Iztapalapa, 09340 Mexico, D.F. J. D. Vergara • Centro de Estudios Nucleares, Universidad Nacional Autónoma de Mexico, Circuito Exterior, C.U., 04510 Mexico, D.F.

confining theories [2]. Very recently it has been at the center of exciting developments in string theories [3].

A different mechanism of making sense of anomalous gauge theories is being studied by several authors, but whether or not it will provide an alternative to the currently used anomaly cancellation paradigm is not known yet [4].

For a long time the perturbative treatment of anomalies was dissociated from a path integral treatment. The work of Fujikawa provided a way to understand the anomaly phenomenon from a path integral formulation and led to a deeper understanding of the problem [5].

Fujikawa realized that the path integral measure is not necessarily invariant under certain transformations even if the classical action is invariant, thus providing space for quantum anomalies. More precisely, his calculation of the Jacobian of the axial transformation does give the right anomaly as it is known from computations of Feynman graphs.

In this chapter we want to call your attention to a complementary aspect of the anomaly phenomenon: the vector anomaly. It is known from the original calculation that the essence of the axial-vector anomaly is the impossibility of the simultaneous conservation of chiral and vector currents. In other words, it is impossible to find a regulator that forces the conservation of both currents. In fact the original calculation uses a point splitting regularization that interpolates continuously between the axial and vector anomalies.*

We ask the question: How do we see the vector anomaly in the path integral? According to Fujikawa, the vector anomaly must show up in the Jacobian of the phase transformation; however, a naive computation of this Jacobian always gives a null result, and in fact such computations have been used by some authors to conclude that there is not a vector anomaly, in obvious contradiction with perturbative calculations. This problem has recently been considered in Refs. 7–9. We present here an alternative solution, which makes use of an Hermitian regulator and which can be in principle generalized to more dimensions, taking properly into account the renormalization problems that will arise.

In Section 2 we review the concept of a family of anomalies in the context of the Schwinger model.

In Section 3 we review the method of Fujikawa.

In Section 4 we reexamine the definition of the Jacobian and show that unless the cyclic property of the trace of products of operators holds, the normal definition of the Jacobian does not apply. In particular we find that the Jacobian of the transformation must depend explicitly on the Dirac operator of the system in order that the integral be the same in any

* See, for example, the first reference in Ref. 1.

(Fermionic) coordinate system. Such Jacobians are called anomalous in Ref. 6.

In Section 5 we present a preliminary interpretation of our regulated Jacobian in terms of "regulated phase transformations."

Section 6 contains a discussion of our results and some open problems.

2. THE SCHWINGER MODEL

The Euclidean classical action for the Schwinger model is given by

$$S[A] = \int dx\, \bar{\psi} i \not{D} \psi \tag{1}$$

ψ and $\bar{\psi}$ are anticommuting variables defined in the space-time point x and

$$\not{D} = \not{\partial} + ie\not{A} \tag{2}$$

is the usual covariant derivative. We assume that $A_\mu(x)$ is a classical external field. Our conventions to go from Minkowski to Euclidean space are those of Ref. 5. The metric is $g^{\mu\nu} = (-1, -1)$ and our gamma matrices are anti-Hermitian, satisfying

$$\gamma^\mu \gamma^\nu = g^{\mu\nu} + \varepsilon^{\mu\nu}\gamma_5 \tag{3}$$

with $\varepsilon^{14} = -i$. With these conventions the Dirac operator $\not{D}$ is Hermitian.

This theory has two classically conserved currents defined by

$$\begin{aligned} j^v_\mu &= \bar{\psi}\gamma_\mu\psi \\ j^A_\mu &= \bar{\psi}\gamma_\mu\gamma_5\psi \end{aligned} \tag{4}$$

They are Noether currents associated with the invariance of the classical action under phase and axial phase transformations.

It is well known that when quantum effects are taken into account there is an axial-vector anomaly that can be parametrized by a parameter b; it contains the regulator dependence of the anomaly. This is what we mean by a family of anomalies.

The explicit result in the Schwinger model is

$$\begin{aligned} \partial_\mu\langle j^V_\mu(x)\rangle &= \frac{e}{\pi}(1-b)\partial_\mu A_\mu(x) \\ \partial_\mu\langle j^A_\mu(x)\rangle &= \frac{e}{\pi} b\varepsilon_{\mu\nu}\partial_\mu A_\nu(x) \end{aligned} \tag{5}$$

We will consider an anomaly calculation using any regulator R to be correct if the results are such that they can be reproduced by a specific choice of b in relations (5).

3. THE METHOD OF FUJIKAWA [5]

Let us write the generating functional of fermionic Green's functions in an external field A_μ corresponding to the action (1):

$$Z[A] = \int \mathscr{D}\psi \mathscr{D}\bar{\psi} \exp \int dx\, \bar{\psi}(x) i \not{D} \psi(x) \tag{6}$$

In order to obtain the Ward identities related to a symmetry of the classical action (1), we follow the standard steps [5]. Let us consider the fermionic part of such an infinitesimal symmetry transformation, which we write as

$$\begin{aligned} \chi &= [1 + K(\alpha)]\psi \\ \bar{\chi} &= \bar{\psi}[1 + L(\alpha)] \end{aligned} \tag{7}$$

where K and L are operators depending upon infinitesimal local parameters, which we call $\alpha = \alpha(x)$ in compact notation. The transformation (7) leaves invariant the action (1) when the parameters are independent of position. Now we use the transformation (7) as a linear change of variables in the generating functional, obtaining

$$Z[A] = J(\alpha) \int \mathscr{D}\bar{\psi} \mathscr{D}\psi \exp \int (dx) \bar{\psi} i (\not{D} + \not{D}K + L\not{D})\psi \tag{8}$$

where we have kept only terms to first order in the parameters in the action and $J(\alpha)$ stands for the Jacobian of the transformation. The explicit calculation of the terms in the exponential leads to the identification of the currents in the form

$$Z[A] = J(\alpha) \int \mathscr{D}\bar{\psi} \mathscr{D}\psi \exp\left[\int (dx) \bar{\psi} i \not{D} \psi - \int (dx) j^\mu(x) \partial_\mu \alpha(x) \right] \tag{9}$$

The Ward identities are obtained by functionally differentiating with respect to $\alpha(x)$ both sides of (9). The result is

$$\frac{1}{Z} \frac{\delta Z}{\delta \alpha(x)} = 0 = \partial^\mu J_\mu(x) + \left. \frac{\delta \ln J(x)}{\delta \alpha(x)} \right|_{\alpha=0} \tag{10}$$

where $J_\mu(x) = \langle j_\mu(x) \rangle$ is the usual average value of $j_\mu(x)$.

Now we use the standard definition of the Jacobian to compute the axial-vector and vector anomalies.

i. Axial-Vector Anomaly. In this case we have that

$$\begin{aligned} \psi' &= e^{i\alpha\gamma_5}\psi \approx (1 + i\alpha\gamma_5)\psi \\ \bar{\psi} &= \bar{\psi}\, e^{i\alpha\gamma_5} \approx \bar{\psi}(1 + i\alpha\gamma_5) \\ K(\alpha) &= i\alpha\gamma_5, \qquad L(\alpha) = i\alpha\gamma_5 \end{aligned} \tag{11}$$

Then $J(\alpha) = \exp(-2 \operatorname{tr} i\alpha\gamma_5)$.

Following Fujikawa [5] we introduce a regulator:

$$\text{tr}\ i\alpha\gamma_5 \to \text{tr}\ i\alpha\gamma_5 \exp(-\not{D}^2/M^2) \tag{12}$$

The regulator M is taken to infinity at the end of the calculation.

In this case we get the right (perturbative) axial anomaly. (See Ref. 5 for more details.)

ii. Vector Anomaly. In this case we have that

$$\begin{aligned} \psi' &= e^{i\alpha}\psi \simeq (1 + i\alpha)\psi \\ \bar{\psi}' &= \bar{\psi}\, e^{-i\alpha} \approx \bar{\psi}(1 - i\alpha) \end{aligned} \tag{13}$$

$$K(\alpha) = i\alpha(x), \qquad L(\alpha) = -i\alpha(x)$$

Then $J(\alpha) \approx \exp[-(i\alpha - i\alpha)] = 1$.

There is no way to get an anomaly. But we know there is one. Moreover, how do we get the family of anomalies discussed in Section 2?

4. REDEFINITION OF THE JACOBIAN [11]

In the previous section we have explicitly reviewed the well-known fact that there is a vector anomaly if one chooses to work with a gauge variant regulator. However, the naive Jacobian of the transformation (13) never produces a vector anomaly. We get no anomaly at all no matter what our regularization procedure is. Clearly something does not work with the standard definition of the Jacobian. Let us see what it is.

We will use the definition

$$\int \mathcal{D}\bar{\psi}\mathcal{D}\psi \exp \int (dx)\bar{\psi}P\psi = \det(P) \tag{14}$$

together with the fact that the Jacobian is precisely the extra factor that produces a coordinate-independent integral.

Since the integral cannot depend on the coordinate system used to evaluate it we must have that (8) is equal to (9).

Therefore the Jacobian must satisfy that

$$J(\alpha) = \frac{\det(\not{D})}{\det(\not{D} + \not{D}K + L\not{D})} \tag{15}$$

which can be rewritten as

$$\begin{aligned} J(\alpha) &= \det[1 - \not{D}^{-1}(\not{D}K + L\not{D})] \\ &= 1 - \text{Tr}[\not{D}^{-1}(\not{D}K + L\not{D})] \end{aligned} \tag{16}$$

where $\not{D}^{-1}$ denotes the Green's function of the operator $\not{D}$ and we recall that K and L are first-order quantities in the infinitesimal parameters. As usual, the trace in (16) is ill defined and needs to be regularized. This is usually achieved by considering an arbitrary Hermitian, positive definite operator S. The regulating operator is subsequently defined by $R = f(-S/M^2)$, where the limit $M \to \infty$ is taken at the end of the calculation. The function $f(x)$ satisfies $f(0) = 1$ and rapidly approaches zero at $x = \infty$. We define our regulated Jacobian by

$$J_R(\alpha) = 1 - \mathrm{Tr}[\not{D}^{-1}(\not{D}K + L\not{D})R] \equiv 1 - T \tag{17}$$

which now is a finite quantity.

Equation (17) is the main result of this work and constitutes our prescription to deal with the Jacobian in anomaly problems. It is appropriate to emphasize that the above construction is independent of the number of dimensions and that it provides the right answer for the anomaly calculation using an arbitrary regulator at least in two dimensions. Before going into some explicit calculations we make some comments upon (17).

The first thing we notice is that $J_R(\alpha)$ depends explicitly upon the system ($\not{D}$), the symmetries (K, L), and the regulator (R). In this way $J_R(\alpha)$ does not coincide in general with the standard calculation where the dependence on $\not{D}$ does not appear. One can recover the currently used expression for the Jacobian provided a regulator R is chosen such that $[R, \not{D}] = 0$. Then (17) reduces to

$$J_R(\alpha) = 1 - \mathrm{Tr}[(K + L)R] \tag{18}$$

which can be directly obtained from (16) using first the cyclic property of the trace and regularizing afterwards. Since we lose information in this way, we infer that such identity can only be used when everything is already regulated. A well-known contradiction in quantum mechanics arising from the naive application of the cyclic property of the trace is obtained when one takes the trace of the commutator $[x, p] = i\hbar$.

Our second comment relates to the Abelian vector anomaly, which is defined by taking $K = -L = i\alpha(x)$. The usual Jacobian (18) produces always a null result independently of the regulator employed. (See Section 3.) We understand this because (18) is correct only when we use a regulator, which might be an arbitrary function of $\not{D}$ and thus preserve gauge invariance. In particular, the expression (18) would be an incorrect starting point for calculating the vector or axial-vector anomaly using, for example, the regulator $R = \exp(-\not{\partial}^2/M^2)$.

Nevertheless, as we will show later, the expression (17) for the Jacobian gives the right answer. Keep in mind that the null result always obtained for the vector anomaly [starting from (18)] is in contradiction with the operator point splitting calculation; for example, some authors have proposed regulating the Jacobian separately in each of its two pieces corresponding to $D\bar{\psi}$ and $D\psi$, respectively [9]. The result would be

$$J(\alpha) = 1 - \mathrm{Tr}(KR_1 + LR_2) \tag{19}$$

which is not identically one for the vector anomaly. Nevertheless, according to our prescription (17) this result would be wrong in general, unless

$$R_2 = \not{D} R_1 \not{D}^{-1} \tag{20}$$

Our last comment is related to expression (20) and has to do with systems whose Dirac operator $\not{D}$ is not Hermitian. This is the case in theories with γ_5 coupling* or in some recent discussions of the vector anomaly in the usual Schwinger model, which is artificially written as a theory with γ_5 coupling. Incidentally, this trick can only be played in two dimensions owing to relation (3) [8]. The particular choice $R_1 = \exp(-\not{D}^\dagger D/M^2)$ in our general expression (17) together with $R_2 = \not{D}\exp(-\not{D}^\dagger D/M^2)\not{D}^{-1} = \exp(-\not{D}\not{D}^\dagger/M^2)$, according to (20), explains the specific choice of regularization used in those references.

Let us apply our definition of the Jacobian to the vector and axial-vector anomaly [11].

In order to perform the calculations it is more convenient to rewrite the corresponding expressions for T in (17) in the following form:

$$T^V = -i\,\mathrm{Tr}(\not{D}^{-1}\beta[\not{D}, R]) \tag{21a}$$

$$T^A = i\,\mathrm{Tr}(2\alpha\gamma_5 R + D^{-1}\alpha\gamma_5[\not{D}, R]) \tag{21b}$$

Here the superscript V refers to the vector case $[K = -L = i\beta(x)]$, while A labels the axial-vector case $[K = L = i\gamma_5\alpha(x)]$. Again we see from (21) that only in the case $[\not{D}, R] = 0$ are the standard expressions for both anomalies recovered, including now the axial-vector case too.

The first regulator we will consider is $R = \exp(-\not{\partial}^2/M^2)$, which was discussed in Ref. 10. In that work the anomalies were calculated by using the naive expression (18) for the Jacobian, which amounts to settting (21a) together with the second term of (21b) equal to zero. A null result was obtained for both anomalies, in obvious contradiction with the family defined in (5). We now indicate how the proper calculation of the terms involving $[\not{D}, R] \neq 0$ leads to the $b = 1/2$ member of family (5). Of course we use the result $\mathrm{Tr}(2\alpha\gamma_5 R) = 0$ obtained previously in Ref. 10. The object of the calculation is

$$X = \mathrm{Tr}(\not{D}^{-1}Q[\not{D}, R]) \tag{22}$$

* See, for example, the last reference in Ref. 5.

where Q stands for either βI or $\alpha\gamma_5$. The space-time trace is conveniently calculated in the plane-wave basis because these vectors are eigenvectors of the regulator. Moreover, the only contribution to the commutator in (22) arises from the external field-dependent part of the operator $\not{D}$. Up to this point the calculation reads

$$X = ie\int (dx)(dy)\frac{dK\,dq}{(2\pi)^4}\operatorname{tr}\{G(x,y)Q(y)\gamma^\mu \times A_\mu(q)e^{-ikx}\,e^{i(k-q)y}[e^{k^2/M^2} - e^{(q-k)^2/M^2}]\} \tag{23}$$

where we have introduced the Fourier transform of the external field. The only divergent integral in expression (23) would come out from the first term in the expansion of $G(x, y)$ in powers of the free Green's function $G_0(x, y)$. The regulated version of this contribution provides the nonzero result in the limit $M \to \infty$. As usual, one rescales the momentum $k \to Mp$, factors out the regulating term $\exp(p^2)$, and expands the remaining square bracket of (23) to leading order in p/M. After taking the spin trace, performing the p integration, substituting in (21), and using (10) we obtain

$$\partial^\mu J^V_\mu = \frac{e}{2\pi}\partial_\mu A^\mu, \qquad \partial^\mu J^A_\mu = \frac{e}{2\pi}\varepsilon^{\mu\nu}\partial_\mu A_\nu \tag{24}$$

which corresponds to the choice $b = 1/2$ in (5).

To conclude we discuss the regulator $R = \exp(-\not{D}_a^2/M)$ with $D_a = \not{D}_a^\dagger = \not{D} + iea\not{A}$ and show that it reproduces the whole family (5). The calculation here is a little more involved and we just give some brief details. The first term in (21b) corresponds to a standard Fujikawa calculation with $A_\mu \to aA_\mu$ and reduces to

$$2\operatorname{Tr}(\alpha\gamma_5 R) = -\frac{iea}{\pi}\int (dx)\alpha(x)\varepsilon^{\mu\nu}\partial_\mu A_\nu \tag{25}$$

The remaining pieces are incorporated in the expression X previously introduced in (22). The commutator is now rewritten as

$$[\not{D}, R] = ie(1-a)[\not{A}, R] \tag{26}$$

and the plane-wave basis is used again. Operating the regulator upon the corresponding plane wave produces the usual shift $\not{D}_a \to \not{D}_a - i\not{k}$ giving

$$X = -ie(a-1)\int \frac{dK}{(2\pi)^2}(dx)(dy)\,e^{iK(x-y)}\,e^{K^2/M^2} \times \operatorname{tr}(G(x,y)Q(y)[\exp(i\{\not{K}, \not{D}_a\} - \not{D}_a^2)/M^2, \not{A}(y)]) \tag{27}$$

Once more the finite contribution to X comes from the free Green's function together with the leading power of k/M in the expansion of the exponential in the commutator of (27). After rescaling $k \to Mp$ and taking the limit

$M \to \infty$ we are left with

$$X = \frac{ie(1-a)}{4\pi} \int (dx)\, \mathrm{Tr}[\gamma^\mu Q(x) \gamma^\nu] \partial_\mu A_\nu(x) \tag{28}$$

from which the vector and axial-vector contributions are easily recovered. The final answer for the anomalies corresponds to (5) with $b = (1+a)/2$.

5. REGULATED PHASE TRANSFORMATIONS

In the previous section we have pointed out that the canonical expression for the Jacobian of the transformation of variables may not hold for systems with an infinite number of degrees of freedom. In addition to this, we were able to find the right generalization of the unregulated Jacobian in such a way that it could be regularized in a standard way, thus avoiding the use of complicated non-Hermitian regulators. We tested our form of the regulated Jacobian by computing the axial-vector and vector anomalies in the Schwinger model. We get directly the most general family of anomalies that is known to exist in the model.

In this section we want to understand our regularization of the Jacobian in terms of more basic concepts. We will discover that our method of regularization can be understood as regularization of the transformation in the following sense: In the classical theory the fermionic action is invariant under constant (i.e., space-time-independent) phase transformations that belong to the representation of the group generated by the identity matrix 1 and γ_5. If we pass to the quantized theory through the Feynman path integral we find that the Jacobian of this transformation is ambiguous (owing to divergences) and needs to be regulated. Instead we propose passing to quantum theory, preserving the group structure of the classical transformation but allowing the generator to be a general operator that acts as a regulator. In other words, passing to quantum theory may change the classical representation of the group but preserve the group structure. The bonuses we get from this change of perspective are as follows:

The Jacobian of the regulated transformation is again "naive" and gives directly the regulated Jacobian we postulated in a previous letter [11]. [See equation (17).]

Imposing that the regulated transformation is still a symmetry of the classical fermionic action, we derive the right relationship between the regularization of the ψ and the $\bar{\psi}$ that was also postulated in our previous letter [11]. [See equation (20).]

We can now impose the group properties of our regulated transformation. This implies a relationship between the regulators of the axial and vector transformations that is sufficient to guarantee the existence of the family of anomalies. [See equation (33).] We believe this last point is an important one.

Let us consider the path integral for a Dirac field in an external electromagnetic potential. The action for such a system is given by

$$S[A] = \int (dx) i\bar{\psi}(x)[\not{\partial} + ie\not{A}]\psi(x) \tag{29}$$

Notice that the action is invariant under the following global (i.e., space-time-independent) transformations of the fermionic variables:

$$\begin{aligned} \psi' &= e^{i\alpha}\psi, \qquad & \psi' &= e^{i\beta\gamma_5}\psi \\ \bar{\psi}' &= \bar{\psi}\, e^{-i\alpha} \qquad & \bar{\psi}' &= \bar{\psi}\, e^{i\beta\gamma_5} \end{aligned} \tag{30}$$

These transformations form a group: $U(1)aXU(1)v$ generated by 1 and γ_5.

As we mentioned above, the Jacobian of these transformations is ambiguous because of divergences. The commonly used strategy is to regulate the Jacobian and from the regulated Jacobian derive the anomalous Ward identities of the theory. We propose instead to regulate the transformation (30) in such a way that we keep the same group but permit the generator of the representation of the group to be an operator that acts as a regulator of the Jacobian.

That is, we replace (30) by (31):

$$\begin{aligned} \psi' &= e^{i\alpha R_1}\psi, \qquad & \psi' &= e^{i\beta\gamma_5 R_2}\psi \\ \bar{\psi}' &= \bar{\psi}\, e^{i\alpha \bar{R}_1}, \qquad & \bar{\psi}' &= \bar{\psi}\, e^{i\beta\gamma_5 \bar{R}_2} \end{aligned} \tag{31}$$

Of course, the action must be invariant under these regulated transformations as well. This implies the following relation:

$$\bar{R}_1 = \not{D} R_1 \not{D}^{-1}, \qquad \bar{R}_2 = \not{D} R_2 \not{D}^{-1} \tag{32}$$

Notice that (32) is precisely the relation among the regulators of the ψ and $\bar{\psi}$ transformation that we need to get the right regulated Jacobian. [See equation (20).]

Since the regulated transformations form the group $U(1)a \times U(1)v$, we must have that

$$[R_1, \gamma_5 R_2] = [1, \gamma_5] = 0 \tag{33}$$

Our previous calculation (see Ref. 13) provides a check of this relationship in the simplest case $R_1 = R_2$, but we have also considered a more general family of regulators:

$$\begin{aligned} R_2 &= e^{-(1-\beta\gamma_5)\not{\partial}^2/M^2} \\ R_1 &= e^{-\not{\partial}^2/M^2} \end{aligned} \tag{34}$$

We get that the anomaly does not depend on beta.

This result is quite interesting. It shows that to get a member of the family of anomalies one can use independent regulators for the axial and vector transformations. Provided that the regulators satisfy (33) we are guaranteed to get a member of the family of anomalies. Notice that (33) is a sufficient, but not a necessary, condition to have a right anomaly. The reason for this is that the original theory as described by (29) is only recovered in the physical limit of the regulators ($M \to \infty$). Therefore we should impose (33) in the neighborhood of this limit. Clearly this is a less restrictive (but more complicated to implement) condition than the actual form of (33).

We want to stress also that the Jacobian of the transformation (31) is now finite, and because of this it can be computed in the standard way.

6. CONCLUSIONS AND OPEN PROBLEMS

To summarize, we have constructed in our expression (17) an extension of the standard definition of the regulated Jacobian in the path integral calculation of anomalies. In two dimensions, this extension directly solves the problem of recovering the whole family of vector and axial-vector anomalies for the Schwinger model in the Fujikawa approach. Notice that we are able to obtain correct results using the regulator $R = \exp(-\not{D}_a^2/M^2)$, thus showing that one is not forced to use non-Hermitian operators as stated in Ref. 8. We expect that for any sensible regulator the anomaly calculation, according to expression (17), in the above-mentioned model will produce a correct member of such families of anomalies. Our prescription is valid for any number of dimensions and could be directly tested in such cases. In particular, we conjecture that the same regulator $R = \exp(-\not{D}_a^2/M^2)$ will reproduce the standard family of anomalies in four-dimensional QED.

In addition to this, we get an interpretation of the regulated Jacobian starting from a "regulated transformation."

There are several open problems. Among them we can mention discussion of the family of anomalies in four-dimensional QED and an application of our regulated Jacobian to chiral theories and non-Abelian anomalies. Moreover, it is very interesting to see whether we can use the concept of a "regulated transformation" to extend our knowledge of the anomaly phenomenon.

ACKNOWLEDGMENTS. J.A. wants to thank Professor Claudio Teitelboim for the invitation to talk at The Second Summer Meeting on Quantum Mechanics of Fundamental Systems. L.F.U. and J.D.V. received partial support from the CONACyT(Mexico) under grant No. PCEXNA-040575, while J.A. was partially supported by FONDECYT 5084/1986, UNESCO 6266, and DIUC 201/86.

REFERENCES

1. R. Jackiw, in *Lectures on Current Algebra and Its Applications* (S. Treiman, R. Jackiw, and D. Gross, eds.), Princeton University Press, Princeton, New Jersey, 1972, and references therein; G. 't Hooft, *Phys. Rev. Lett. 37*, 8 (1976).
2. C. Bouchiat, J. Iliopoulos, and Ph. Meyer, *Phys. Lett. 38B*, 519 (1972); G. 't Hooft, in *Recent Development in Gauge Theories, Cargèse 1979* (G. 't Hooft *et al.*, eds.), Plenum Press, New York, 1980.
3. M. B. Green and J. H. Schwarz, *Nucl. Phys. 243B*, 475 (1984).
4. L. D. Faddeev and S. L. Shatashvili, *Phys. Lett. 167B*, 225 (1986); O. Babelon, F. A. Schaposnik, and C. M. Viallet, *Phys. Lett. 177B*, 385 (1986).
5. K. Fujikawa, *Phys. Rev. Lett. 42*, 1195 (1979); *Phys. Rev. D 21*, 2848 (1980); *29*, 285 (1984).
6. R. E. Gamboa-Saravi, M. A. Muschietti, and J. R. Solomin, *Commun. Math. Phys. 89*, 363 (1983).
7. A. Das and V. S. Mathur, *Phys. Rev. D 33*, 489 (1986).
8. A. Smailagic and R. E. Gamboa-Saravi, *Phys. Lett. 192B*, 145 (1987).
9. J. M. Gibson, *Phys. Rev. D 33*, 1061 (1986).
10. G. A. Christos, *Z. Phys. C 18*, 155 (1983).
11. J. Alfaro, L. F. Urrutia, and J. D. Vergara, Extended definition of the regulated Jacobian in the path integral calculation of anomalies, UNAM(Mexico)preprint, October 1987; it appears in *Phys. Lett. B 202*, 121 (1988).

Chapter 2

String Phenomenology

Thomas Banks

1. INTRODUCTION

The two most common questions people ask me about string theory are: What is it? and What does it have to do with the real world? Sadly, I cannot claim to have more than a preliminary and partial understanding of the answers to these questions. Let me try nonetheless to communicate what I think I know about this difficult subject.

String theory is supposed to be a generalization of Einstein's general theory of relativity. If this is true, there should be an analog of the space-time metric tensor $G_{\mu\nu}$, an elegant Lagrangian, and a deep set of principles like the principle of general covariance, which lie at the base of it all. The analog of the metric tensor is a two-dimensional quantum field theory. To understand this, write the Lagrangian for a quantum mechanical string propagating through a space-time with metric $G_{\mu\nu}(x)$

$$\mathscr{L} = \int d^2\xi \frac{\partial}{\partial \xi^\alpha} x^\mu(\xi) \frac{\partial}{\partial \xi^\alpha} x^\nu(\xi) G_{\mu\nu}(x(\xi)) \tag{1}$$

ξ^α is a two-dimensional parameter for the world sheet of the string. In this Lagrangian the metric $G_{\mu\nu}$ plays the role of a set of coupling constants.

THOMAS BANKS • Santa Cruz Institute for Particle Physics, University of California, Santa Cruz, California 95064.

The dynamical variable of string theory is simply the set of all possible coupling constants that one can write for the variables x^μ. Clearly, this rule endows string theory with an infinite number of space-time fields. Remarkably, once gauge invariance is properly treated, this infinite set of fields is in one-to-one correspondence with the states of a single string.

In (1) we have chosen a particular metric (the flat one) and a Cartesian coordinate system on the world sheet. It seems reasonable to require that physical quantities be independent of such arbitrary choices. Thus physical observables in string theory should be independent of the world sheet metric. The requirement that a quantum field theory not depend on the world sheet coordinates is easily implemented in the BRST [1] formalism. We choose conformal coordinates in which the two-dimensional metric has the form $g_{\alpha\beta} = e^\phi \delta_{\alpha\beta}$ and introduce Fadeev Popov ghosts.

Independence of the conformal factor e^ϕ is achieved only by restricting the coupling constants $G_{\mu\nu}$, etc., to be fixed points of the renormalization group. As first shown by Friedan [2], this leads to differential equations for the metric that are equivalent to Einstein's equations (to lowest order in the σ model loop expansion). Friedan and Shenker [3] and Lovelace [4] conjectured that this remarkable fact was no coincidence. Rather, the equations of motion of string theory are precisely the requirement of conformal invariance. It has also been shown that these equations follow from an action, but the true geometrical significance of this construction is not yet understood.

A familiar example of a theory in which the equations of motion follow from an invariance principle is general relativity (GR). In the Arnowitt–Deser–Misner [5]-Wheeler–deWitt [6] (ADMWDW) formulation of this theory, coordinate transformations are divided into two classes. Those transformations that act within a given spacelike surface are easily implementable and carry no dynamical information. Those that move the spacelike slice generate the equations of motion. It is tempting to make the analogy between world sheet metric independence in string theory and coordinate independence of GR. Conformal invariance would be the analog of local time translations. One puzzling aspect of this idea is that the ADMWDW approach to GR is intrinsically Hamiltonian, while conformal invariance leads directly to the Lagrange equations of motion. From the two-dimensional point of view it is quite natural to generalize the requirements of reparametrization and conformal invariance to those of local supersymmetry and superconformal invariance. We do not yet have a deep understanding of what we are doing here. Drawing an analogy with what happens for supersymmetric particle mechanics, I would conjecture that string theories with world sheet SUSY have variables that are differential forms of arbitrary rank on whatever passes here for space-time. By contrast, bosonic string theories involve only zero forms on this semimythical space.

While there are many types of local SUSY in two dimensions, it appears that only strings based on (0, 0), (0, 1) [or, equivalently, (1, 0)] or (1, 1) local SUSY can live consistently in more than two flat space-time dimensions. These have come to be called bosonic, heterotic, and superstring theories (this last category includes both the type IIA and IIB superstrings). They are all closed string theories. If we allow open strings, there is one other consistent theory. We will ignore it in hopes that it will soon go away. It will be a fundamental assumption of this chapter that all closed string models can be thought of as perturbation expansions around a classical vacuum state [(p, q) superconformal field theory] of one of these three basic theories. We will concentrate on the heterotic string. The bosonic string probably does not have stable vacua, while the superstring has none that are compatible with low-energy phenomenology [7]. We will be dealing with classical solutions of heterotic string theory, so this chapter should be entitled "Phenomenology and (0, 1) Superconformal Field Theory."

Despite early claims of almost unique predictions of the low-energy spectrum, heterotic string theory has many perturbatively stable vacua that have particle spectra compatible with observations. It also has many that are not. We do not yet have the tools to decide which (if any) of these myriad candidate vacua *the theory* prefers to sit in. The best we can hope to do at present is to find a vacuum that resembles the real world. It seems that the best way to put some order in this search is to prove general results like the superstring No-Go theorem of Ref. 7. This is what we will attempt to describe in this chapter.

We will sketch the proofs of the following results:

(1) A criterion will be formulated for a heterotic vacuum to satisfy the usual SU_5 relation between tree-level gauge couplings. Most heterotic vacua seem to belong to continuously connected families of vacuum states. If the tree-level value of the Weinberg angle were to vary continuously along these flat directions in the potential, one could not claim that string theory made any tree-level prediction at all for this parameter. We will show that for vacua whose excitation spectra contain chiral space-time fermions this does not happen: the Weinberg angle is constant along all flat directions.

(2) We will show that the only continuous global symmetries of string theory are translation symmetries of axionlike fields (under which no particle carries a charge) and the Lorentz groups of uncompactified flat space-time directions. In particular, global baryon- or lepton-number symmetries are nonexistent. Note, however, that a discrete conservation law like "baryon-number modulo N" is not ruled out by our theorem, so we have not shown that proton decay, neutrino masses or neutron–antineutron oscillations are inevitable consequences of string theory.

(3) We complete the arguments of Ref. 8 that space-time SUSY is equivalent to (0, 2) *global* world sheet SUSY in heterotic string theory.

Furthermore, we show that $N = 2$ space-time SUSY puts strong restrictions on the world sheet theory. It must be a minimal (0, 4) theory tensored with two free (0, 1) superfields [or, equivalently, one free (0, 2) superfield]. $N = 4$ space-time SUSY can appear only in models based entirely on free (0, 1) superfields.

(4) We show that classical string theory does not allow small breaking of space-time SUSY. That is, every vacuum near a supersymmetric one is supersymmetric. The options for SUSY breaking in string theory thus appear to be

(a) Classical breaking: $M_{\text{SUSY}} \sim M_{\text{Planck}}$.
(b) D term breaking: $M_{\text{SUSY}} \sim \alpha M_{\text{Planck}}$
(c) Non-perturbative: $M_{\text{SUSY}} \sim e^{-k/\alpha} M_{\text{Planck}}$

A fourth possibility, that SUSY breaking is related to the Kaluza–Klein scale, which is much less than M_{Planck}, appears to be ruled out within the semiclassical approximation by experimental bounds on couplings [10]. Since all the material covered either has appeared in the literature [9] or soon will, I will give only a brief outline of the arguments for points 3 and 4.

2. GAUGE SYMMETRIES AND GLOBAL SYMMETRIES

In Kaluza–Klein theories, gauge bosons in four-dimensional space-time arise from continuous global symmetries (isometries) of the metric of the compactified dimensions. In accordance with our analogy between the metric of GR and the two-dimensional quantum field theory of string theory, we expect global symmetries of the two-dimensional field theory to lead to gauge bosons. Let us see how this happens.

A Kaluza–Klein vacuum for the heterotic string consists of four left-moving free scalars (x_L^μ), four right-moving free superfields ($x_R^\mu + \theta\psi_R^\mu$), and an "internal" (0, 1) superconformal field theory with Virasoro central charge $(C_L, C_R) = (22, 9)$. The internal theory has a positive metric Hilbert space.

Suppose the internal theory has a global symmetry with generator I. In a local two-dimensional field theory, I will be the line integral of a local conserved current

$$I = \int d\Sigma^\alpha J_\alpha$$
$$\partial_\alpha J^\alpha = \partial \bar{J} + \bar{\partial} J = 0 \tag{2}$$

The second version of the conservation equation is written in complex coordinates: $z = \xi_1 + i\xi_2$, $\bar{z} = \xi_1 - i\xi_2$.

If a symmetry is to map the space of physical states into itself, it must commute with the BRST charge. This implies that

$$
\begin{aligned}
T(z)J(w) &\sim \frac{1}{(z-w)^2} J(w) + \frac{1}{(z-w)} \partial J(w) \\
\bar{T}(\bar{z})\bar{J}(\bar{w}) &\sim \frac{1}{(\bar{z}-\bar{w})^2} \bar{J}(\bar{w}) + \frac{1}{(\bar{z}-\bar{w})} \bar{\partial}\bar{J}(\bar{w}) \\
\bar{T}_F(\bar{z})\bar{J}(\bar{w}) &\sim \frac{\bar{\psi}(\bar{w})}{(\bar{z}-\bar{w})^2} + \frac{\bar{\partial}\bar{\psi}(\bar{w})}{(\bar{z}-\bar{w})} \\
\bar{T}_F(\bar{z})\bar{\psi}(\bar{w}) &\sim \frac{\bar{J}(\bar{w})}{(\bar{z}-\bar{w})}
\end{aligned} \tag{3}
$$

where ψ has dimension 1/2. These equations imply that $J(z)$ is a $(1, 0)$ conformal field, while $\bar{J}$ is the highest component of a $(0, 1/2)$ conformal superfield. The usual rules of conformal field theory then imply

$$
\langle J(z,\bar{z})J(w,\bar{w})\rangle = \frac{a}{(z-w)^2} \tag{4}
$$

which implies that the two-point function of $\bar{\partial}J$ is a derivative of a δ function. In a positive metric theory, this means that $\bar{\partial}J = 0$ as an operator. Similarly $\partial\bar{J} = 0$. Thus continuous symmetries of a heterotic vacuum state lead to holomorphically conserved currents and/or antiholomorphically conserved supercurrents. In either case, we can construct a BRST invariant vertex operator for the gauge boson associated with the symmetry

$$
V = \varepsilon^{\mu}(k)J(z)\, e^{ikx}\psi_{\mu}(\bar{z})\, e^{-\phi}
$$

or (5)

$$
\begin{aligned}
V &= \varepsilon^{\mu}(k)\bar{\psi}(\bar{z})\, e^{ikx}\, \partial x^{\mu}\, e^{-\phi} \\
k^2 &= k\cdot\varepsilon = 0
\end{aligned}
$$

where $\bar{\psi}$ is the dimension 1/2 superpartner of $\bar{J}$ and ϕ the bosonized superconformal ghost.

This is the proof that there are no global continuous symmetries in string theory. It is worth pausing for a moment to understand the major exception to this theorem: the Lorentz symmetry of the uncompactified dimensions. The currents of this symmetry have the form

$$
\begin{aligned}
J^{\mu\nu} &= x^{\mu}\, \partial x^{\nu} - x^{\nu}\, \partial x^{\mu} \\
\bar{J}^{\mu\nu} &= x^{\mu}\, \bar{\partial} x^{\nu} - x^{\nu}\, \bar{\partial} x^{\mu} + \bar{\psi}^{\mu}\bar{\psi}^{\nu}
\end{aligned} \tag{6}
$$

They satisfy (3) but not (4) and are conserved

$$
\bar{\partial}J^{\mu\nu} + \partial\bar{J}^{\mu\nu} = 0 \tag{7}
$$

without being holomorphically conserved. There are logarithms in the correlation functions of these operators, and vertex operators of the form

$$(\bar{\partial}x^\lambda + k \cdot \psi\psi^\lambda)J_{\mu\nu}\, e^{ikx} \quad \text{etc.} \tag{8}$$

are not BRST invariant. There are no gauge bosons associated with Lorentz invariance in string theory.

It is clear that the dimension 0 field x^μ is responsible for all of these curiosities. In conventional conformal field theory it is easy to prove that there are no dimension 0 conformal fields beside the identity. One simply takes the expectation value of the commutator

$$[L_{-1}, L_1] = 2L_0 \tag{9}$$

in the highest weight state $|h\rangle$ obtained by applying the dimension 0 field V to the vacuum. Positivity implies $L_{-1}|h\rangle = 0$ so that $|h\rangle$ is the SL_2 vacuum and $V = 1$. The existence* of x^μ in free field theory is really a consequence of the continuous spectrum of vertex operators e^{ikx} with arbitrarily small dimensions. Clearly we do not want such an occurrence for the *internal* conformal field theory (otherwise a four-dimensional observer would see a continuum of particle states with arbitrarily small masses). Let us define a compact conformal field theory to be one with a discrete spectrum of dimensions. If the internal part of the vacuum state is compact, then all continuous symmetries will be gauge symmetries.

An important question in heterotic string theory is whether the observed gauge interactions arise from currents on the world sheet which transform under the world sheet SUSY or from left-moving, purely "bosonic" currents. The remarkably simple answer to this question is that in phenomenologically acceptable vacua, gauge symmetries arise from bosonic currents. I will briefly outline the argument for this, which first appeared in Ref. 7.

The basic empirical input to the argument is the existence of chiral fermions in complex representations of the low-energy gauge group. If "mirror" partners to the quarks and leptons are found, this question will have to be reexamined. A set of currents belonging to dimension 1/2 superfields form a super Kac–Moody algebra (SKM). The superpartners of the currents are free fermions, and the whole algebra can be written in terms of these fermions and a set of currents that commute with them [7, 11].

Space-time fermions come from the Ramond sector of the theory. In this sector all fermionic components of superfields, and in particular the free fermions of the SKM have periodic boundary conditions on the cylinder. This has two simple consequences:

* Actually a truly careful field theorist would claim that x^μ does not exist.

(1) It is easy to obtain a lower bound on the dimensions of operators which shows that all massless space-time fermions are singlets of any non-Abelian SKM. This result, first proven by Friedan and Shenker [11], implies that the $SU_3 \times SU_2$ algebra of the low-energy world must be associated with the left-moving, bosonic currents.

(2) The existence of a $U(1)$ SKM in the spectrum does not preclude the existence of massless fermions. However, since it is simply a free superfield, the zero mode of its fermionic components enlarges the Dirac algebra of the four space-time fermions. The GSO projection no longer projects onto a fixed space-time chirality. Each fermion in a given representation of $SU_3 \times SU_2$ (which as we have seen is generated by left-moving currents) must come in both chiralities.

We conclude that in a heterotic vacuum with chiral fermions, all gauge symmetries are associated with left-moving Kac–Moody currents.

Within the semiclassical approximation to string theory, grand unification, if it can be said to occur at all, occurs only at the Planck scale (~string scale ~scale of compactification). Generally, Planck scale breaking of a GUT gives rather different results for the tree-level relation between the $SU(3, 2, 1)$ gauge couplings from those obtained by Georgi, Quinn, and Weinberg (GQW) [12]. The Lagrangian can have nonrenormalizable operators of the form

$$f^{ab}(\phi/M_p)F^a_{\mu\nu}F^b_{\mu\nu} \tag{10}$$

Here f^{ab} is a function of the Higgs fields ϕ. If $\langle\phi\rangle \sim M_p$, these terms generically give corrections of $O(1)$ to the GQW relations.

Let us try to understand the magnitude of ratios between gauge couplings in a general heterotic string vacuum. The vertex operators for the gauge bosons of some simple group (in the zero ghost charge picture) are

$$Nj^a(z)[\bar{\partial}x^\mu(\bar{z}) + k^\mu k \cdot \bar{\psi}(\bar{z})]\, e^{ik\cdot x} \tag{11}$$

where the currents $j^a(z)$ form a Kac–Moody algebra

$$j^a(z)j^b(w) \sim \frac{k\delta^{ab}}{(z-w)^2} + \frac{if^{abc}}{(z-w)}j^c(w) \tag{12}$$

If f^{abc} is normalized so that the roots of the algebra have length squared $=2$, then k, which is called the level of the algebra, must be an integer. For a $U(1)$ current k need not be quantized.

The normalization factor N for any one simple gauge group can be absorbed into the overall string coupling (dilaton expectation value). Let us choose to do this by setting $N = 1$ for the $SU(3)$ color group. How then do we fix the normalization of the other gauge-boson vertex operators?

Consider graviton emission from a single gauge-boson

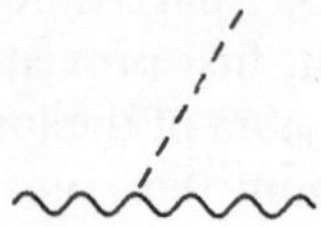

In string theory this process is governed by the coefficient of the graviton vertex operator in the product of two gauge-boson vertices. It is easy to see that this is proportional to $N_i^2 k_i$ for the ith gauge group. So we must choose $N^i = (k_3/k_i)^{1/2}$ in order to have universal gauge-boson–graviton couplings. It is then clear that the ith set of non-Abelian gauge bosons has a self-coupling $(k_3/k_i)^{1/2} f^{abc}$. For a $U(1)$ boson the situation is somewhat more subtle. The current has the form $i\sqrt{k_1}\partial\varphi$, where φ is a canonically normalized left-moving free scalar

$$\langle\varphi(z)\varphi(w)\rangle = -\ln(z-w)$$
$$\varphi = p\ln z + q + i\sum_n \frac{\alpha_n}{n} x^{-n} \tag{13}$$

If our internal conformal field theory is compact, p must be quantized

$$p = \frac{n}{R} \tag{14}$$

and the allowed charged vertex operators have the form

$$e^{i\pi mR\varphi}\tilde{V} \tag{15}$$

where $\tilde{V}$ is φ independent. The operator product expansion of the properly normalized gauge-boson vertex operator with an exponential

$$i\sqrt{k_3}\partial\varphi(z)\, e^{2\pi imR\varphi(w)} \sim \frac{2\pi mR\sqrt{k_3}}{z-w}\, e^{2\pi imR\varphi(w)} \tag{16}$$

tells us that the amplitude for emission of a gauge boson from a charged particle goes like $2\pi mR\sqrt{k_3}$; so this is the charge of the particle.

In order to have the tree-level GQW relations between the gauge couplings we must require that $k_2 = k_3$ and that $2\pi R\sqrt{k_3} = 1$. The charges of known particles are obtained if we assign m values to their $U(1)$ vertex operators as follows:

$$\begin{aligned} &qL, && m - 2 \\ &\bar{u}_R, && m = 8 \\ &d_R, && m = -4 \\ &l_L, && m = -3 \\ &e_R, && m = -6 \end{aligned} \tag{17}$$

There are a number of interesting aspects of these constraints. The $U(1)$ charges are quantized as they are in Kaluza–Klein theories. More surprisingly the ratio of SU_2 and SU_3 "fine structure constants" is always a rational number. This can happen in Kaluza–Klein theory [13], but as far as I know, it is not guaranteed. Note also that the possibility of a color singlet hypercharge 1/6 particle is not ruled out at this point. In GUT containing $SU(5)$, such a particle is not allowed. It may be that world sheet locality provides the same constraint in string theory, though we know too little of the structure of quark and lepton vertex operators to prove this. If not, there may be realistic string vacua that contain such exotic particles.

To obtain the GQW values for tree-level couplings we must "tune" the rational number k_2/k_3 and the continuous parameter R. Witten has argued that it is possible to obtain these values naturally by introducing Wilson line-breaking in a vacuum with grand unified symmetry. Actually, since k_2 and k_3 are integers, we only have a discrete set of choices. It is "technically natural" to set them equal. The most attractive choice from the phenomenological point of view is to set $k_2 = k_3 = 1$ since this automatically rules out light particles which are not in the fundamental representation of $SU(3) \times SU(2)$.

The possibility of continuous variation of R is more dangerous, even given Witten's demonstration of a natural set of vacua with the correct value. All known superstring vacua have many continuously variable parameters in them. From the low-energy point of view these are exact flat direction in the scalar potential. k_2 and k_3 cannot vary along these directions since they are integers. If R does, then string theory does not naturally explain the GQW relations.

To lowest order in perturbation theory a flat direction is represented by perturbing the vacuum conformal field theory by a $(1, 1)$ operator V which is the highest component of a $(0, 1)$ superfield. In order to represent an exact flat direction, the operator must be "truly marginal," i.e., the quantum field theory defined by adding $\lambda \int V$ to the action must be $(0, 1)$ superconformally invariant for all λ. We will not need to invoke this further constraint to prove our results.

How, in two-dimensional language, can the gauge coupling change? The perturbation V does not affect the space-time coordinates, so it can only affect the $U(1)$ current $j(z) = i\sqrt{k}\partial\varphi$ (φ is a canonically normalized holomorphic scalar field). The conformal fields that do not commute with $j(z)$ are $j(z)$ itself and $e^{iq\varphi}$ for $q = 2\pi mR$. The latter all carry $U(1)$ charge. If V contains them, it represents a vacuum in which $U(1)$ is spontaneously broken. We are discussing the possibility of a $U(1)$ preserving vacuum in which the gauge coupling varies. The only possible form for the operator V is thus

$$j(z)\bar{j}(z) \tag{18}$$

where $\bar{j}(\bar{z})$ is a $(0, 1)$ conformal field which is the highest component of a superfield. This is the definition of a $U(1)$ super-Kac–Moody algebra. The fermionic partner of $\bar{j}$ has a zero mode in the Ramond sector that enables both space-time chiralities of fermions to survive the GSO projection. As mentioned previously, the existence of chiral space-time fermions in the original vacuum precludes the existence of an SKM. We conclude that vacua with space-time chiral fermions have no perturbations which can change the gauge coupling. This, combined with Witten's discovery of vacua which naturally give the tree-level GQW relations, shows that string theory, like GUTS, can predict the Weinberg angle.

3. SPACE-TIME SUPERSYMMETRY IN STRING THEORY

The existence of space-time SUSY in string theory, like that of internal gauge symmetries, is a property of particular vacua. A vacuum will have SUSY if there exists (0, 1), BRST-invariant operators $q_\alpha(\bar{z})$, $q_{\dot{\alpha}}(\bar{z})$ such that $\oint q_{\dot{\alpha}}$ satisfy the SUSY algebra. In the FMS formalism for fermion vertex operators we must choose a "picture" or ground state for the superconformal ghosts. We will use the $-1/2$ picture for fermion vertices and the 0 and -1 pictures for bosons. The OPE for the q_α required by the SUSY algebra are

$$\begin{aligned} q_\alpha(\bar{z})q_\beta(\bar{w}) &\sim \text{nonsingular} \\ q_\alpha(\bar{z})q_{\dot{\beta}}(\bar{w}) &\sim \frac{1}{z-\bar{w}}\,\sigma^\mu_{\alpha\dot{\beta}}\psi_\mu(\bar{w})\,e^{-\phi} \end{aligned} \tag{19}$$

(ψ_μ are the four-dimensional NSR fermion fields; ϕ is the bosonized superconformal ghost; and the right-hand side is the -1 picture version of the momentum density).

In order to have the right ghost number and Lorentz transformation properties q_α must have the form

$$\begin{aligned} q_\alpha &= \Sigma(\bar{z})S_\alpha\, e^{-\phi/2} \\ q_{\dot{\alpha}} &= \Sigma^+(\bar{z})S_{\dot{\alpha}}\, e^{-\phi/2} \end{aligned} \tag{20}$$

where S_α and $S_{\dot{\alpha}}$ are dimension 1/4 spin field built from the four ψ_μ. Σ and Σ^+ must have dimension 3/8 and (19) constrains the singular terms in their OPE

$$\begin{aligned} \Sigma(\bar{z})\Sigma^+(\bar{w}) &\sim \frac{1}{(\bar{z}-\bar{w})^{3/4}} + \tfrac{1}{2}(\bar{z}-\bar{w})^{1/4}J(\bar{w}) \\ \Sigma(\bar{z})\Sigma(\bar{w}) &\sim (\bar{z}-\bar{w})^{3/4}O(w) \end{aligned} \tag{21}$$

The dimension of J is 1 and that of O is 3/2. In principle either or both of them could be zero. The singularity in the $\Sigma\Sigma^+$ OPE is there to ensure the correct OPE of SUSY generators of opposite chirality. The powers in the subleading terms differ from this by positive integers. Otherwise the SUSY generators would not be local. The OPE of the space-time part of the equal chirality SUSY generators has a singularity $1/(z-w)^{3/4}$ and the $\Sigma\Sigma$ OPE must cancel it to protect the SUSY algebra.

The heterotic string has a dimension 3/2 holomorphic field $T_F(z)$ that generates (0, 1) SUSY and must exist in any consistent vacuum. Our aim is to show that T_F combines with the $U(1)$ current J to form an $N=2$ SUSY algebra. We must first show that J exists. This is easy. The singularity structure of the OPE (21), combined with analyticity, enable us to calculate the four-point function of Σ. Expanding it as $\Sigma \to \Sigma^+$ we find a nonzero term with a power of the separation indicating a dimension 1 operator. The coefficient of this term determines the three-point function

$$\langle \Sigma(z_1)\Sigma^+(z_2)J(z_3)\rangle = \tfrac{3}{2}z_{12}^{1/4}z_{13}^{-1}z_{23}^{-1} \tag{22}$$

and the two-point function

$$\langle J(z)J(w)\rangle = \frac{3}{(z-w)^2} \tag{23}$$

J can then be written as $i\sqrt{3}\,\bar{\partial}H(\bar{w})$, with H a canonically normalized free field, and

$$\Sigma = e^{i\sqrt{3}/2H}V \tag{24}$$

where V commutes with H. Since Σ has dimension 3/8, V has dimension 0, and so is equal to 1 [the OPE (21) normalizes Σ] since V lives in a compact conformal field theory.

The next crucial point is that the SUSY currents must be BRST invariant. It is easy to see that there are no poles in any terms in the OPE of the BRST current with the SUSY generators except perhaps

$$\gamma(\bar{z})T_F(\bar{z})\begin{Bmatrix}\Sigma(\bar{w})\,e^{-\phi/2(\bar{w})}S_\alpha\\ \Sigma^+(\bar{w})\,e^{-\phi/2(\bar{w})}S_{\dot{\alpha}}\end{Bmatrix} \tag{25}$$

Σ creates states in the Ramond sector of the internal superconformal field theory. Its OPE with T_F has only 1/2 integer powers. Thus

$$T_F = \sum_{q=-\infty}^{\infty} e^{i(q/\sqrt{3})H}\tilde{T}_F^q = \sum T_F^q \tag{26}$$

where $\tilde{T}_F^q$ commutes with H. The SUSY charges will be BRST-invariant only if $q=\pm 1$. It is now easy to show that $T_F^\pm$, $\bar{T}$, and J close to form the $N=2$ superconformal algebra. This completes the argument of Ref. 2 that space-time SUSY is equivalent to $N=2$ world sheet SUSY.

Now assume that we perturb the vacuum along some flat direction. The perturbation is described by a $(1, 1)$ operator P, which must be the highest component of an $N = 1$ superfield. This is the zero momentum vertex operator for some massless scalar particle. It is written in the "picture" with zero ghost number where it only depends on matter fields. The corresponding vertex operator in the -1 picture is a $(1, 1/2)$ operator. This operator must be local with respect to the space-time SUSY generator in the $-1/2$ picture. Since it commutes with S_α and has the form $e^{-\phi X}$, it must be a sum of terms

$$\sum_{q=-\infty}^{\infty} e^{i(2q+1/\sqrt{3})H} X_q \tag{27}$$

But its antiholomorphic dimension is $1/2$, so $(2q+1)^2/6 < 1/2$, which means $2q + 1 = \pm 1$. This field is obtained by commuting the BRST charge with ξP (ξ is the "fermionized ghost field" [14]). The relevant term in the BRST operator is again

$$\int \gamma T_F \tag{28}$$

It carries $q = \pm 1$, so P has $q = 0$. Conformal invariance now restricts P to be

$$P_0(z, \bar{z}) + O(z)J(\bar{z}) \tag{29}$$

where P_0 commutes with J. But J is not the lowest component of an $N = 1$ superfield (its superpartner is the other superstress tensor), so the second term is absent.

Thus any acceptable perturbation of a space-time supersymmetric vacuum commutes with the $U(1)$ current of the $N = 2$ algebra and so preserves the space-time SUSY generators. Clearly, it also preserves the $U(1)$ charges of all operators in the theory. Thus the vertex operators remain local with respect to the SUSY charges and scattering amplitudes are invariant under SUSY. There is no way in heterotic string theory to obtain a "small" spontaneous breakdown of SUSY at the classical level, and hierarchical SUSY breaking must be a nonperturbative quantum effect.

The final topic mentioned in my introduction was the constraints on the world sheet theory that follow from extended space-time SUSY. Briefly, introduction of $N = 2$ or 4 SUSY generators implies 2 or 4 spin fields Σ_I. The same sort of analysis that we performed for $N = 1$ gives us several different $U(1)$ currents and some free world sheet fermions. For $N = 2$ we get two free superfields (one $N = 2$ world sheet superfield) direct product with an $SU(2)$ level 1 current algebra which completes the superstress tensor into a representation of the world sheet $N = 4$ superconformal algebra. With $N = 4$ in space-time we prove that the right-moving part of the theory is simply six free superfields. Modular invariance probably implies that we have a toroidal compactification. The details of these arguments will appear in Ref. 9.

4. CONCLUSIONS

We have seen an intricate and beautiful interplay between the world sheet and space-time properties of superstrings. Clearly, the key to these results was holomorphic symmetries of the world sheet theory. I expect that most of the nontrivial results that have been obtained about stringy phenomenology can be recast in this language. It is to be hoped that these kinds of general constraints will help us to understand the plethora of ground states with which string theory provides us.

REFERENCES

1. C. Becchi, A. Rouet, and R. Stora, *Phys. Lett. 58B*, 344 (1974); I. V. Tyutin, Lebedev preprint FIAN-39, 1975, unpublished.
2. D. Friedan, *Phys. Rev. Lett. 45*, 1057 (1980); *Ann. Phys.* (*N.Y.*) *163*, 318 (1985).
3. D. Friedan and S. Shenker, Talk at Aspen Summer Institute 1984.
4. C. Lovelace, *Phys. Lett. 135B*, 75 (1984); and talk at Princeton University, December, 1985.
5. R. Arnowitt, S. Deser, and C. W. Misner, in *Gravitation* (E. Witten, ed.), Wiley, New York, 1982, p. 227.
6. B. S. de Witt, *Phys. Rev. 160*, 1113 (1967).
7. D. Friedan and S. Shenker, unpublished; L. Dixon, V. Kaplunovsky, and C. Vafa, *Nucl. Phys. B 294*, 43 (1987).
8. D. Friedan, A. Kent, S. Shenker, and E. Witten, unpublished; A. Sen, *Nucl. Phys. B 278*, 289 (1986); L. Dixon, D. Friedan, E. Martinec, and S. Shenker, *Nucl. Phys. B 282*, 13 (1987).
9. T. Banks, L. Dixon, D. Friedan, and E. Martinec, SLAC-Pub-4377 (1987), *Nucl. Phys. B* (in press); T. Banks and L. Dixon, SCIPP Preprint 88/05 (1988).
10. V. Kaplunovsky, *Phys. Rev. Lett. 55*, 1033 (1985); M. Dine and N. Seiberg, *Phys. Rev. Lett. 55*, 366 (1985).
11. D. Friedan and S. Shenker, unpublished.
12. H. Georgi, H. Quinn, and W. Weinberg, in *Gauge Theories of Fundamental Interactions* (R. N. Mohapatra and C. H. Lai, ed.), World Scientific, Singapore (1974), p. 428; *Phys. Rev. Lett. 33*, 451 (1974).
13. S. Weinberg, in Proceedings of the Workship on Grand Unification, Philadelphia, 1984, p. 380.
14. D. Friedan, E. Martinec, and S. Shenker, *Nucl. Phys. B 271*, 93 (1986).

Chapter 3

Open Gauge Algebra and Ghost Unification

Laurent Baulieu

In this chapter, I will explain how the notion of an antifield allows one to solve the problem of quantizing the gauge invariant actions of charged 2-forms coupled to a Yang–Mills field in a four-dimensional space-time. I will also prove on this example that antifields can be geometrically unified in the same enlarged space as ordinary gauge fields and their ghosts.

The enlarged space in which BRST quantization is harmoniously described is a double complex $\mathscr{C}$ whose vertical direction is determined by the ghost number and the horizontal one by the usual form degree. In the BRST scheme a grading exists that is defined as the sum of the ghost number and of the form degree [1]. As an example, a Yang–Mills field valued in a given Lie algebra $\mathscr{G}$ is a 1-form with ghost number zero; The corresponding Faddeev–Popov ghost is a $\mathscr{G}$-valued 0-form with ghost number one. Therefore a Yang–Mills field and its Faddeev–Popov ghost are objects of a similar nature, which are separated only after a choice of coordinates within $\mathscr{C}$. In a double complex notation one can write the Yang–Mills field as A_1^0 and the Faddeev–Popov ghost as A_0^1, and it is consistent to unify A_1^0 and A_0^1 into the geometrical object $A_1^0 + A_0^1$. Similarly, the exterior differential $d = dx^\mu \partial \mu$ with form degree 1 and ghost number 0 and the BRST operator s with form degree 0 and ghost number 1 can be unified into the operator $d + s$ acting in $\mathscr{C}$.

LAURENT BAULIEU • Université Pierre et Marie Curie, 75252 Paris, Cedex 05, France.

If one generalizes the usual BRST structure of the Yang–Mills theory by allowing the introduction of new fields with negative ghost numbers, one must consider the existence of a $\mathcal{G}$-valued 2-form with ghost number -1, of a $\mathcal{G}$-valued 3-form with ghost number -2, and of a $\mathcal{G}$-valued 4-form with ghost number -3, denoted as A_2^{-1}, A_3^{-2}, and A_4^{-3}, respectively, since these objects can be consistently unified into the following generalized $\mathcal{G}$-valued 1-form in $\mathscr{C}$:

$$\tilde{A} = A_0^1 + A_1^0 + A_2^{-1} + A_3^{-2} + A_4^{-3} \tag{1}$$

We call the new fields with negative ghost number antifields: Indeed, they can be identified with the additional fields that Batalin and Vilkoviski have introduced for quantizing actions invariant under systems of gauge transformations that close only up to classical equations of motion [2]. The antifields occurring in (1) will turn out to be useful for building the gauge symmetry characterizing a charged 2-form.

Consider now a classical $\mathcal{G}$-valued 2-form gauge field $B_2 = \frac{1}{2}B_{\mu\nu}\, dx^\mu\, dx^\nu$. The field-strength of B_2 is

$$G_3 = \tfrac{1}{6}G_{\mu\nu\rho}\, dx^\mu\, dx^\nu\, dx^\rho = dB_2 + [A, B_2] = DB_2 \tag{2}$$

In order to couple B_2 with the Yang–Mills field A, one introduces the following action:

$$I_{\rm cl} = \int (B_2 F + F^* F) = \int d^4x\, (\varepsilon_{\mu\nu\rho\sigma} B^{\mu\nu} F^{\rho\sigma} + F_{\mu\nu} F^{\mu\nu}) \tag{3}$$

where $F = dA + AA$. In our notation all products are graded by the sum of the ghost number and of the form degree. The asterisk stands for the duality operation. $I_{\rm cl}$ is invariant under the infinitesimal gauge transformations:

$$\begin{aligned} \delta A_\mu &= D_\mu \varepsilon \\ \delta B_{\mu\nu} &= D_{[\mu}\varepsilon_{\nu]} + [B_{\mu\nu}, \varepsilon] \end{aligned} \tag{4}$$

In these equations ε is the known $\mathcal{G}$-valued local 1-form parameter for an infinitesimal Yang–Mills transformation. The quantization of the action (3) is troublesome because the infinitesimal 1-form parameter ε_ν occurring in the transformation law of B_2 is degenerate modulo the field equation of B_2, $F = 0$:

$$\varepsilon_\nu \sim \varepsilon_\nu + D_\nu \alpha \tag{5}$$

This degeneracy has the following consequence. In the usual BRST scheme one introduces a 1-form ghost B_1^1 in correspondence with the local infinitesimal 1-form parameter ε_ν and a 0-form ghost of ghost B_0^2 in correspondence with the local infinitesimal 0-form parameter α. Introducing the ghost of ghost B_0^2 is a necessity since the primary ghost B_1^1 is itself a gauge field due to the degeneracy (5). One discovers then that there is no way to build a BRST operator s acting only on the fields A, c, B_2, B_1^1, and B_0^2 with the requirement of nilpotency independently of any restriction on A [2]. As a matter of fact, starting from the condition that the BRST transformations of A and B_2 reproduce the gauge transformations (4) up to the changes $\varepsilon \to c$ and $\varepsilon_\mu \to B_\mu^1$, one finds the following expression for the action of s:

$$\begin{aligned} sA &= -Dc \\ sc &= -\tfrac{1}{2}[c, c] \\ sB_2 &= -DB_1^1 - [c, B_2] \\ sB_1^1 &= -DB_0^2 - [c, B_1^1] \\ sB_0^2 &= -[c, B_0] \end{aligned} \tag{6}$$

A straightforward computation shows that the nilpotency of s is broken on B_0^2:

$$s^2 B_0^2 = [B_0^2, F] \tag{7}$$

Notice that the nilpotency breaking is proportional to the Yang–Mills curvature F, i.e., to the equation of motion of B_2 stemming from the action (3).

The origin of this problem can be recognized as an incompatibility between the BRST equations (6) and their Bianchi identities: Equation (6) can be equivalently written as

$$(s + d)(A + c) + \tfrac{1}{2}[A + c, A + c] = F \tag{8a}$$

$$(s + d)(B_2 + B_1^1 + B_0^2) + [A + c, B_2 + B_1^1 + B_0^2] = G_3 \tag{8b}$$

The property $(d + s)^2 = s^2 = 0$ amounts to the fulfilment of the Bianchi identities of (8):

$$(d + s)F + [A + c, F] = 0 \tag{9a}$$

$$(d + s)G_3 + [A + c, G_3] = [F, B_2 + B_1^1 + B_0^2] \tag{9b}$$

It is in fact obvious by expansion in ghost number that equation (9b) is true if and only if $F = 0$.

Having a BRST operator that is nilpotent only up to a field equation as seen either in (7) or in (9) is geometrically absurd. Moreover, this forbids the construction of a consistent gauge fixed action.

The root of the problem is the fact that we have considered a set of fields that is not large enough. We now allow for the existence of geometrical objects with negative ghost numbers, i.e., of antifields, and define the following generalized gauge fields:

$$\tilde{A} = A_0^1 + A_1^0 + A_2^{-1} + A_3^{-2} + A_4^{-3}$$
$$\tilde{\mathbf{A}}_2 = \mathbf{A}_3^{-1} + \mathbf{A}_4^{-2} \qquad (10)$$
$$\tilde{B}_2 = B_0^2 + B_1^1 + B_2$$

The field strengths are defined as follows:

$$\tilde{F} = (s + d)\tilde{A} + \tfrac{1}{2}[\tilde{A}, \tilde{A}]$$
$$\tilde{\mathbf{F}}_3 = (s + d)\tilde{\mathbf{A}}_2 + [\tilde{A}, \tilde{\mathbf{A}}_2] \qquad (11)$$
$$\tilde{G}_3 = (s + d)\tilde{B}_2 + [\tilde{A}, \tilde{B}_2]$$

The property $s^2 = (d + s)^2 = 0$ on all the field components of $\tilde{A}, \tilde{\mathbf{A}}_2$, and $\tilde{B}_2$ which are displayed in (10) is equivalent to the fulfillment of the Bianchi identities of $\tilde{F}, \tilde{\mathbf{F}}_3$, and $\tilde{G}_3$:

$$(d + s)\tilde{F}_3 + [\tilde{A}, \tilde{F}] = 0$$
$$(d + s)\tilde{\mathbf{F}}_3 + [\tilde{A}, \tilde{\mathbf{F}}_3] = [\tilde{F}, \tilde{\mathbf{A}}_2] \qquad (12)$$
$$(d + s)\tilde{G}_3 + [\tilde{A}, \tilde{G}_3] = [\tilde{F}, \tilde{B}_2]$$

The knowledge of the action of s on the various fields amounts to that of a set of constraints on the components with different ghost numbers in $\tilde{F}, \tilde{\mathbf{F}}_3$, and $\tilde{G}_3$. In order to obtain a BRST operator that is nilpotent independently of any restriction on the field components in $\tilde{\mathrm{A}}, \tilde{\mathbf{A}}_2$, and $\tilde{B}_2$ it is necessary and sufficient that these constraints be compatible with the Bianchi identities (12). A straightforward inspection determines the following constraints on $\tilde{F}$ and $\tilde{G}_3$:

$$\tilde{F} = 0$$
$$\tilde{G}_3 = \tilde{G}_3^0 + \tilde{G}_4^{-1} \qquad (13)$$

Besides, $\tilde{\mathbf{F}}_3$ must be constrained as follows:

$$\tilde{\mathbf{F}}_3 = \mathcal{F}_3^0 + \tilde{\mathbf{F}}_4^{-1} \qquad (14)$$

where $\mathcal{F}_3^0[A]$ is any given gauge covariant quantity, for instance:

$$\tilde{\mathbf{F}}_3 = \delta/\delta A(F^*F) + \tilde{\mathbf{F}}_4^{-1} = D^*F + \tilde{\mathbf{F}}_4^{-1} \qquad (15)$$

Notice that $\tilde{\mathbf{A}}_2$ and $\tilde{B}_2$ can be further unified into $\tilde{\mathbf{B}}_2 = \tilde{\mathbf{A}}_2 + \tilde{B}_2$. One can then define

$$\tilde{\mathbf{B}}_2 = \mathbf{A}_4^{-2} + \mathbf{A}_3^{-1} + B_2 + B_1^1 + B_0^2$$
$$\tilde{\mathbf{G}}_3 = (s + d)\tilde{\mathbf{B}}_2 + [\tilde{A}, \tilde{\mathbf{B}}_2] \qquad (16)$$

At this level, the unification of fields and antifields has been completed.

The Bianchi identity (12) is now $(d+s)\tilde{F}+[\tilde{A},\tilde{F}]=0$ and $(d+s)\tilde{\mathbf{G}}_3+[\tilde{A},\tilde{\mathbf{G}}_3]=[\tilde{F},\tilde{\mathbf{B}}_2]$ and the BRST equations (13–15) can be thus written under the following compact form

$$\tilde{F}=0$$
$$\tilde{\mathbf{G}}_3=\delta/\delta A(FB_2+F^*F)+\tilde{\mathbf{G}}_4^{-1}=G_3+D^*F+\tilde{\mathbf{G}}_4^{-1} \tag{17}$$

By expanding equations (13)–(15) or (17) in ghost number, one finds the following expressions for the action of s on the fields:

$$\begin{aligned}
sA &= -Dc\\
sc &= -\tfrac{1}{2}[c,c]\\
sB_2 &= -DB_1^1-[c,B_2]+[A_2^{-1},B_0^2]\\
sB_1^1 &= -DB_0^2-[c,B_1^1]\\
sB_0^2 &= -[c,B_0^2]\\
sA_2^{-1} &= F-[c,A_2^{-1}]\\
sA_3^{-2} &= -DA_2^{-1}-[c,A_3^{-2}]\\
sA_4^{-3} &= -DA_3^{-2}-\tfrac{1}{2}[A_2^{-1},A_2^{-1}]-[c,A_4^{-3}]\\
s\tilde{\mathbf{A}}_3^{-1} &= D^*F-[c,\tilde{\mathbf{A}}_3^{-1}]\\
s\tilde{\mathbf{A}}_4^{-2} &= -D\tilde{\mathbf{A}}_3^{-1}-[c,\tilde{\mathbf{A}}_4^{-2}]
\end{aligned} \tag{18}$$

The nilpotency of s is obvious by construction from the compatibility of the Bianchi identities (12) with the BRST equations (13), (15). It can be also checked from equations (18).

The most interesting equation in (18) is the one defining the action of s on the classical 2-form gauge field B_2. Indeed, the BRST transformation law of B_2 contains the term $[A_2^{-1},B_0^2]$. This term is necessary for consistency, i.e., for ensuring the nilpotency of the BRST operator. However, there is no classical interpretation for such a term, owing to its nonlinear dependence on the ghost fields.

Because of the necessary modification of the gauge symmetry, the original action (3) is not invariant under the BRST symmetry (16). The full s-invariant action is in fact the following one:

$$\begin{aligned}
\mathcal{I}-\int \mathrm{Tr}\,\{&B_2F+F^*F+\tilde{\mathbf{A}}_3^{-1}Dc-\tfrac{1}{2}\tilde{\mathbf{A}}_4^{-2}[c,c]\\
&-A_2^{-1}(DB_1^1+[c,B_2]-\tfrac{1}{2}[A_2^{-1},B_0^2])\\
&-A_3^{-2}(DB_0^2+[c,B_1^1])-A_4^{-3}[c,B_0^2]\}
\end{aligned} \tag{19}$$

This action is ghost and antifield dependent. $\mathscr{I}$ represents the solution modulo d-exact terms of the equation $s\Delta_4^0 = 0$, where Δ_4^0 is a 4-form with ghost number 0. Therefore, it should be considered as the classical invariant action for a charged 2-form gauge field B_2, although it is ghost dependent.

Since we have at our disposal a nilpotent BRST operator it is straightforward to compute the quantized version of the classical action (3). By using the standard BRST technology, i.e., by introducing the relevant Lagrange multipliers and antighosts, one can easily derive the following BRST invariant and gauge fixed action:

$$\begin{aligned} I_{\text{GF}} = \int d^4x\{ & \varepsilon_{\mu\nu\rho\sigma} B^{\mu\nu} F^{\rho\sigma} + F_{\mu\nu}F^{\mu\nu} + \alpha(\partial^\mu A_\mu)^2 + \bar{c}\partial^\mu D_\mu c \\ & \times \beta((\partial^\mu B_{[\mu\nu]})^2 + (\partial^\mu B_0^0)^2) + D_{[\mu} B^1_{\nu]} \partial_{[\mu} \bar{B}^{-1}_{\nu]} + \gamma \partial^\mu \bar{B}^{-1}_\mu \partial^\nu B^1_\nu \\ & + \partial^\mu \bar{B}_0^{-2} D_\mu B_0^2 + \varepsilon_{\mu\nu\rho\sigma} B_0^2 [\partial_\mu \bar{B}_\nu^{-1}, \partial_\rho \bar{B}_\sigma^{-1}]\} \end{aligned} \tag{20}$$

α, β, and γ are arbitrarily chosen real gauge parameters. The interesting characteristic of the action (20), which is the simplest consistent gauge fixed version of the action (19), is the presence of a cubic ghost interaction which forbids the elimination of ghost fields under the form of a determinant in the partition function. More details on the derivation of the action (20) can be found in Ref. 3.

To conclude this talk let us stress that more complicated open gauge algebras than that of the non-Abelian 2-form gauge field can be analyzed with techniques similar to the ones that I have presented. It is in particular interesting to apply these techniques to supergravity since this permits one to bypass the method of auxiliary fields for defining the partition function. Thereby recent progress has been made in the cases of $N = 1$ supergravity in 4 and 11 dimensions [4]. In string field theory, the notion of an antifield is also important in order to interpret consistently the ghost structure [3, 5]. Finally, trying to include in a geometrical framework these fields with negative ghost number that permit systematically the closure of open gauge algebra seems an interesting challenge.

REFERENCES

1. L. Baulieu, Cargese lectures 1983, in *Perspectives in Fields and Particles* (J. L. Basdevant and M. Levy, eds.), Plenum Press, New York, 1984.
2. I. A. Batalin and G. A. Vilkovisky, *Phys. Rev. D 28*, 2567 (1983).
3. L. Baulieu, I. Sezgin, and E. Bergshoeff, *Nucl. Phys. B 307*, 348 (1988).
4. L. Baulieu and M. Bellon, in preparation.
5. R. Thorn, *Nucl. Phys. B 257*, 61 (1987); M. Bocchichio, *Phys. Lett. 188B*, 332 (1987); *193B*, 31 (1987); S. P. De Alvis, M. Grisaru, and L. Mezincescu, *ibid. 190B*, 122 (1987).

Chapter 4

Algebras of the Virasoro, Neveu–Schwarz, and Ramond Types on Genus g Riemann Surfaces

L. Bonora, M. Rinaldi, and J. Russo

1. INTRODUCTION

During the last few years much work has been devoted to trying to reconcile the operator formalism in conformal field theory and string theory, which was originally formulated in the complex plane, with the fact that in Polyakov string theory Riemann surfaces of arbitrary genus must be taken into account. Also, independently of any string interpretation, it is interesting to know the features of a conformal field theory on a Riemann surface of genus g, and therefore it is important to have a manageable operator formalism for any genus. In two recent papers [1, 2] Krichever and Novikov have introduced a new general formalism that may prove very important in this sense. The basic ingredient in their approach is a discrete basis for the algebra of meromorphic vector fields over a Riemann surface that are holomorphic outside two distinguished points. The basis elements form a closed algebra, which is referred to as the Krichever–Novikov (KN) algebra.

What is new and remarkable in Refs. 1 and 2 is first of all that the existence of a unique discrete basis suggests immediately what the operator

---

L. Bonora and M. Rinaldi • S.I.S.S.A., Trieste and I.N.F.N., Sezione di Trieste, Italy. J. Russo • S.I.S.S.A., Trieste, Italy.

formalism should be. Moreover, the KN algebra has a universal aspect, but simultaneously preserves the information of the genus in the structure constants. Thirdly, it permits us to treat diffeomorphisms and deformations on the same footing.

In this chapter we will review some work done using the Krichever–Novikov approach [3, 4]. In particular we show that one can explicitly construct in the bosonic string case [3] a BRST charge over any Riemann surface and show that it is nilpotent in the critical ($D = 26$) dimension. Then we show that one can extend the construction of Krichever and Novikov and of Ref. 3 so as to generalize the Neveu–Schwarz and Ramond algebras to Riemann surfaces of arbitrary genus, construct a corresponding BRST charge, and recover the expected ($D = 10$) critical dimension in the superstring case.

In more detail, following the procedure of Krichever and Novikov, we will construct bases of suitably defined mathematical objects over an arbitrary Riemann surface, as well as natural binary operations among them that define two types of super-Virasoro algebras. In the same way as in the genus 0 case the KN algebra boils down to the Virasoro algebra, these superalgebras reduce either to the Neveu–Schwarz or the Ramond algebras. For this reason we will call them NS–KN algebras and R–KN algebras, respectively. Next we will define the relevant central extensions. Once this is done, we will realize the above superalgebras as algebras of the "momenta" of the energy momentum tensor and of the supercurrent ensuing from the classical Poisson brackets in a superstring theory. Finally, we will consider the expansion coefficients as operators acting on suitable Fock spaces. Such realizations give rise, via normal ordering, to central extensions. In order to define a nilpotent BRST charge we will need to introduce suitable ghosts. The matching of the ghost contribution with the "matter" contribution to yield a nilpotent charge will be, as usual, the origin of the critical ($D = 10$) dimension.

2. THE BASES

Given a Riemann surface Σ of genus g, let us consider two distinguished (but generic) points P_+ and P_- and local coordinates z_+ and z_- around them such that $z_\pm(P_\pm) = 0$. On Σ there exists a whole family of tensors $f_j^{(\lambda,x)}$, parametrized by two real numbers λ (the conformal weight) and x. The label j is discrete (see below). The $f_j^{(\lambda,x)}$ are holomorphic everywhere on Σ except possibly for poles or branch points in P_+ and P_- and a branch cut from P_+ to P_-. The limiting values at the cut satisfy

$$f_j^{(\lambda,x)+} = e^{2\pi i x} f_j^{(\lambda,x)-} \tag{1}$$

Moreover, the expansion of $f_j^{(\lambda,x)}$ near $P_\pm$ is of the form

$$f_j^{(\lambda,x)} = a_j^{(\lambda,x)\pm} z_\pm^{\pm j \pm x - S(\lambda)}[1 + O(z_\pm)](dz_\pm)^\lambda \tag{2}$$

where $S(\lambda) = g/2 - \lambda(g-1)$ and $a_i^{(\lambda,x)\pm}$ are constants to be specified below. In particular [1, 2], for $\lambda = -1$, $x = 0$ we obtain meromorphic vector fields $e_j \equiv f_j^{(-1,0)}$, holomorphic outside $P_\pm$. In this case j is integer, $j = \ldots, -1, 0, 1, \ldots$; or half integer, $j = \ldots, -\frac{1}{2}, \frac{1}{2}, \ldots$, according to whether g is even or odd.

For later use we recall that for $x = 0$, $\lambda = 0, 1, 2$ we obtain meromorphic functions $A_j \equiv f_j^{(0,0)}$, differentials $\omega_j \equiv f_{-j}^{(1,0)}$, and quadratic differentials $\Omega_j \equiv f_{-j}^{(2,0)}$, respectively, which are holomorphic except at $P_\pm$. The definition of A_j and ω_j must be slightly modified for $|j| \leq g/2$ owing to the Weierstrass theorem. In all these cases the Riemann-Roch theorem tells us that e_j, A_j, ω_j, and Ω_j are uniquely determined up to an arbitrary constant. So we normalize them by setting $a_i^{(\lambda,x)+} = 1$; $a_i^{(\lambda,x)-}$ will then be uniquely determined, as well as all the coefficients appearing in the tails $O(z_\pm)$. These coefficients contain the dependence on the genus.

The other objects we need are those with half-integer weight λ and $x = \frac{1}{2}$ or $x = 0$ (the latter characterization will define, respectively, the Neveu-Schwarz and Ramond sector). In particular let us consider the $\lambda = -\frac{1}{2}$ case. Let us set $g_\alpha \equiv f_j^{(-\frac{1}{2},0)}$ with $\alpha = j$ integer and $g_\alpha \equiv f_{j+\frac{1}{2}}^{(-\frac{1}{2},\frac{1}{2})}$ with $\alpha = j + \frac{1}{2}$ half integer. That is, collectively,

$$g_\alpha(z_\pm) = a_\alpha^\pm z_\pm^{\pm\alpha - g + \frac{1}{2}}[1 + O(z_\pm)](dz_\pm)^{-\frac{1}{2}} \tag{3}$$

With α integer g_α is holomorphic outside $P_\pm$ (Ramond sector), while when α is half integer g_α has a branch cut from P_+ to P_- (Neveu–Schwarz sector). For later use, we define also $k_\alpha \equiv f_j^{(\frac{3}{2},0)}$ with $\alpha = -j$ integer and $K_\alpha \equiv f_{j+\frac{1}{2}}^{(\frac{3}{2},\frac{1}{2})}$ with $\alpha = -j - \frac{1}{2}$ half integer, and $h_{-\alpha} = f_j^{(\frac{1}{2},0)}$ with $\alpha = j$ integer and $h_\alpha = f_{j+\frac{1}{2}}^{(\frac{1}{2},\frac{1}{2})}$ with $\alpha = j + \frac{1}{2}$ half integer.

Let us come now to the binary operations, which will allow us to define a superalgebra. Let us concentrate on the e_i's and the g_α's. As for the first we take the Lie bracket [1, 2] $[e_i, e_j]$, for the latter we have the tensor product of sections and set $\{g_\alpha, g_\beta\} \equiv g_\alpha g_\beta + g_\beta g_\alpha$. Finally we set $[e_i, g_\alpha] \equiv L_{e_i} g_\alpha$, where $L_e g = [\varepsilon(z)\partial\gamma(z) + \lambda\gamma(z)\partial\varepsilon(z)](dz)^\lambda$ in a local patch where $e = \varepsilon(z)\partial/\partial z$ and $g = \gamma(z)(dz)^\lambda$. For integer λ, L_e reduces to the Lie derivative along the vector field e. Then from an analysis of the singularities in $P_\pm$ we obtain

$$[e_i, e_j] = \sum_{s=-g_0}^{g_0} C_{ij}^s e_{i+j-s}, \qquad g_0 \equiv \tfrac{3}{2}g \tag{4a}$$

$$[e_i, g_\alpha] = \sum_{s=-g_0}^{g_0} H_{i\alpha}^s g_{i+\alpha-s} \tag{4b}$$

$$\{g_\alpha, g_\beta\} = \sum_{p=-g}^{g} B^p_{\alpha\beta} e_{\alpha+\beta-p/2} \tag{4c}$$

The coefficients C^s_{ij}, $H^s_{i\alpha}$, $B^p_{\alpha\beta}$ can be calculated from the constants appearing in the expansion of e_i and g_α near $P_\pm$. For example, in the simplest case, we have $C^{g_0}_{ij} = j - i$, $H^{g_0}_{i\alpha} = \alpha - i/2 - g + g_0/2$, $B^g_{\alpha\beta} = 2$.

Equation (4a) defines the KN algebra, while equations (4) together define the NS-KN superalgebra or the R-KN superalgebra for $\alpha, \beta, \gamma, \ldots$ integer or half integer, respectively. We will denote by $\mathcal{M}_\Sigma$ the algebra generated by the e_i's through equation (4a), and by $\mathcal{A}_\Sigma$ the superalgebra generated by the e_i's and the g_α's through equations (4). The algebra $\mathcal{M}_\Sigma$ splits according to

$$\mathcal{M}_\Sigma = \mathcal{M}^+_\Sigma + \mathcal{M}^0_\Sigma + \mathcal{M}^-_\Sigma$$

where $\mathcal{M}^\pm_\Sigma$ are the subalgebras generated by the e_i's with $\pm i \geq g_0 - 1$, and generate diffeomorphisms. The complement $\mathcal{M}^0_\Sigma$ generated by the e_i with $|i| \leq g_0 - 2$ corresponds to deformations that change the conformal structure; its complex dimension is $3g - 3$ and it is naturally identified with the tangent space to the moduli space.

Similarly the algebra $\mathcal{A}_\Sigma$ splits according to

$$\mathcal{A}_\Sigma = \mathcal{A}^+_\Sigma + \mathcal{A}^0_\Sigma + \mathcal{A}^-_\Sigma$$

where $\mathcal{A}^\pm_\Sigma$ are the superalgebras generated by e_i with $\pm i \geq g_0 - 1$ and g_α with $\pm\alpha \geq g - \frac{1}{2}$. These generate superconformal transformations. The complement $\mathcal{A}^0_\Sigma$ generated by e_i with $|i| \leq g_0 - 2$ and g_α with $|\alpha| < g - \frac{1}{2}$ correspond to deformations that change the superconformal structure. $\mathcal{A}^0_\Sigma$ is naturally identified with the tangent space to the supermoduli space. One can easily see that the complex dimension of $\mathcal{A}^0_\Sigma$ is $3g - 3 + 2g - 2$, the dimension of the supermoduli space.

3. THE CENTRAL EXTENSIONS

In order to define the central extensions of the KN algebra and of the GS-KN and R-KN superalgebras, let us introduce the following cocycles:

$$\chi(e_i, e_j) = \frac{1}{24\pi i} \oint \tilde{\chi}(e_i, e_j) \tag{5}$$

$$\varphi(g_\alpha, g_\beta) = \frac{1}{6\pi i} \oint \tilde{\varphi}(g_\alpha, g_\beta) \tag{6}$$

where the integral is over a contour surrounding P_+ and $\tilde{\chi}$ and $\tilde{\varphi}$ are defined as follows.

Let f and g be meromorphic vector fields which are holomorphic outside $P_\pm$, and let $f = f(z_+)\partial/\partial z_+$ and $g = g(z_+)\partial/\partial z_+$ near P_+; then

$$\tilde{\chi}(f, g) = [\tfrac{1}{2}(f'''g - g'''f) - R(f'g - fg')]\, dz_+ \tag{7}$$

where R is a Schwarzian connection. Likewise let ρ and σ have weight $-1/2$ and be holomorphic on Σ except possibly for poles or branch points in $P_\pm$ (with associated branch cut), and let $\rho = \rho(z_+)(dz_+)^{-1/2}$, $\sigma = \sigma(z_+)(dz_+)^{-1/2}$. Then

$$\tilde{\varphi}(\rho, \sigma) = \rho'\sigma'\, dz_+ \tag{8}$$

It is immediate to see that they verify the following properties:

i. $\chi(e_i, e_j) = -\chi(e_j, e_i)$, $\varphi(g_\alpha, g_\beta) = \varphi(g_\beta, g_\alpha)$.

ii. They are independent of the coordinate system [for $\chi(f, g)$ this follows from the properties of R].

iii. They satisfy the following cocycle conditions:

$$\chi(f, [g, h]) + \chi(g, [h, f]) + \chi(h, [f, g]) = 0$$
$$\varphi(\rho, [\sigma, f]) - \varphi(\sigma, [f, \rho]) + \chi(f, \{\rho, \sigma\}) = 0$$

iv. They are "local," in the sense that

$$\chi(e_i, e_j) = 0 \qquad \text{for } |i + j| > 3g \tag{9a}$$
$$\varphi(g_\alpha, g_\beta) = 0 \qquad \text{for } |\alpha + \beta| > 2g \tag{9b}$$

as follows from an elementary computation of the zeros and poles in $P_\pm$.

So finally we can centrally extend both NS–KN and R–KN superalgebras as follows:

$$[e_i, e_j] = \sum_{s=-g_0}^{g_0} C_{ij}^s e_{i+j-s} + t\chi(e_i, e_j) \tag{10a}$$

$$[e_i, g_\alpha] = \sum_{s=-g_0}^{g_0} H_{i\alpha}^s g_{i+\alpha-s} \tag{10b}$$

$$\{g_\alpha, g_\beta\} = \sum_{p=-g}^{g} B_{\alpha\beta}^p e_{\alpha+\beta-p/2} + t\varphi(g_\alpha, g_\beta) \tag{10c}$$

$$[e_i, t] = [g_\alpha, t] = 0 \tag{10d}$$

Of course equation (10a) defines the central extension of the KN algebra. A few final remarks:

- The cocycles χ and φ are easily calculated in a few cases. For example, for $R = 0$,

$$\chi(e_i, e_{3g-i}) = \tfrac{1}{12}[(i - g_0)^3 - (i - g_0)] \tag{11a}$$
$$\varphi(g_\alpha, g_{2g-\alpha}) = -\tfrac{1}{3}(\alpha - g)^2 + \tfrac{1}{12} \tag{11b}$$

- It has been shown by Krichever and Novikov [1] that up to trivial cocycles there is only one cocycle satisfying the "locality" condition (9a).
- The above superalgebras reduce to the usual Virasoro, Neveu-Schwarz, and Ramond superalgebras in the genus 0 case.

4. A STRING REALIZATION

Now we want to realize the above algebras as intrinsic algebras of (super)string theories, first from a classical and then from a quantum point of view. In the following the superstring case will be treated explicitly. The purely bosonic case can be easily recovered by setting to zero the fermionic variables and the corresponding ghosts. We start from the energy momentum tensor, which, in local coordinates, is given by $T = T^{X\psi} + T^{\mathrm{gh}}$

$$T^{X\psi} \equiv -\partial X^\mu \partial X_\mu - \tfrac{1}{2}\partial\psi^\mu\psi_\mu, \qquad T^{\mathrm{gh}} \equiv c\partial b + 2\partial c b - \tfrac{1}{2}\gamma\partial\beta - \tfrac{3}{2}\partial\gamma\beta \tag{12}$$

and the supersymmetric current $J = J^{X\psi} + J^{\mathrm{gh}}$

$$J^{X\psi} = \psi_\mu \partial X^\mu, \qquad J^{\mathrm{gh}} = 2c\partial\beta + 3\partial c\beta - \gamma b \tag{13}$$

$X^\mu(Q)$ and $\psi^\mu(Q)$ are fields of weight 0 and $\frac{1}{2}$, respectively. $b(Q)$ and $c(Q)[\beta(Q)$ and $\gamma(Q)]$ are anticommuting (commuting) ghost fields of weight 2 and -1 ($\frac{3}{2}$ and $-\frac{1}{2}$), respectively. T and J have weight 2 and $\frac{3}{2}$.

We can use the bases $\{e_i\}$, $\{\omega_i\}$, etc., $\{g_\alpha\}$, $\{h_\alpha\}$, etc., introduced above in order to expand these fields. The coefficients will be later interpreted as creation and annihilation operators in suitable Fock spaces:

$$\lambda = -1: \qquad c(Q) = \sum_i c_i e_i(Q), \qquad c_i = \frac{1}{2\pi i}\oint_{C_\tau} c(Q)\Omega_i(Q) \tag{14a}$$

$$\lambda = 2: \qquad b(Q) = \sum_i b_i \Omega_i(Q), \qquad b_i = \frac{1}{2\pi i}\oint_{C_\tau} b(Q) e_i(Q) \tag{14b}$$

$$\lambda = -\tfrac{1}{2}: \qquad \gamma(Q) = \sum_\alpha \gamma_\alpha g_\alpha(Q), \qquad \gamma_\alpha = \frac{1}{2\pi i}\oint_{C_\tau} \gamma(Q) k_\alpha(Q) \tag{14c}$$

$$\lambda = \tfrac{1}{2}: \qquad \psi^\mu(Q) = \sum_\alpha d^\mu_\alpha h_\alpha(Q), \qquad d^\mu_\alpha = \frac{1}{2\pi i}\oint_{C_\tau} \psi^\mu(Q) h^+_\alpha(Q) \tag{14d}$$

$$\lambda = \tfrac{3}{2}: \qquad \beta(Q) = \sum_\alpha \beta_\alpha k_\alpha(Q), \qquad \beta_\alpha = \frac{1}{2\pi i}\oint_{C_\tau} \beta(Q) g_\alpha(Q) \tag{14e}$$

$$\begin{gathered}\lambda = 1: \qquad dX^\mu(Q) + P^\mu(Q) = \sqrt{2}\sum_i \alpha^\mu_i \omega_i(Q) \\ \sqrt{2}\alpha^\mu_i = \frac{1}{2\pi i}\oint_{C_\tau} [dX^\mu(Q) + P^\mu(Q)] A_i(Q)\end{gathered} \tag{14f}$$

where P^μ is the conjugate momentum of X^μ and C_τ are the level curves of a suitable univalent function $\tau(Q)$ over Σ. These level curves can be interpreted as representing closed string configurations on the Riemann surface and τ as a proper time. As τ tends to $\pm\infty$, C_τ tends to a circle around $P_\pm$. Now we introduce the Poisson brackets

$$[X^\mu(Q), P^\nu(Q')] = 2\pi\eta^{\mu\nu}\Delta_\tau(Q, Q'), \qquad Q, Q' \in C_\tau \tag{15a}$$

$$\{\psi^\mu(Q), \psi^\nu(Q')\} = 2\pi\eta^{\mu\nu}\delta_\tau(Q, Q') \tag{15b}$$

$$\{c(Q), b(Q')\} = 2\pi D_\tau(Q, Q') \tag{15c}$$

$$[\gamma(Q), \beta(Q')] = 2\pi d_\tau(Q, Q') \tag{15d}$$

The symbols in the right-hand side play the role of δ functions over C_τ for smooth tensors of weight $0, \frac{1}{2}, -1, -\frac{1}{2}$, respectively. For example, for a generic smooth function $f(Q)$ over C_τ we have

$$f(Q) = \oint_{C_\tau} \Delta_\tau(Q, Q') f(Q'), \qquad Q, Q' \in C_\tau$$

As a consequence of equations (15) we have the following Poisson brackets for the coefficients of the expansion (14):

$$[\alpha_i^\mu, \alpha_j^\nu] = -i\gamma_{ij}\eta^{\mu\nu} \tag{16a}$$

$$\{d_\alpha^\mu, d_\beta^\nu\} = -i\eta^{\mu\nu}\delta_{\alpha+\beta} \tag{16b}$$

$$\{b_i, c_j\} = -i\delta_{ij} \tag{16c}$$

$$[\gamma_\alpha, \beta_\beta] = -i\delta_{\alpha\beta} \tag{16d}$$

where $\gamma_{ij} = (1/2\pi i)\oint A_i\, dA_j$.

Now let us consider $L_i = L_i^{X\psi} + L_i^{\text{gh}}$ and $G_\alpha = G_\alpha^{X\psi} + G_\alpha^{\text{gh}}$ defined by

$$T(Q) = \sum_i L_i\Omega_i(Q), \qquad J(Q) = \sum_\alpha G_\alpha k_\alpha(Q) \tag{17}$$

$$\begin{aligned} L_i^{X\psi} &= -\tfrac{1}{2}\sum_{jk} l_{jk}^i \alpha_j \cdot \alpha_k + \tfrac{1}{4}\sum_{\alpha\beta} d_\alpha \cdot d_\beta F^i{}_{\alpha\beta} \\ L_i^{\text{gh}} &= \sum_j \sum_{s=-g_0}^{g_0} C_{ij}^s c_j b_{i+j-s} - \sum_\alpha \sum_{s=-g_0}^{g_0} H_{i\alpha}^s \gamma_\alpha \beta_{i+\alpha-s} \end{aligned} \tag{18}$$

and

$$\begin{aligned} G_\alpha^{X\psi} &= \sum_{\beta j} d_\beta \cdot \alpha_j D_{\beta j}^\alpha \\ G_\alpha^{\text{gh}} &= -2\sum_j \sum_{s=-g_0}^{g_0} c_j \beta_{j+\alpha-s} H_{j\alpha}^s - \tfrac{1}{2}\sum_\beta \sum_{p=-g}^{g} B_{\alpha\beta}^p \gamma_\beta b_{\alpha+\beta-p/2} \end{aligned} \tag{19}$$

where

$$l^i_{jk} = \frac{1}{2\pi i}\oint_{C_\tau} \omega_i\omega_k e_i, \qquad F^i_{\alpha\beta} = \frac{1}{2\pi i}\oint_{C_\tau} (h_\beta\partial h_\alpha - h_\alpha\partial h_\beta)e_i$$

$$D^\alpha_{\beta j} = \frac{1}{2\pi i}\oint_{C_\tau} h_\beta\omega_j g_\alpha$$

Then the Poisson brackets for L_i and G_α are

$$[L_i, L_j] = -i \sum_{s=-g_0}^{g_0} C^s_{ji} L_{i+j-s} \tag{20a}$$

$$[L_i, G_\alpha] = -i \sum_{s=-g_0}^{g_0} H^s_{i\alpha} G_{i+\alpha-s} \tag{20b}$$

$$\{G_\alpha, G_\beta\} = -i \sum_{p=-g}^{g} B^p_{\alpha\beta} L_{\alpha+\beta-p/2} \tag{20c}$$

These are a realization of equations (4), apart from the opposite sign in the first equation and the $-i$ factor. Of course, equation (20a) alone defines a realization of the KN algebra.

5. QUANTIZATION

All the classical quantities considered so far are promoted to operators acting in a Fock space. The Poisson brackets are replaced by quantum commutators according to the recipe: $[\,,]_{\mathrm{PB}} \to -i[\,,]_{\mathrm{quantum}}$. In order to avoid ambiguities we have to define normal ordering. As for the α_i's, the normal ordering prescription is any one given in Ref. 2. For the other relevant operators it is defined by considering as annihilation operators b_i for $i > 0$ and c_i for $i \le 0$, d_α and γ_α for $\alpha \le 0$, and β_α for $\alpha > 0$, and as creation operators the complementary ones (choosing another discriminating value for the normal ordering instead of zero would amount to modifying the central charges by trivial cocycles). With this prescription we have calculated the algebra of $:L_i:$ and $:G_\alpha:$ and we have obtained

$$[:L_i:, :L_j:] = \sum_{s=-g_0}^{g_0} C^s_{ji} :L_{i+j-s}: + \hat{\chi}_{ij} \tag{21a}$$

$$[:G_\alpha:, :L_i:] = \sum_{s=-g_0}^{g_0} H^s_{i\alpha} :G_{i+\alpha-s}: \tag{21b}$$

$$\{:G_\alpha:, :G_\beta:\} = \sum_{p=-g}^{g} B^p_{\alpha\beta} :L_{\alpha+\beta-p/2}: + \hat{\varphi}_{\alpha\beta} \tag{21c}$$

This again is a replica of equation (10). Of course the crucial quantities are the central charges $\hat{\chi}_{ij}$ and $\hat{\varphi}_{\alpha\beta}$. In order to give an idea of the problems involved without introducing too many technicalities, from now on we will limit ourselves to the purely bosonic string case. The relevant central charge can be written as

$$\hat{\chi}_{ij} = D\chi^{\Lambda}_{ij} + \tilde{\chi}_{ij} \tag{22}$$

D is the target space dimension, χ^{Λ}_{ij} is given in Ref. 2, and $\tilde{\chi}_{ij}$ is equation (21) of Ref. 3. That this central charge is a cocycle is a rather nontrivial fact. One can easily prove that it is antisymmetric and satisfies the locality condition (9a). But the Jacobi identity is more complicated to deal with. By using an explicit construction of the KN algebra by means of semi-infinite forms we have been able to prove that both χ^{Λ}_{ij} and $\tilde{\chi}_{ij}$ are indeed cocycles and are proportional to $\chi(e_i, e_j)$. Therefore it is enough to calculate them for a particular value of the indices in order to know the proportionality constant. We have calculated $\hat{\chi}_{ij}$ for $i+j=3g$ and found

$$\chi^{\Lambda}_{i,3g-i} = \tfrac{1}{12}(i-g_0)^3 + (i-g_0)A(\Lambda) \tag{23a}$$

$$\tilde{\chi}_{i,3g-i} = -\tfrac{13}{6}(i-g_0)^3 + (i-g_0)(\tfrac{1}{6} + g_0^2 - g_0) \tag{23b}$$

$A(\Lambda)$ is a number depending on the normal ordering prescription chosen for the α operators. Equation (23) should be compared with equation (11a). The trivial parts, which depend on the normal ordering or on the Schwarzian connection, can be taken care of by a suitable redefinition of the generators. The nontrivial parts allow us to calculate the proportionality constant. Up to trivial cocycles we have

$$\hat{\chi}_{ij} = (D-26)\chi(e_i, e_j) \tag{24}$$

Following an analogous procedure, in the superstring case we find

$$\hat{\chi}_{ij} = (\tfrac{3}{2}D - 15)\chi(e_i, e_j), \qquad \hat{\varphi}_{\alpha\beta} = -(\tfrac{3}{2}D - 15)\varphi(g_\alpha, g_\beta) \tag{25}$$

6. THE BRST OPERATOR

It is now easy to define a BRST operator on Σ corresponding to the NS-KN and R-KN superalgebras. We define

$$Q = \frac{1}{2\pi i}\oint_{C_\tau} (T^{X,\psi}(Q)c(Q) + J^{X,\psi}(Q)\gamma(Q) + \tfrac{1}{2}B(Q)[C(Q), C(Q)] - \beta(Q)[C(Q), \gamma(Q)] - \tfrac{1}{2}\{\gamma(Q), \gamma(Q)\}b(Q)) \tag{26}$$

The integrand in equation (26) is a global expression and the commutators are geometrical commutators [in the sense of equation (4)].

After quantization we have to consider $\hat{Q} =: Q:$. We obtain

$$\hat{Q}^2 = \tfrac{1}{2}\{\hat{Q}, \hat{Q}\} = \sum_{i,j} \hat{\chi}_{i,j} :c_i c_j: + \sum_{\alpha,\beta} \hat{\varphi}_{\alpha\beta} :\gamma_\alpha \gamma_\beta: \tag{27}$$

From equation (25) we have that, up to trivial cocycles, $\hat{Q}^2 = 0$ for $D = 10$. The BRST operator for the purely bosonic case is obtained from equation (26) by setting to zero the fermionic fields and relevant ghosts. Because of equation (25), nilpotence holds for $D = 26$.

REFERENCES

1. I. M. Krichever and S. P. Novikov, *Funk. Anal. Pril. 21*(2), 46 (1987).
2. I. M. Krichever and S. P. Novikov, *Funk. Anal. Pril. 21*(4), 47 (1987).
3. L. Bonora, M. Bregola, P. Cotta-Ramusino, and M. Martellini, *Phys. Lett. 205B*, 53 (1988).
4. L. Bonora, M. Martellini, M. Rinaldi, and J. Russo, *Phys. Lett. 206B*, 444 (1988).

Chapter 5

Quantum Groups, Integrable Theories, and Conformal Models

H. J. de Vega

The construction of exact solutions of two-dimensional integrable theories has made impressive progress in recent years. By integrable theories we mean models possessing as many commuting and conserved physical magnitudes as degrees of freedom. That is an infinite number for field theories or statistical models.

Integrable theories are interesting since they are usually exactly solvable to a large extent. One obtains detailed information about the physics of the models without relying on any approximation. In addition, integrable models happen to be realistic for different phenomena in condensed matter physics. In particle physics, two-dimensional QFT are interesting laboratories to understand four-dimensional physics and they can probably be used to build string models.

Eigenvalues and eigenvectors of the Hamiltonian, the momentum, and higher conserved magnitudes are usually explicitly calculable in an integrable model. In this way exact mass spectra and S matrices for QFT and the free energy in statistical models are derived. Moreover, form factors and one-point functions (order parameters) can also be computed explicitly.

These results are obtained through the use of the Bethe Ansatz (BA) and its different generalizations [1-3]. Actually the BA is not merely an

H. J. DE VEGA • Laboratoire de Physique Théorique et Hautes Energies (Laboratoire Associé au CNRS), Université Pierre et Marie Curie, 75252 Paris, Cedex 05, France.

Ansatz but it provides the exact solution of the models. To be more precise, the investigations of these last years show that the structure underlying all integrable theories (both QFT and statistical models) is the so-called Yang-Baxter-Zamolodchikov-Faddeev algebra (YBZF) (sometimes also called quantum groups). These structures are new in mathematics and their ultimate mathematical meaning and scope is still under investigation.

Integrable theories in more than two dimensions are known. As classical field theories one can mention self-dual Yang-Mills in $(4+0)$ [4] or $(3+1)$ dimensions [5], supersymmetric Yang-Mills in 10 dimensions [6], and $N = 3$ and 4 extended SUSY Yang-Mills in four dimensions [7]. The integrability structure of these models seems to be deeply connected with twistors and supertwistors [8]. In three-dimensional statistical mechanics, the tetrahedron equations and Zamolodchikov's solution [9] are three-dimensional extensions of the two-dimensional YBZF equations.

A YBZF algebra can be defined as follows. Let $T_{ab}(\Theta)$ be a set of quantum operators for $1 \leq a, b \leq q$ and $\Theta \in \mathbb{C}$ acting on a quantum space $\mathcal{V}$. $q \geq 2$ is a given number that defines the dimensionality of the auxiliary space (A) where T_{ab} acts as a $q \times q$ matrix. Θ is called the spectral parameter owing to the connection with the spectral problems in the inverse scattering method. Now the YBZF algebra is defined by the set of relations

$$R(\Theta - \Theta')[T(\Theta) \otimes T(\Theta')] = [T(\Theta') \otimes T(\Theta)]R(\Theta - \Theta') \tag{1}$$

where the tensor product notation

$$(A \otimes B)_{ab,cd} = A_{ac}B_{bd}$$

in the auxiliary space is used. More explicitly equation (1) reads

$$R(\Theta - \Theta')_{ab,ef}T_{ec}(\Theta)T_{fd}(\Theta') = T_{ae}(\Theta')T_{bf}(\Theta)R_{ef,cd}(\Theta - \Theta') \tag{2}$$

where we sum over repeated indices. Here $R_{ab,cd}(\Theta - \Theta')$ are c numbers in the quantum space $\mathcal{V}$. That is, R is a matrix in $A \otimes A$. In equations (1) and (2), it is understood an operational product for the $T_{ab}(\Theta)$ acting on $\mathcal{V}$. The $R_{ab,cd}(\Theta - \Theta')$ form the so-called R matrix that defines the YBZF algebra. They can be thought of as "structure constants" and the $T_{ab}(\otimes)$ play the role of "generators."

The link between a YBZF algebra (1) and integrability (in the sense defined above) is immediate. Multiplying (1) by $R(\Theta - \Theta')^{-1}$ yields

$$T(\Theta) \otimes T(\Theta') = R(\Theta - \Theta')^{-1}[T(\Theta') \otimes T(\Theta)]R(\Theta - \Theta')$$

Taking now the trace on $A \otimes A$ gives

$$[\tau(\Theta), \tau(\Theta')] = 0 \tag{3}$$

where

$$\tau(\Theta) \equiv \sum_{a=1}^{q} T_{aa}(\Theta) \tag{4}$$

and we used the property

$$\operatorname{tr}(A \otimes B) = \operatorname{tr} A \cdot \operatorname{tr} B$$

Therefore we have a family of commuting transfer matrixes $\tau(\Theta)$ and an infinite number of operators C_n defined by

$$\log \tau(\Theta) = \sum_n C_n \Theta^n, \qquad [C_n, C_m] = 0 \tag{5}$$

These C_n are usually conserved charges since the Hamiltonian and the momentum are connected with the lower C_n [1-3] or with $\log \tau(\Theta)$ at some special Θ. In this way, the connection of YBZF with integrable theories is straightforward.

The associativity requirement for the product of operators $T_{ab}(\Theta)$ as defined by equation (1) puts constraints on the "structure constants" $R_{ab,cd}(\Theta - \Theta')$. Taking a product of three operators $T(\Theta_1) \otimes T(\Theta_2) \otimes T(\Theta_3)$ and reordering it with the help of equation (1) in two inequivalent ways leads to the triangle relations or Yang-Baxter equations

$$\begin{aligned} &S^{(12)}(\Theta_1 - \Theta_2) S^{(13)}(\Theta_1 - \Theta_3) S^{(23)}(\Theta_2 - \Theta_3) \\ &\quad = S^{(23)}(\Theta_2 - \Theta_3) S^{(13)}(\Theta_1 - \Theta_3) S^{(12)}(\Theta_1 - \Theta_2) \end{aligned} \tag{6}$$

where the matrix $S^{(ij)} (1 \leq i < j \leq 3)$ acts in the tensor product of spaces $A_i \otimes A_j$ as $\mathrm{PR}(\Theta_i - \Theta_j)$ and it is the unit matrix in the remaining space. Here

$$P_{ab,cd} = \delta_{ad} \delta_{bc} \tag{7}$$

The algebraic equations (6) are a sufficient condition for the associativity of the YBZF algebra (see Refs. 1-3 for more details). It must be stressed that the Yang-Baxter equations (6) (or factorization equations) are a heavily overdetermined set since they contain a priori q^6 equations and only q^4 unknowns. Despite this fact, a rich set of solutions is known. All of them possess at least a discrete symmetry $Z_q \otimes Z_q$ that reduces the number of independent equations and probably permits the very existence of nontrivial solutions.

The main property of the YBZF algebras is the following reproduction property. If the operators $T_{ab}(\Theta)$ acting on $\mathscr{V}$ obey equations (1)-(2), so does

$$\begin{aligned} T_{ab}^{(K)}(\Theta, \{\mu, g\}) = &\sum_{a_1 \cdots a_{k-1}} [g_1 T(\Theta - \mu_1)]_{aa_1} \otimes [g_2 T(\Theta - \mu_2)]_{a_1 a_2} \\ &\otimes \cdots \otimes [g_K T(\Theta - \mu_K)]_{a_{K-1} b} \end{aligned} \tag{8}$$

on $\mathscr{V} \otimes \mathscr{V} \otimes \cdots \otimes \mathscr{V}$ (K times) with the *same* R matrix. Here $g_i \in \mathscr{G}$ $(1 \leq i \leq K)$ and the μ_i are arbitrary parameters. We have

$$\begin{aligned} &R_{(\Theta-\Theta')}[T^{(K)}(\Theta, \{\mu, g\}) \otimes T^{(K)}(\Theta', \{\mu, g\})] \\ &\quad = [T^{(K)}(\Theta', \{\mu, g\}) \otimes T^{(K)}(\Theta, \{\mu, g\}) R_{(\Theta-\Theta')} \end{aligned} \tag{9}$$

Equation (7) follows inserting equation (6) and using equation (1) repeatedly. So, starting from a given YBZF algebra on $\mathscr{V}$, it is always possible to construct other representations on $(\mathscr{V})^K$. One can also construct still other representations $\tilde{T}_{ab}^{(K)}(\Theta, \{\mu, g\})$ multiplying the generators $g_i T(\Theta - \mu_i)$ from right to left as

$$\tilde{T}_{ab}^{(K)}(\Theta, \{\mu, g\}) = \sum_{a_1 \cdots a_{K-1}} [g_1 T(\Theta - \mu_1)]_{a_1 b} \otimes [g_2 T(\Theta - \mu_2)]_{a_2 a_n}$$

$$\otimes \cdots \otimes [g_K T(\Theta - \mu_K)]_{a_n K-1} \tag{10}$$

It must be stressed that any solution $R(\Theta)$ of the YB equations (1)-(6) yields a YBZF algebra. Setting

$$[t_{ab}(\Theta)]_{cd} \equiv R_{ca}^{bd}(\Theta)$$

one finds from equation (6)

$$R(\Theta - \Theta')[t(\Theta) \otimes t(\Theta')] = [t(\Theta') \otimes t(\Theta)]R(\Theta - \Theta') \tag{11}$$

As we see, the representation of YBZF algebras sounds like a natural generalization of Lie Algebras. This assertion is actually correct in the sense that YBZF algebras are deformations of Lie algebras just as quantum mechanics is a generalization of classical mechanics [10, 11]. It can be noticed that the reproduction property (8) in the particular case $K = 2$ defines a comultiplication of the generators leading to the structure of a Hopf algebra.

Let us now consider physical applications of the YBZF algebra concepts first in two-dimensional statistical mechanics and then in field theory.

The matrix elements of the generators $t_{ab}(\Theta)$ can be defined as statistical weights of a vertex configuration (Fig. 1). The indices a, b $(1 \leq a, b \leq q)$ label the states of the horizontal bonds and the indices α, β those of the vertical bonds $(1 \leq \alpha, \beta \leq \dim \mathscr{V})$. For meaningful statistical models one

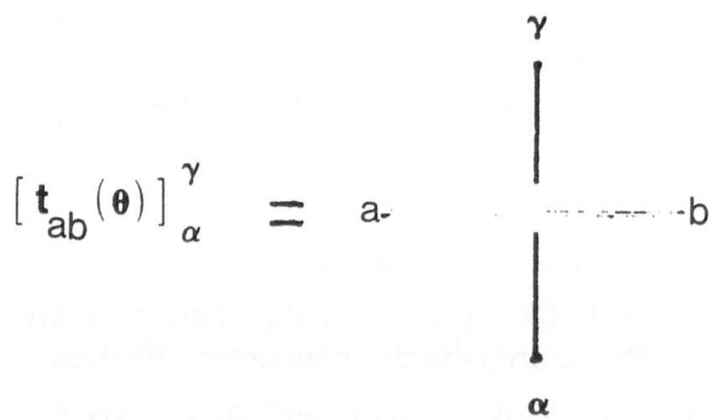

Figure 1. The local statistical weight depends on the states of the four bonds joining at the vertex.

needs $[t_{ab}]_\alpha^\gamma(\Theta) \geq 0$. However, models with some negative and complex weights have interesting mathematical structure too [3, 12, 13]. For a rectangular $N \times L$ lattice, the partition function follows by the usual definition:

$$Z = \sum_{\substack{\text{all}\\ \text{configurations}}} \prod_{\substack{\text{whole}\\ \text{lattice}\\ 1\leq i\leq N, 1\leq l\leq L}} t_{a_i b_{i+1}}(\Theta)_{\alpha_e}^{\beta_{e+1}} \tag{12}$$

This Z can be written in terms of the transfer matrix $\tau(\Theta)$ associated by equation (4) to the YBZF generator

$$T_{ab}(\Theta) = \sum_{a_1\cdots a_{L-1}} t_{aa_1}(\Theta) \otimes t_{a_1 a_2}(\Theta) \otimes \cdots \otimes t_{a_{L-1}b}(\Theta) \tag{13}$$

One finds

$$Z = \mathrm{Tr}[\tau(\Theta)^N] \tag{14}$$

where

$$\tau(\Theta) = \sum_{a=1}^{q} T_{an}(\Theta)$$

for periodic boundary conditions in both directions. Therefore, the free energy in the thermodynamic limit is given by the maximum eigenvalue of $\tau(\Theta)$, $\lambda_{\max}(\Theta)$

$$f = -\lim_{\substack{N\to\infty\\ L\to\infty}} \frac{1}{NL} \log Z = -\lim_{L\to\infty} \frac{1}{L} \log \lambda_{\max(\Theta)} \tag{15}$$

This relation shows that the physical properties of the system are directly related to the YBZF algebra. The fact that $\tau(\Theta)$ is a commuting family [equation (3)] allows us to diagonalize it. This can be done just using the YBZF algebra as a tool to build the eigenvectors and eigenvalues [1-3].

Besides two-dimensional vertex models, two-dimensional classical spin models (IRF models) and solid-on-solid models possess YBZF algebras. In one dimension, quantum magnetic Hamiltonians like the Heisenberg model and generalizations can be built from these algebras [1-3, 13, 14].

Implicit and explicit connections between YBZF algebras and two-dimensional QFT have been known for some time [1-3]. Let us discuss here the light-cone approach [15, 16]. This approach is the more general and precise way of constructing integrable QFT and conformal invariant theories. One starts from integrable lattice models like vertex models (Fig. 1) on a diagonal lattice [16]. This diagonal lattice is a discretization of Minkowski space-time in light-cone coordinate $X_\pm = X \pm T$. The matrix elements $[t_{ab}(\Theta)]_\alpha^\beta$ are now interpreted as quantum mechanical transition amplitudes of bare particles, propagating to the right or to the left by the bonds at the speed of light. In the simplest case dim $\mathcal{V} = q = 2$ and we interpret these particles as bare fermions without internal degrees of freedom. The allowed microscopical amplitudes assuming a $U(1)$ charge conservation are depicted in Fig. 2. This corresponds to the six-vertex model in the statistical mechanical language of Fig. 1. We have only three independent amplitudes because we assumed parity invariance. We can organize the microscopic amplitudes at a site into a unitary scattering matrix

$$R^{\alpha\beta}_{\alpha'\beta'} = \quad \text{[diagram]} \quad = \begin{pmatrix} 1 & 0 & 0 & 0 \\ 0 & c & b & 0 \\ 0 & b & c & 0 \\ 0 & 0 & 0 & \omega \end{pmatrix} \tag{16}$$

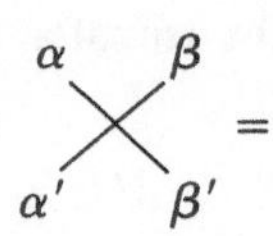

We can build now the operators describing the evolution by one lattice step in the diagonal directions

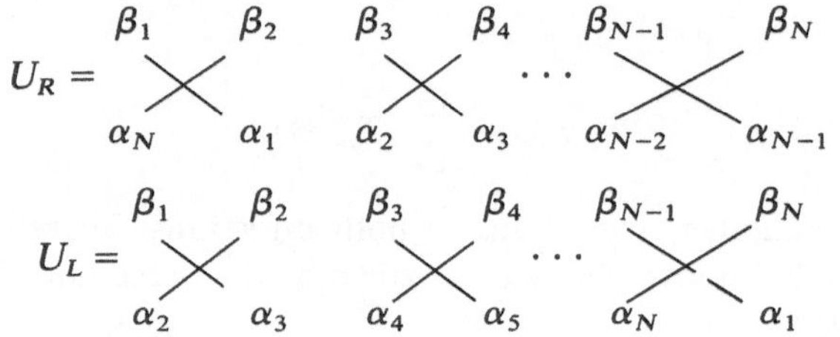

(17)

where the numbers $1, \ldots, 2N$ label the sites.

A second quantized formalism can be introduced defining fermion operators

$$\psi_{R,n} \quad \text{and} \quad \psi_{L,n}$$

They are associated with the links stemming upward from each site to the right and left, respectively (Fig. 3). They fulfill canonical anticommutations rules:

$$\{\psi_{A,n}, \psi_{B,m}\} = 0, \{\psi_{A,B}, \psi^{+}_{B,m}\} = \delta_{AB}\delta_{nm}$$
$$1 \leq n, m \leq N, \qquad A, B = R, L \tag{18}$$

$\psi_{R,n}$ and $\psi_{L,n}$ can be assembled into a two-component spinor. This provides a diagonal representation for γ_5; it is the chiral representation. We avoid

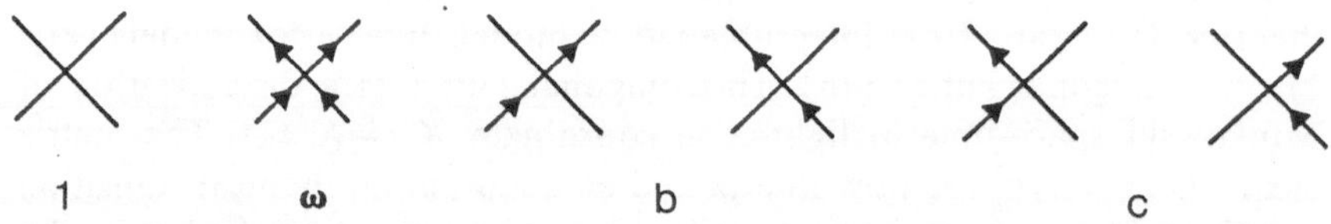

Figure 2. The six nonzero microscopic amplitudes. They coincide with the statistical weights of the six-vertex model.

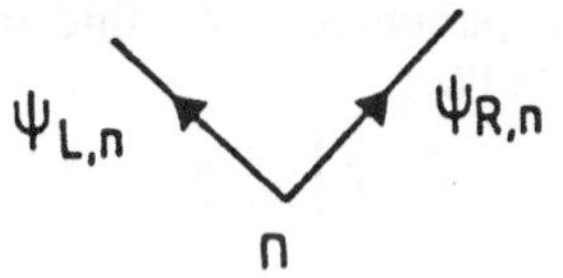

Figure 3. Fermion lattice operators associated with the links stemming upward from each site.

species doubling since our lattice Hamiltonian and momentum defined here from U_R and U_L are nonlocal operators. It is convenient to write

$$\psi_{R,n} = \psi_{2n}, \qquad \psi_{L,n} = \psi_{2n-1}, \qquad 1 \le n \le N$$

Now the R matrix (16) can be expressed in second-quantized language as [16]

$$R_{n,n+1} = 1 + bK_{n,n+1} + (c-1)K^2_{n,n+1} + (\omega - 1)\psi^+_n\psi_n\psi^+_{n+1}\psi_{n+1} \tag{19}$$

where

$$K_{n,m} = \psi^+_n\psi_m + \psi^+_m\psi_n \tag{20}$$

It is now possible to derive the lattice equations of motion for the fermion operators ψ_n. We find using equations (17)-(20) [16]

$$U_R\psi_{2n-2}U^+_R = U_L\psi_{2n}U^+_L = b^*\psi_{2n} + c^2\psi_{2n-1} + (c/\omega - c^*)\psi^+_{2n}\psi_{2n}\psi_{2n-1} - (b/\omega + b^*)\psi^+_{2n-1}\psi_{2n-1}\psi_{2n} \tag{21}$$

$$U_L\psi_{2n+1}U^+_L = U_R\psi_{2n-1}U^+_R = b^*\psi_{2n-1} + c^*\psi_{2n} + (c/\omega - c^*)\psi^+_{2n-1}\psi_{2n-1}\psi_{2n} - (b/\omega + b^*)\psi^+_{2n}\psi_{2n}\psi_{2n-1} \tag{22}$$

We can interpret in equations (21) and (22) the b terms as kinetic energy, the c terms as mass terms, and the trilinear pieces as interactions. It must be remarked that the equations of motion are *local* on the lattice, although the lattice Hamiltonian is *totally* nonlocal.

The continuum limit equations (21) and (22) are carefully analyzed in Ref. 16, where the continuum field Hamiltonian H and momentum P are derived. We find that as the lattice spacing tends to zero

$$U_{\substack{R\\L}} = e^{2i\mu Q}\left[1 - \frac{ia}{2}(H \pm P) + O(a^2)\right] \tag{23}$$

provided the weights b and c behave for $a \to 0$ as [16]

$$b \underset{a\to 0}{=} e^{i\mu}[1 + O(a^2)]$$
$$c \underset{a\to 0}{=} -\frac{i}{2}e^{i\mu}m_0a[1 + O(a^2)] \tag{24}$$

where μ and m_0 are fixed parameters. We find for H and P those of the massive Thirring model (MTM).

$$P = -i \int dx\, \psi^+ \partial_x \psi$$

$$H = \int dx \left[-i\psi^+ \gamma_5 \partial_x \psi + m_0 \bar{\psi}\psi + \frac{g}{2} (\bar{\psi}\gamma_\mu \psi)^2 \right] \tag{25}$$

with

$$g = -2 \operatorname{cotan}(\mu - \mu_0), \qquad \omega = e^{2i\mu_0}, \qquad \psi = \begin{pmatrix} \psi_R \\ \psi_L \end{pmatrix}$$

Equation (25) gives the bare operators and equation (24) defines the bare scaling limit. This is different from the renormalized scaling limit giving the physical sector of the Fock space [15, 16]. Both the particle spectrum and physical S matrices follow rigorously in the renormalized scaling limit computed in this light-cone approach. The bare limit, (24) and (25), tells us which model one is actually solving. We use the word "rigorous" since in this approach we solve a lattice model exactly, then we take the infinite volume limit and finally the $a \to 0$ (scaling) limit. In other words, here one solves (exactly) a model with both UV and volume cutoffs and then lets the cutoffs to infinity. This is clearly better than coordinate Bethe Ansatz (CBA) where the UV cutoff is introduced *after* obtaining the solution. For the MTM and the chiral Gross–Neven model the results of the CBA coincide with the light-cone approach for on-shell magnitudes. Hence the CBA works well in these cases. This is not the case of the multiflavor chiral model treated in Ref. 17 by CBA. As is shown in Ref. 18, the results of Ref. 17 are not correct.

The lattice light-cone approach was extended in Ref. 18 to chiral fermionic models with any simple Lie group of symmetry and Lagrangian

$$\mathscr{L} = i\bar{\psi}\partial\psi - g(\bar{\psi}\gamma_\mu T^\alpha \psi)(\bar{\psi}\gamma^\mu T_\alpha \psi) \tag{26}$$

Here ψ transforms under an irreducible representation ρ of G and T_α are the G generators in that representation. The light-cone lattice approach works here as follows. The lattices H and P are defined as

$$H \pm P = \frac{2i}{a} \log U_{\substack{R \\ L}}(\Theta) \tag{27}$$

where a is the lattice spacing and $U_{R,L}(\Theta)$ are given by equation (13) from the rational R-matrix invariant under G taken in the ρ representation. For large Θ we have

$$R(\Theta) = P\left[1 + \frac{\Pi + \lambda}{i\Theta} + O\left(\frac{1}{\Theta^2}\right)\right] \tag{28}$$

where λ is a numerical constant, the exchange operator P was defined in equation (7), and

$$\Pi = \sum_{\alpha=1}^{\dim G} T_\alpha \otimes T^\alpha \tag{29}$$

We then introduce the lattice operator

$$T_n^\alpha \equiv 1 \otimes \cdots \otimes \overbrace{T^\alpha}^{n\text{th site}} \otimes \cdots \otimes 1 \tag{30}$$

Using equations (13), (28), and (29) and the Lie algebra commutators

$$[T_\alpha, T_\beta] = i f_{\alpha\beta}{}^\gamma T_\gamma$$

we can show that T_n^α obey local equation of motion on the lattice [18]

$$U_R T_{2n-2}^\alpha U_R^+ = U_L T_{2n}^\alpha U_L^+ = T_{2n}^\alpha + \frac{2i}{\Theta} f_{\beta\gamma}^\alpha T_{2n-1}^\beta T_{2n}^\gamma + O\left(\frac{1}{\Theta^2}\right) \tag{31}$$

$$U_R T_{2n-1}^\alpha U_R^+ = U_L T_{2n+1}^\alpha U_L^+ = T_{2n-1}^\alpha - \frac{2i}{\Theta} f_{\beta\gamma}^\alpha T_{2n-1}^\beta T_{2n}^\gamma + O\left(\frac{1}{\Theta^2}\right)$$

The bare scaling limit is now defined as $a \to 0$, $\Theta \to \infty$, $x = na$ fixed. We get

$$\begin{aligned} &\partial_\mu J^{\mu\alpha} = 0 \\ &\partial_0 J_1^\alpha - \partial_1 J_0^\alpha + ig f_{\beta\gamma}^\alpha [J_0^\beta, J_1^\gamma] = 0 \end{aligned} \tag{32}$$

where

$$J_R^\alpha(x) = \frac{1}{ga\Theta} T_{2n}^\alpha, \qquad J_L^\alpha(x) = \frac{1}{ga\Theta} T_{2n-1}^\alpha$$

Therefore we have a lattice version of the G-algebra currents $J_\mu^\alpha(x)$ associated to an exactly solvable discretization of the field models.

Let us now discuss the renormalized scaling limit. The operators U_R and U_L, as explained before, are the light-cone evolution operators in a discretized Minkowski space-time. The Hamiltonian and momentum in the continuum theory are defined by the $a = 0$ limit of equation (27). This limit is to be taken such that the physical masses are finite. We derive the form of $\Theta = \Theta(a)$ from the spectrum of $U_R(\Theta)$ and $U_L(\Theta)$ on the lattice in order to obtain such a finite mass spectrum for $a \to 0$.

The light-cone transfer matrices U_R and U_L express in terms of the row-to-row transfer matrix

$$\tau(\Theta, \{\mu\}) = \sum_{a=1}^{q} T_{aa}^{(N)}(\Theta, \{\mu, 1\})$$

that follows from equation (8) when $g_1 = \cdots = g_K = 1$ and $\mu_k = (-1)^{k+1}\Theta$. Then we showed [18] that

$$\begin{aligned} U_L(\Theta) &= \tau(\Theta, \{\mu_k = (-1)^{k+1}\Theta\}) \\ U_R(\Theta) &= \tau(-\Theta, \{\mu_k = (-1)^{k+1}\Theta\})^+ \end{aligned} \tag{33}$$

Then, the spectrum of $\tau(\Theta_0; \{\Theta_k\})$ provides that of $U_R(\Theta_0)$ and $U_L(\Theta_0)$. The eigenvectors and eigenvalues can be exactly computed by the Bethe Ansatz and its nested generalizations [3, 14, 19]. The antiferromagnetic ground state eigenvector provides in the scaling limit the vacuum state of the QFT. The particle states follow from the lowest-lying excitations of the lattice vertex model. Since there is a factor a^{-1} in equation (27), only gapless lattice models may yield finite energy states in this scaling limit.

Let us take the G-invariant fundamental vertex models defined by equations (28)–(32). The low-lying excitation over the antiferromagnetic ground state are holes with large rapidity ϕ. The large ϕ behavior of the $\tau(\Theta_0, \{\Theta_k\})$ eigenvalues is independent of Θ_j and given by [20]

$$f_l(\phi) = -\lim_{N\to\infty} \frac{\lambda_l(\phi, \Theta_0, \{\Theta_K\})}{\lambda_{\max}(\Theta_0, \{\Theta_K\})} = \frac{im_l}{\pi} e^{\mp\kappa(\phi+\Theta_0)} + O(e^{\mp 2\kappa(\phi+\Theta_0)})$$

$$\phi + \Theta_0 \to \pm\infty \tag{34}$$

Here $\lambda_l(\phi, \Theta_0, \{\Theta_K\})$ is the contribution of a hole in the lth branch ($1 \leq l \leq \text{rank } G$) to the eigenvalue of $\tau(\Theta_0, \{\Theta_K\})$. The dimensionless parameters κ and m_l are given in Table I.

Combining equation (34) with equations (33) and (27) yields the dispersion relation for $\Theta \to +\infty$

$$\begin{aligned} E_l(\phi) &= \frac{e^{-\kappa\Theta}}{\pi a} m_l \cosh(\kappa\phi) + O(e^{-2\kappa\Theta}) \\ P_l(\phi) &= \frac{e^{-\kappa\Theta}}{\pi a} m_l \sinh(\kappa\phi) + O(e^{-2\kappa\Theta}) \end{aligned} \tag{35}$$

It is then natural to define the scaling limit according to

$$a \to 0, \qquad \Theta \to \infty, \qquad \mu = \frac{e^{-\kappa\Theta}}{\pi a} = \text{fixed} \tag{36}$$

μ is the renormalized or physical mass, and the particle mass spectrum of these integrable QFT is given by

$$M_l = \mu m_l \tag{37}$$

We recognize in equation (35) $\kappa\phi$ as the physical particle rapidity.

This is a very general way of constructing integrable QFTs. The operators H and P given by equation (27) are well defined on the lattice as well as all the higher conserved charges. In the continuum limit $a \to 0$, they provide the energy and momentum of a relativistic invariant QFT, as long as the spectrum of the original vertex model is gapless. This is usually the case for statistical weights $[t_{ab}(\Theta)]_c^d$ that are rational or trigonometric functions of the spectral parameter Θ. In addition to the particle spectrum,

Table I. Dimensionless Parameters as Defined in Equation (34)

Lie algebra	Dynkin's diagram	κ	m_l
A_n	O-O-O-···-O (1 2 3 … n)	$2\pi/n+1$	$\sin(\pi k/n+1), 1 \le k \le n$
B_n	O-O-O-···-O⇒O (1 2 3 … $n-1$ n)	$\pi/2n-1$	$\sin(\pi k/2n-1), 1 \le k \le n-1$; $m_n = \frac{1}{2}$
C_n	O-O-O-···-O⇐O (1 2 3 … $n-1$ n)	$\pi/n+1$	$\sin[\pi k/2(n+1)], 1 \le k \le n$
D_n	O-O-O-···-O< O(−), O(+) (1 2 3 … $n-2$)	$\pi/n-1$	$\sin[\pi k/2(n-1)], 1 \le k \le n-2$; $m_\pm = \frac{1}{2}$
E_6		$\pi/6$	$m_1 = m_5 = \dfrac{m_6}{2} = \dfrac{\sqrt{3}}{2}$; $m_2 = m_4 = \dfrac{3+\sqrt{3}}{2}$; $m_3 = \dfrac{3+\sqrt{3}}{\sqrt{2}}$
E_7	O—O—O—O—O—O (with O attached above the fourth node)	$\pi/9$	[a]
E_8	O—O—O—O—O—O—O (with O attached above the fifth node)	$\pi/15$	[a]
F_4	O—O⇒O—O	$\pi/9$	[a]
G_2	O⇚O	$\pi/6$	[a]

[a] These values can be (very laboriously) extracted from Ref. 27.

the S matrix is exactly calculable from the BA equation by standard methods [1-3].

The field theoretic models discussed up to here correspond to finite values of q, namely, a finite-dimensional vector space for each link in the light-cone lattice. This is clearly appropriate for fermionic fields. Since there exist representations of the YB algebra for $q = \infty$, also bosonic QFTs may be described in this framework.

The $S = \infty$ representation of the XXX magnet relates to the $SU(2)$ principal chiral σ model (PCM), as was developed in Ref. 21. Let us recall that the physical particle states of this model transform under the $SU(2)_L \times SU(2)_R$ group. The counting of states in the BA equations [21] and our derivation [22] show that only the $SU(2)_L$ singlet sector of the model is described by the H and P associated through equation (27) to the infinite S limit of the R matrix [23]

$$R_{12}(\Theta) = \frac{\Gamma(2S+1+i\Theta)\Gamma(\mathbb{J}+1-i\Theta)}{\Gamma(2S+1-i\Theta)\Gamma(\mathbb{J}+1-i\Theta)} \tag{38}$$

Here the operator $\mathbb{J}$ is defined by

$$\mathbb{J}(\mathbb{J}+1) = 2S(S+1) + 2\mathbf{S}_1 \cdot \mathbf{S}_2 \tag{39}$$

$\mathbf{S}_1$ and $\mathbf{S}_2$ are spin-S operators $[\mathbf{S}_1^2 = \mathbf{S}_2^2 = S(S+1)]$ acting on the horizontal and vertical spaces, respectively.

In other words, the Hamiltonian of the quantum PCM *does not follow* from the vertex constructions (27) and (38), even in the scaling limit. Only at the classical level can an equivalence be established between the respective classical analogs [21]. Although (27) and (38) do not provide at $S = \infty$ the full PCM Hamiltonian, they correctly reproduce its restriction to the $SU(2)_L$ singlets, and this is enough to calculate all particle masses as well as the invariant S-matrix amplitudes.

The lattice current construction, equations (30)–(32), also applies to the PCM. For large Θ the R matrix (38) admits a semiclassical expansion of the type (28). Therefore the whole construction holds. It must be noted that we have once again only one conserved and curvatureless matrix current: either the one associated with $SU(2)_R$ or that with $SU(2)_L$.

In conclusion, the light-cone transfer matrices U_R and U_L associated to each integrable gapless vertex model yield integrable and massive QFTs in the continuum limit. Since the scaling limit can sometimes be performed in several inequivalent ways, one can construct different QFTs from a single vertex model.

Besides the scaling limits leading to massive QFT such as equation (36), there exists the scaling limit

$$a \to 0, \qquad \Theta = \text{fixed} \tag{40}$$

yielding a conformal invariant theory.

The conformal invariant theory describes only the long-range properties of the integrable lattice model. In this sense integrable models have much more structure than conformal field theories.

The finite size resolution of Bethe Ansatz equations of the lattice model following the methods of Ref. 24 gives the values of the central charge C and of the conformal dimensions $(h, \bar{h})$ of operators for a large class of models [20, 25]. Branching coefficients can be related to one-point functions.*

The large size behavior of the logarithm of the transfer matrix eigenvalues close to the ground state λ_n takes the form (for periodic boundary conditions)

$$\frac{f_{n,N}(\Theta) - f_{n,\infty}(\Theta)}{v} = -\frac{\pi}{6N^2}\,c + \frac{2\pi}{N^2}\,x_n + o\left(\frac{1}{N^2}\right) \tag{41}$$

* For a review, see [26].

Here $f_{n,N} \equiv -(\operatorname{Re}\log \Lambda_n)/N$ and v is the speed of sound. That is, the dispersion relation for the excitations behave as

$$E \underset{p\to 0}{=} vp + o(p^3) \tag{42}$$

where p stands for the momentum.

The values of c and x_n follow from the BA equations, and comparison with conformal field theory tells us that c is the center of Virasoro algebra and $x_n = h_n + \bar{h}_n$ (for the ground state $x_n = 0$).

The value of c is known for all fundamental vertex models with YB algebras associated to simply laced Lie algebras [5, 6] and for the spin-S $SU(2)$ model [28]. These results indicate the general formula for c [18]:

$$c = \frac{x \dim G}{x + \tilde{h}} \tag{43}$$

Here $\dim G$ in the dimension of the Lie algebra and $\tilde{h}$ the dual Coxeter number. The Sugawara construction of Ref. 29 also fulfills equation (43). We believe that the gapless integrable theories associated to a Lie algebra G provide through their long-range behavior an alternative realization of the conformal algebra.

The conformal dimensions for the six-vertex model result [25]

$$x_{p,q} = \Delta + \bar{\Delta} = \frac{1}{2}\left(1 - \frac{\gamma}{\pi}\right)p^2 + \frac{q^2}{2(1-\gamma/\pi)}, \qquad p, q \in z$$

$$S_{p,q} = \Delta - \bar{\Delta} = pq$$

Here S_{pq} is the spin of operator A associated with the excitated state (that is, $\langle \phi_n | A | 0 \rangle \neq 0$). The value of S_{pq} can be extracted from the finite size correction to the momentum

$$P_N = P_\infty + \frac{2\pi S_{pq}}{N^2} + o\left(\frac{1}{N^2}\right)$$

and

$$P_N = -\operatorname{Im}\log \tau(\Theta)|_{\Theta=0}$$

For the critical q states Potts model we find [25]

$$c = 1 - 6/[m(m+1)]$$

where $q = 4\cos^2[\pi/(m+1)]$; $m = 2, 3, 4, \ldots$. The conformal dimension fulfills Kac's formula [25]. The conformal dimensions for models with many states per bond like those considered in Ref. 3, 13 indicate the presence of an extended conformal algebra (in the sense of Ref. 30) for long distances [31].

More precisely, we have computed exactly the transfer matrix eigenvalues for a general class of q-state vertex models (q possible different states per bond, $q \geq 2$) from their nested Bethe Ansatz equations at finite but large size [31].

Comparing with equation (41) gave for the conformal weight of an excited state

$$\Delta = \frac{1}{(1-\gamma/\pi)} \sum_{l,l'=1}^{q-1} \left(B_l^+ - \frac{\gamma}{2\pi} S_l \right) (M^{-1})_{ll'} \left(B_{l'}^+ - \frac{\gamma}{2\pi} S_{l'} \right)$$

and a similar formula for $\bar{\Delta}$ with $B_e^+ \leftrightarrow B_e^-$. Here $B_l^\pm$ are the number of holes near the end points in the lth branch, S_l is the lth "spin" of the state, M is the Cartan matrix of the underlying Lie algebra, and γ is the anisotropy parameter ($0 \leq \gamma \leq \pi$). Notice that Δ varies continuously with γ. When $\gamma = \pi/(m+1)$, $m = q+1, q+2, \ldots$ one recovers the conformal dimension of theory possessing extended Virasoro invariance [30]. More precisely, one must consider the RSOS version of the vertex model.

In this way the central charge takes on the values

$$c = (q-1)\left[1 - \frac{q(q+1)}{m(m+1)}\right], \qquad m \geq q+1$$

These integrable lattice models provide explicit realizations of the extended Virasoro algebra through their long-range behavior. They may be a very useful framework to uncover the physical meaning of the extended conformal symmetries.

REFERENCES

1. L. D. Faddeev and L. A. Takhtadzhyan, *Russ. Math. Surv. 34*, 11 (1979); L. D. Faddeev, *les Houches Lectures*, North-Holland, Amsterdam, 1982.
2. P. P. Kulish and E. K. Sklyanin, in: *Springer Lectures in Physics*, Vol. 151, Springer-Verlag, New York, 1981.
3. H. J. de Vega, *Lectures in the Sâo Paulo School* (E. Abdalla and M. C. Abdalla, eds.), World Scientific, Singapore (1988) and *Int. J. Mod. Phys.* (to be published).
4. R. Ward, *Phys. Lett. 61A*, 81 (1977); A. A. Belavin and V. E. Zakharov, *Phys. Lett. 73B*, 53 (1978); M. F. Atiyah, *Geometry of Yang–Mills Fields*, Lezione Fermiane, Pisa, 1979.
5. H. J. de Vega, CERN preprint TH 4817; *Commun. Math. Phys. 116*, 659 (1988).
6. E. Witten, *Nucl. Phys. B 266*, 245 (1985).
7. E. Witten, *Phys. Lett. B 77*, 394 (1978).
8. J. Avan and H. J. de Vega, *Int. J. Mod. Phys. 3*, 1263 (1988).
9. A. B. Zamolodchikov, *Sov. Phys. JETP 52*, 325 (1980); *Commun. Math. Phys. 79*, 489 (1981); R. J. Baxter, *ibid. 88*, 185 (1983).
10. V. G. Drinfeld, *Dokl. Akad. Nauk SSSR 32*, 254 (1985).
11. H. J. de Vega, *Phys. Lett. B 177*, 171 (1986).
12. A. A. Belavin, *Nucl. Phys. B 180*, 189 (1981).
13. O. Babelon, H. J. de Vega, and C. M. Viallet, *Nucl. Phys. B 190*, 542 (1981).
14. O. Babelon, H. J. de Vega, and C. M. Viallet, *Nucl. Phys. B 200*, 266 (1982).

15. T. T. Truong and K. D. Schotte, *Nucl. Phys. B 220,* 77 (1983); *230,* 1 (1984); M. F. Weiss and K. D. Schotte, *Nucl. Phys. B 225,* 247 (1983).
16. C. Destri and H. J. de Vega, *Nucl. Phys. B 290,* 363 (1987).
17. L. V. Avdeev and M. V. Chizov, *Phys. Lett. B 184,* 363 (1987).
18. C. Destri and H. J. de Vega, *J. Phys. A* (to be published).
19. L. A. Takhtadzhyan, *J. Sov. Math.* 2470 (1983); N. Y. Reshetikhin, *Theor. Math. Phys. 63,* 555 (1986); H. J. de Vega and M. Karowski, *Nucl. Phys. B 280,* 225 (1987).
20. H. J. de Vega, *J. Phys. A 20,* 6023 (1987).
21. L. D. Faddeev and N. Y. Reshetikhin, *Ann. Phys.* (N.Y.) *167,* 227 (1986).
22. C. Destri and H. J. de Vega, *Phys. Lett. 201B,* 261 (1988).
23. P. P. Kulish, N. Y. Reshetikhin, and E. K. Sklyanin, *Lett. Math. Phys. 5,* 393 (1981).
24. H. J. de Vega and F. Woynarovich, *Nucl. Phys. B 251,* 439 (1985); H. P. Eckle and F. Woynarovich, *J. Phys. A 20,* L97 (1987).
25. H. J. de Vega and M. Karowski, *Nucl. Phys. B 285,* 619 (1987); F. Woynarovich, *Phys. Rev. Lett. 59,* 259 (1987); M. Karowski, Brasov Lectures (1987); C. J. Hamer, G. R. W. Quispel and M. J. Batchelor, Canberra preprint 9/87; see also Ref. 31.
26. E. Date, M. Jimbo, T. Miwa, and M. Okado, *RIMS 590,* Kyoto preprint, 1987.
27. E. Ogievetsky and P. B. Wiegmann, *Phys. Lett. B 168,* 360 (1986)
28. H. M. Babujian, *Nucl. Phys. B 215,* 317 (1983); I. Affleck, *Phys. Rev. Lett. 56,* 746 (1986).
29. P. Goddard and D. Olive, *Int. J. Mod. Phys. A1,* 303 (1987).
30. V. A. Fateev and A. B. Zamolodchikov, *Nucl. Phys. B 280,* 644 (1987).
31. H. J. de Vega, *J. Phys. A 21,* L1089 (1988).

Chapter 6

Small Handles and Auxiliary Fields

M. Dine, W. Fischler, and N. Seiberg

In the superconformal field theory approach to superstrings, the space-time supersymmetry is not manifest. In contrast, conventional four-dimensional $N = 1$ supersymmetric field theories admit a superspace formulation that easily reveals consequences of supersymmetry for these theories, such as nonrenormalization theorems. The existence of auxiliary fields is crucial to this formulation. It would be useful to introduce such space-time auxiliary fields in the superconformal formulation of strings. However, in string theory we usually study only very specific correlation functions, i.e., S-matrix elements. These correlation functions involve BRS invariant vertex operators which create on-shell physical states. This seems to preclude the description of the auxiliary field by vertex operators. On the other hand, a set of candidate auxiliary fields was introduced in Ref. 1. Because they do not correspond to physical states, these operators are not BRS invariant. The connection of these operators to contact terms required to maintain two-dimensional supersymmetry was explained in Ref. 2. In fact, as explained there and in Ref. 3, these terms are properly taken into account if one works

M. DINE • Institute for Advanced Study, Princeton, New Jersey 08540 and Physics Department, City College of CUNY, New York, New York 10031. W. FISCHLER • Theory Group, Physics Department, University of Texas, Austin, Texas 78712. N. SEIBERG • Institute for Advanced Study, Princeton, New Jersey 08540; on leave of absence from the Department of Physics, Weizmann Institute of Science, Rehovot 76100, Israel.

in superspace (or with two-dimensional auxiliary fields) and does not drop terms which formally vanish by equations of motion. These terms only contribute for special values of the momenta [2]; correct results can always be obtained by staying away from these points and then analytically continuing. On the other hand, it is often convenient to use these operators. That such operators could be used in practical calculations was shown in Ref. 3. In this short talk we will review this work from a slightly different viewpoint. In particular, we will show that these space-time auxiliary fields satisfy the usual four-dimensional equations of motion, including loop effects.

In the language of two-dimensional field theory the space-time fields can be thought of as an infinite set of coupling constants. These coupling constants flow according to the renormalization group [4]. The requirement of conformal invariance leads to coupled equations for these coupling constants. These are partial differential equations which are the equations of motion for the space-time fields.

We are familiar with the fact that Einstein's equations for gravity involve not only physical fields, namely, the transverse traceless component of the metric (in the weak field approximation) but also nonpropagating, longitudinal modes. In the same way the renormalization group flow involves such longitudinal degrees of freedom. Let us show how, for example, integrating out a small handle renormalizes an operator that creates a state that is not in the physical spectrum [5]. What do we mean by a "small handle" in a conformally invariant theory? Take a sphere (for simplicity), attach to it a handle of size a, and place two probes on this surface separated by a distance $|z| \gg a$. These probes can be any operators on the world sheet; let us take the space-time coordinates $X_\mu(z_1)$ and $X_\nu(z_2)$. From the vantage point of these X's the effect of a small handle should be represented by a series of operators on the sphere located at the "center of mass" of the handle. In other words there is an operator product expansion (O.P.E.) that mimics the effect of the handle. How do we determine this O.P.E. and especially the marginal operators in this expansion?

Consider the correlation function $\langle X_\mu(z_1)X_\nu(z_2)\rangle$ in the presence of a handle:

$$\langle X_\mu(z_1)X_\nu(z_2)\rangle_{1\text{ handle}} = \langle X_\mu(z_1)X_\nu(z_2)\rangle_{\text{no handle}} + \sum_n a^n f_n(z_1, z_2; z_c) \quad (1)$$

where $\langle X_\mu(z_1)X_\nu(z_2)\rangle_{\text{no handle}} = \delta_{\mu\nu}\log|z_{12}|$ and z_c is the center of the handle. What can be said about the f_n's? First they satisfy $\nabla^2_{z_{1,2}} f_n = 0$ since $\nabla^2_z \log|z| = \delta^2(z)$. Also since one integrates over the relative orientations of the handle with respect to the probes, f_n has to be isotropic. Therefore up to $O(a^2)$ equation (1) becomes

$$\langle X_\mu(z_1)X_\nu(z_2)\rangle_{1\text{ handle}} = \log|z_{12}| + A\frac{a^2}{(z_1 - z_c)(z_2^* - z_c^*)}$$

$$+ A\frac{a^2}{(z_1^* - z_c^*)(z_2 - z_c)} \tag{2}$$

where A is a number that can be calculated. It is now straightforward to determine the marginal operator which appears in the O.P.E. for one handle. Indeed one has to determine $O(z_c)$ such that

$$\langle X_\mu(z_1)X_\nu(z_2)O(z_c)\rangle_{\text{sphere}} = \frac{Aa^2\delta_{\mu\nu}}{(z_1 - z_c)(z_2^* - z_c^*)} + \frac{Aa^2\delta_{\mu\nu}}{(z_1^* - z_c^*)(z_2 - z_c)} \tag{3}$$

The operator that has this property is $O(z_c) = Aa^2 :\partial X\,\bar\partial X: (z_c)$. Since we have to integrate over locations z_c and size a of the handle we obtain

$$B\int_{\Lambda^{-1}} d^2z_c(da/a^3)Aa^2 :\partial X\,\bar\partial X: (z_c)$$

where B is a number, part of the scale invariant measure. Then (1) becomes

$$\langle X_\mu(z_1)X_\nu(z_2)\rangle_{1\text{ handle}} \approx \langle X_\mu(z_1)X_\nu(z_2)\rangle_{S_2} + \langle X_\mu(z_1)X_\nu(z_2)\rangle_{S_2}AB\int d^2z_c :\partial X\,\bar\partial X: \log\Lambda$$

Therefore

$$\langle X_\mu(z_1)X_\nu(z_2)\rangle_{1\text{ handle}} \approx \langle X_\mu(z_1)X_\nu(z_2) \times \exp AB\log\Lambda\int d^2z_c :\partial X\,\bar\partial X(z_c):\rangle_{S_2} \tag{4}$$

So there is a renormalization of the coupling constant associated to $:\partial X\,\bar\partial X:$ by a factor $1 - AB\log\Lambda$. This coupling however is not a physical space-time field.

For the rest of this talk we focus on the auxiliary field for the gauge supermultiplet, D. Similar considerations apply to the F terms. More specifically, consider the $SO(32)/Z_2$ heterotic string compactified to four dimensions on an orbifold or Calabi-Yau manifold. The surviving low-energy gauge group is $O(26)\times U(1)$. In such a compactification the X's divide into six interacting fields X^i and $X^{\bar i}$, $i, \bar i = 1, 2, 3$ and four free fields X^μ, $\mu = 0, \ldots, 3$. The right moving fermions ψ^M decompose similarly.* The left moving fermions λ^A decompose into six interacting fermions, λ^i and $\lambda^{\bar i}$, and 26 free fermions, λ^a, $a = 1, \ldots, 26$. The $O(26)$ and $U(1)$ Kac–Moody currents are, respectively,

$$j^{ab} = \lambda^a\lambda^b, \qquad a, b = 1, \ldots, 26$$

and

$$j = \sum_{i\bar i} g_{i\bar i}\lambda^i\lambda^{\bar i}, \qquad i, \bar i = 1, 2, 3$$

* In the case of an orbifold, of course, all of the fields are free, but the "internal" fields are subject to nontrivial projections and boundary conditions [6].

The $U(1)$ appears to have gravitational and gauge anomalies; these can be canceled by assigning the space-time axion a transformation law under the gauge symmetry [7]. By supersymmetry, this gives rise to a Fayet–Iliopoulos D term [8].

We will illustrate the appearance and the use of the auxiliary D field by focusing in this chapter on the potential $V(A)$ for the scalars charged under $U(1)$. More precisely we will calculate $\partial V/\partial A$ using the renormalization group flow. To evaluate $\partial V/\partial A$ we will consider an arbitrary S-matrix element with n vertex operators with nonvanishing background scalars. The action with the background scalars is obtained from the original one by adding the scalar vertex operators with momentum-dependent coefficients, which play the role of coupling constants. To simplify the writing, we consider the case of compactification on a Z_3 orbifold, with the "standard embedding" [6]. In this case, in the untwisted sector, there are nine scalar fields, $A^{ai\bar{j}}$ $(i, \bar{j} = 1, 2, 3)$ in the 26 of $O(26)$; their vertex operators are given simply by

$$V_A = \int d^2z\, d\theta\, e^{ik\cdot x}\Lambda^a\Lambda^i DX^{\bar{j}}$$

The vertex operators for their CPT conjugates, $A^{a\bar{i}j}$ are obtained by complex conjugating this expression. For states in the twisted sectors, the vertex operators can be constructed using the methods of Ref. 9. The construction of the vertex operators for more general cases is described in the papers in Ref. 1. For this case, the action of the string in the scalar background is

$$S = S_0 + \int d^2z\, d\theta \sum_k A^{ai\bar{j}}(k)\, e^{ikX}\Lambda^a\Lambda^i DX^{\bar{j}} + \text{c.c.} \tag{5}$$

where

$$X^M = x^M + \theta\psi^M, \qquad \Lambda^A = \lambda^A + \theta F^A$$

F^A is a two-dimensional auxiliary field, and

$$D = \frac{\partial}{\partial\theta} + \theta\frac{\partial}{\partial z}$$

Let us first evaluate $\partial V/\partial A$ at string tree level. Consider therefore the following S-matrix element

$$\left\langle \int d^2z_1 V_{-1}(z_1) \int d^2z_2 V_{-1}(z_2) \int d^2z_3 V_0(z_3) \cdots \int d^2z_n V_0(z_n) \right\rangle_{\text{sphere}} \tag{6}$$

where the subscripts -1 and 0 refer to the different ghost pictures [10]. Since we are considering the effective potential it is sufficient to calculate this correlation function retaining only the zero-momentum components of the charged scalars in equation (5). Therefore assuming weak couplings (weak fields):

$$(6) = \left\langle \int V_{-1}(z_1) \int V_{-1}(z_2) \cdots \int V_0(z_n) \right.$$
$$\left. \times \left\{ 1 + \sum_{p \neq 0} \left[\frac{(\int d\theta\, d^2 w A^{ai\bar{j}}_{k=0} \Lambda^a \Lambda^i DX^{\bar{j}} + \text{c.c.})^p}{p!} \right] \right\} \right\rangle_{S_2} \tag{7}$$

where this correlation function is calculated with the action S_0. In general, the zero-momentum limit is a delicate one. As stressed in Ref. 2, at zero momentum we must be careful to include δ functions arising from contractions of the two-dimensional auxiliary fields, F.

The first nontrivial renormalization of the operators $\Lambda^a \Lambda^i DX^{\bar{j}}$, which have $U(1)$ charge $+1$, occurs at order $(A_{k=0})^3$ and comes from the region of integration where the three vertex operators for the scalar background fields are close to each other (see Fig. 1). Indeed consider for example the S-matrix element:

$$\left\langle \int V_{-1}(z_1)\, d^2 z_1 \int V_{-1}(z_2)\, d^2 z_2 \int V_0(z_3)\, d^2 z_3 \cdots \right.$$
$$\times \int V_0(z_n) d^2 z_n \int A^{ai\bar{j}} \Lambda^a \Lambda^i DX^{\bar{j}}\, d\theta\, d^2 w_1$$
$$\left. \times \int A^{*bk\bar{l}} \Lambda^b \Lambda^{\bar{k}} DX^l\, d\theta\, d^2 w_2 \int A^{cp\bar{q}} \Lambda^c \Lambda^p DX^{\bar{q}}\, d\theta\, d^2 w_3 \right\rangle_{S_2} \tag{8}$$

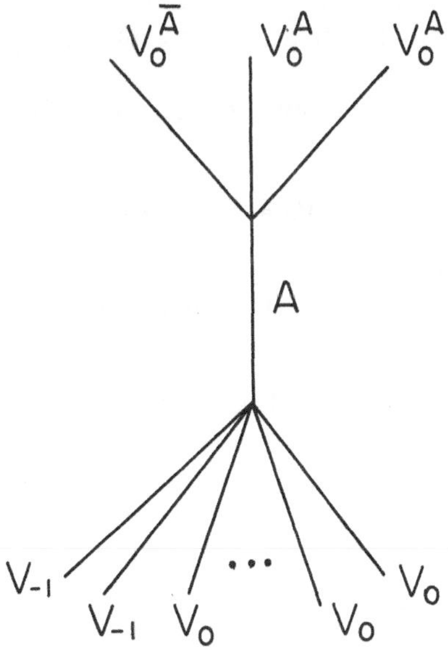

Figure 1. Factorization of the amplitude of n vertex operators in the presence of three charged backgrounds.

where the operators for the background are in the zero ghost picture and the sum of their $U(1)$ charges is +1. For simplicity in evaluating this O.P.E. we will first picture change the vertex operator with $U(1)$ charge +1 from the 0 ghost picture to the −1 ghost picture.

$$(8) = \left\langle \int V_0(z_1) \int V_{-1}(z_2) \int V_0(z_3) \cdots \int V_0(z_n) \right.$$
$$\times \int A^{ai\bar{j}} \lambda^a \lambda^i \Psi^{\bar{j}}\, d^2w_1\, e^{-\varphi} \int A^{*b\bar{k}l} \Lambda^b \Lambda^{\bar{k}} DX^l\, d\theta\, d^2w_2$$
$$\left. \times \int A^{cp\bar{q}} \Lambda^c \Lambda^p DX^{\bar{q}}\, d\theta\, d^2w_3 \right\rangle_{S_2} \tag{9}$$

Next, consider the product of the two vertex operators:

$$\int d\theta\, \Lambda^b \Lambda^{\bar{k}} DX^l(w_2) \int d\theta\, \Lambda^c \Lambda^p DX^{\bar{q}}(w_3)$$
$$= (\lambda^b F^{\bar{k}} \psi^l + F^b \lambda^{\bar{k}} \psi^l + \lambda^b \lambda^{\bar{k}} \partial X^l)_{w_2} (\lambda^c F^p \psi^{\bar{q}} + F^c \lambda^p \psi^{\bar{q}} + \lambda^c \lambda^p \partial X^{\bar{q}})_{w_3}$$
$$= \delta^2(w_2 - w_3)[\lambda^{\bar{k}} \psi^l \lambda^p \psi^{\bar{q}} \delta^{bc} + \lambda^b \psi^l \lambda^c \psi^{\bar{q}} \delta^{\bar{k}p}](w_2) + \cdots$$

The dots here denote other operators which will not be relevant to our discussion. As discussed in Refs. 1-3, the operators appearing here (if the indices on the right-moving fields, ψ, are contracted together) are naturally interpreted as the vertex operators for the auxiliary fields in the gauge multiplets. In particular, the operator for the D field associated with the $U(1)$ symmetry is

$$V_D = \lambda^{\bar{i}} \lambda^i \psi^{\bar{j}} \psi^j$$

We focus on this term in the following. For the region $w_2 \to w_1$, we can use the O.P.E. of this operator with the scalar vertex operator:

$$e^{-\varphi} \lambda^a \lambda^i \psi^{\bar{j}}(w_1) V_D(w_2) \sim \lambda^a \lambda^i \psi^{\bar{j}} \frac{e^{-\varphi}}{|w_1 - w_2|^2} + \cdots$$

to obtain from this integration region

$$(9) = \left\langle \int V_0(z_1) \int V_{-1}(z_2) \int V_0(z_3) \cdots \right.$$
$$\left. \times \int V_0(z_n) \int d^2w\, \lambda^a \lambda^p \psi^{\bar{q}}\, e^{-\varphi} \tilde{A}^{ap\bar{q}} \log \Lambda \right\rangle_{S_2} \tag{10}$$

where Λ^{-1} is a short distance cutoff on the world sheet and $\tilde{A}^{ap\bar{q}} = A^{ap\bar{j}} A^{*c\bar{i}j} A^{ci\bar{j}}$. Picture changing once more, we recover

$$(10) = \left\langle \int V_{-1}(z_1) \int V_{-1}(z_2) \int V_0(z_3) \cdots \right.$$
$$\left. \times \int V_0(z_n) \int d^2w \, d\theta \, \Lambda^a \Lambda^p D X^{\bar{q}} \tilde{A}^{ap\bar{q}} \log \Lambda \right\rangle_{S_2} \tag{11}$$

By virtue of equations (7) and (11) we have

$$(6) = \left\langle \int V_{-1}(z_1) \int V_{-1}(z_2) \cdots \int V_0(z_n) \right.$$
$$\left. \times \exp \int d^2w \, d\theta \, \Lambda^a \Lambda^p D X^{\bar{q}} \tilde{A}^{ap\bar{q}} \log \Lambda \right\rangle$$

Thus there is a renormalization of $A^{ai\bar{j}}$, the β function

$$\beta^{ai\bar{j}} = dA^{ai\bar{j}} / d \log \Lambda = \tilde{A}^{ai\bar{j}} = A^{ai\bar{j}} A^{c\bar{p}q} A^{cp\bar{q}} \tag{12}$$

The right-hand side of equation (12) is a derivative of part of the usual potential for charged scalars A_i:

$$V(A_i) = \tfrac{1}{2} \left(\sum_i e_i A_i^* A_i \right)^2$$

where e_i are the $U(1)$ charges of the scalar fields A_i. Once all the charged fields are taken into account we indeed recover:

$$\beta(A_i) = \frac{\partial V}{\partial A_i^*} = e_i A_i \left(\sum_j e_j A_j^* A_j \right)$$

We have seen that this term can be thought of as arising from the "exchange" of a D term at zero momentum. At slightly nonzero momentum, things would have worked somewhat differently [1–3]. Taking, again, two scalars in the zero ghost picture and two in the -1 picture, the operator product of the two -1 picture operators yields

$$V_{-1}(z, k_1) V_{-1}(w, k_2) = |z - w|^{-2} e^{-2\phi} V_D + \cdots$$

while that of the two in the zero ghost picture gives

$$V_0(z, k_1) V_0(w, k_2) = k_1 \cdot k_2 |z - w|^{-2} V_D + \cdots$$

Thus the zero ghost picture operators "couple" to the D term with a factor of k^2, while the -1 picture operators couple with no such factor. In other words, while the operator V_D in the -2 picture is naturally identified with the usual auxiliary field, that in the 0 picture is identified with k^2 times this field. In the four-point function, after Möbius gauge fixing, the remaining integral yields a factor of $1/k^2$, which cancels the k^2 in the zero ghost picture coupling. One can think of this as the "propagator" of the D term, which is $k^2/k^2 = 1$ in momentum space.* We will see this phenomenon again when we consider the β functions for the auxiliary fields. Note that had we performed the computation with all of the fields in the zero ghost number picture, the calculations would not have resembled the exchange of a D term. It is straightforward to show, however, that the same result is obtained.

There is also as expected a string loop contribution to first order in the couplings A to the renormalization of A. This is the contribution to the equation of motion for the charged scalar coming from the Fayet–Iliopoulos D term (8). Indeed, consider the following S-matrix element

$$\left\langle \int V_0(z_1)\cdots\int V_0(z_n)\left[1+\int AV_0^A(w)\right]\right\rangle_{\text{Torus}} \tag{13}$$

evaluated in the free two-dimensional action, where

$$AV_0^A(w) = A^{ai\bar{j}}[F^a\lambda^i\psi^{\bar{j}} + \lambda^a F^i\psi^{\bar{j}} + \lambda^a\lambda^i\partial X^{\bar{j}}](w)$$

The renormalization that we are considering comes from the O.P.E. of the vertex operator $V_0^A(w)$ and a small handle. This region of moduli space is equivalent to the region of integration where all the $V_0(z_i)$ coalesce. Then by factorization one obtains a contribution from the second term in (13)

$$\left\langle \int V_0(z_1)\int V_0(z_2)\cdots\int V_0(z_n)\int V_{-2}^A(w)\right.$$
$$\left.\times\log\Lambda\left\langle\int V_0^{\bar{A}}(w_1)\int AV_0^A(w_2)\right\rangle_{T_2}\right\rangle_{S_2} \tag{14}$$

where $V_{-2}^A = e^{-2\varphi}V_0^A$ is the vertex operator for the charged scalar in the -2 ghost picture. Also note that

$$\left\langle \int V_0(z_1)\int V_0(z_2)\cdots\int V_0(z_n)\int V_{-2}^A(w)\right\rangle_{S_2}$$

is the coefficient of V_0^A in the O.P.E. of $V_0(z_1)\cdots V_0(z_n)$. The term $\langle\int V_0^{\bar{A}}(w_1)\int V_0^A(w_2)\rangle_{T_2}$ in equation (14) has been calculated in Ref. 1 and was shown to be equal to

$$\langle\lambda^j\lambda^{\bar{j}}\psi^i\psi^{\bar{i}}\rangle_{T_2} = \langle j_L j_R\rangle_{T_2} = \mu^2$$

* One might think that this phenomenon is special to auxiliary fields, but a similar cancellation occurs when one factorizes an amplitude on a fermion pole. As here, the sum of the ghost numbers of the two operators is -2. If one of the fermions is in the $-\frac{1}{2}$ picture, the other is in the $-\frac{3}{2}$ picture. The correlation function of the latter has a factor of $k\!\!\!/$ which cannot be set to zero. It is canceled by the factor of $1/k^2$ arising from the integral over the modular parameter associated with the pinched cycle to yield the fermion propagator, $k\!\!\!/ /k^2 = 1/k\!\!\!/$.

By picture changing the vertex operator $V_{-2}^{A}(w)$ appearing in (14) back to the zero ghost picture we obtain

$$(14) = \left\langle \int V_{-1}(z_1) \int V_{-1}(z_2) \int V_0(z_3) \cdots \int V_0(z_n) \times \left(1 + \mu^2 A \log \Lambda \int V_0^A(w)\right)\right\rangle_{S_2}$$

$$= \left\langle \int V_{-1}(z_1) \int V_{-1}(z_2) \int V_0(z_3) \cdots \times \int V_0(z_n) \exp \mu^2 A \log \Lambda \int V_0^A(w)\right\rangle_{S_2}$$

This corresponds to a renormalization of the couplings A_i:

$$\beta_{A_i} = \sum_j e_j A_j^* A_j e_i A_i + \mu^2 e_i A_i$$

I will conclude this talk by showing that the renormalization group flow for D_k indeed agrees with the known four-dimensional equations:

$$D_k = \sum_{i,k'} e_i A_i^*(k') A_i(k-k') + \mu^2$$

Consider

$$D_k \lambda^i \lambda^{\bar{i}} \psi^j \psi^{\bar{j}} e^{ikX} \tag{15}$$

where D_k are again to be thought of as coupling constants. (As explained in Refs. 1 and 3, it is only this contraction of indices that leads to an operator that is marginal in these compactifications.) By normal ordering (15) we obtain

$$D_k :\lambda^i \lambda^{\bar{i}} \psi^j \psi^{\bar{j}} e^{ikX}: e^{-k^2 \log \Lambda}$$

which contributes $k^2 D_k$ to the β-function associated to $D_k : \beta_{D_k} = k^2 D_k + \cdots$. To fill in the ellipsis we will next consider the O.P.E. for two charged scalars in the zero ghost picture

$$\int d\theta\, \Lambda^a \Lambda^i D X^{\bar{j}} e^{ik_1 X}(z_1) \int d\theta\, \Lambda^b \Lambda^{\bar{i}} D X^j e^{ik_2 X}(z_2) \tag{16}$$

This operator product contains a term

$$\frac{\lambda^i \lambda^{\bar{i}} \psi^{\bar{j}} \psi^j e^{i(k_1+k_2)X}}{|z_1 - z_2|^{2-k_1 k_2}}$$

From the region of integration where $z_1 \approx z_2$ we obtain an additional contribution to β_{D_k}:

$$\beta_{D_k} = k^2 D_k + k^2 \sum_i e_i A_i^* A_i$$

Before continuing to the computation on the torus, it is instructive to see how the scalar potential emerges; this parallels closely our discussion of the correlation function in equation (8). Working with the scalars in the -1 picture, these obey (schematically) a renormalization group equation of the form

$$\frac{\partial \phi}{\partial t} = k^2 \phi + \phi D$$

On the other hand, in the 0 ghost picture, the D term obeys

$$k^2 \frac{\partial D}{\partial t} = k^2 \phi^2 + k^2 D$$

Here $t = \ln(\Lambda)$. Integrating this second equation with boundary condition $D_0 = -\phi^2(t_0)$ at t_0, some reference scale, gives $D(t) = -\phi^2$; substituting back in the first equation and setting $\partial\phi/\partial t = 0$ gives

$$k^2 \phi + \phi^3 = 0$$

the desired equation of motion. Note that in this analysis it was important that the auxiliary field operators differed by a factor of k^2 in the two pictures. This is, of course, just the renormalization group description of our analysis of the four-point function above.

The only missing piece is the contribution of small handles to the renormalization of D_k. Consider two probes on a torus, $V_0^{\bar{A}}(k_1)$ and $V_0^{A}(k_2)$, where

$$V_0^{A}(k_i) = [F^a \lambda^i \psi^{\bar{j}} + \lambda^a F^i \psi^{\bar{j}} + \lambda^a \lambda^i \partial X^{\bar{j}} (ik_i \cdot \psi)]\, e^{ik_i X}$$

The small handle contribution can be evaluated by looking at the region of integration where $z_1 \approx z_2$. The O.P.E. of these two operators contains

$$(k_1 \cdot k_2) \frac{\lambda^i \lambda^{\bar{i}} \psi^j \psi^{\bar{j}}\, e^{i(k_1+k_2)X}}{|z_1 - z_2|^{2-k_1 k_2}}$$

Therefore the S-matrix element

$$\left\langle \int d^2 z_1\, V_0^{\bar{A}}(k_1) \int d^2 z_2\, V_0^{A}(k_2) \right\rangle_{T_2} \tag{17}$$

becomes by factorization

$$\left\langle \int d\theta\, d^2 z_1\, \Lambda^a \Lambda^i D X^{\bar{j}}\, e^{ik_1 X} \int d\theta\, d^2 z_2\, \Lambda^b \Lambda^{\bar{i}} D X^j\, e^{ik_2 X} \right.$$

$$\left. \times \int d^2 w\, \lambda^i \lambda^{\bar{i}} \psi^j \psi^{\bar{j}}\, e^{-2\varphi} \right\rangle_{S_2} (k_1 + k_2)^2 \log \Lambda \left\langle \int d^2 w\, \lambda^i \lambda^{\bar{i}} \psi^j \psi^{\bar{j}} \right\rangle_{T_2}$$

which after picture changing and exponentiation leads to an additional contribution to β_{D_k} (see Fig. 2)

$$\beta_{D_k} = k^2 D_k + k^2\left(\sum_i e_i A_i^* A_i\right) + k^2 \mu^2$$

where $\mu^2 = \langle V_0^{\bar{A}} V_0^{A} \rangle_{T_2}$.

At the conformal fixed point we recover

$$D_k + \sum_i \sum_{k'} e_i A_i^*(k') A_i(k-k') + \mu^2 = 0$$

So the coupling to vertex operators for auxiliary fields does indeed satisfy the proper equations of motion derived from the renormalization group flow.

There are a number of applications that one can envisage for the ideas presented here. First, using the auxiliary fields, it may be possible to provide a simplified proof of the nonrenormalization theorem in string theory. The subject is complicated, not only by the intricacies of higher-order string perturbation theory, but also by the need to separate out wave function renormalizations in each order. Still, it may be possible to make general statements.

Another possible application arises in theories in which Fayet–Iliopoulos terms are generated. In these models, it is frequently possible to find new supersymmetric vacua by giving small expectation values to some scalar fields [11]. To prove this, one examines the low-energy effective field theory, and notes that there are usually combinations of fields that are forbidden by the gauge symmetries to appear in the superpotential, and

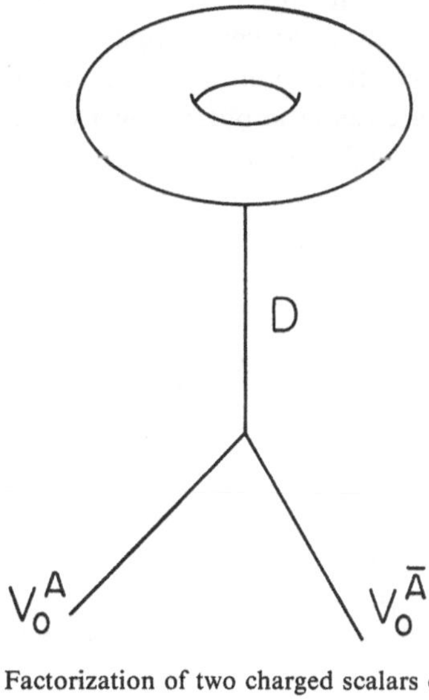

Figure 2. Factorization of two charged scalars on a torus.

with appropriate $U(1)$ charges to cancel the D term. At first sight, however, it might appear difficult to describe these new vacua directly in string theory. For example, one might imagine that one has to excite Kaluza–Klein or string modes, and that one has to solve a nontrivial two-dimensional field theory. This is not the case, however. First, at string tree level, we are accustomed to the idea that it is only necessary to consider marginal and relevant operators as perturbations of conformal field theories. This should be the case in loops as well. Thus it should be necessary to add to the conformal field theory only the vertex operator for the scalar, with a small coefficient of order g (where g^2 counts string loops). Moreover, we only need to consider this term perturbatively. At order g^{2l} in the loop expansion, one needs to expand the path integral on the sphere to order g^{2l}, on the torus to order g^{2l-1}, and so forth. Thus it is entirely straightforward to study these vacua in string perturbation theory.

The "Fischler–Susskind" program [4] is easily implemented for these models. For example, at one loop, the equations we have studied for the scalar field beta functions have two types of solutions, corresponding to vanishing and nonvanishing expectation values for the scalars. (Nonvanishing scalar field expectation value leads to a nonvanishing dilaton beta function, but the error is of two-loop order and can be ignored.) It is easy to show that the various fields have the correct masses in this order as well [12]. The auxiliary field vertex operators should be very useful in the study of these vacua. For example, in the field theory analysis, one shows that the superpotential vanishes in suitable directions by using the gauge symmetry. On the world sheet, this argument translates into the statement that global symmetries yield vanishing correlations between certain combinations of auxiliary field and scalar vertex operators.

Because of the lack of a manifestly supersymmetric formulation of string theory, many statements about string theory have been proven by examining low-energy effective actions. Hopefully, the auxiliary field vertex operators will allow us to short circuit this intermediate step, yielding the same sorts of simplifications that we are accustomed to in supersymmetric field theories.

ACKNOWLEDGMENTS. One of us (W.F.) would like to thank Claudio Teitelboim, his colleagues, and the staff at C.E.C.S. for the unforgettable hospitality during the second summer meeting on "Quantum Mechanics of Fundamental Systems." The work of M. Dine is supported by the DOE under contract No. DE-ACO2-83ER40107 and the NSF under grant No. PHY 8620266 and by an A. P. Sloan Foundation Fellowship. The work of W. Fischler is supported in part by the Robert A. Welch Foundation by the NSF under grant No. PHY 8605978. The work of N. Seiberg is supported by the NSF under grant No. PHY 8620266.

REFERENCES

1. M. Dine, I. Ichinose, and N. Seiberg, *Nucl. Phys. B 293*, 253 (1987); J. Atick, L. Dixon, and A. Sen, *Nucl. Phys. B 292*, 109 (1987).
2. M. Green and N. Seiberg, *Nucl. Phys. B 299*, 559 (1988).
3. M. Dine and N. Seiberg, IAS preprint HEP-87/50 (1987).
4. C. Lovelace, *Phys. Lett. 135B*, 75 (1984); *Nucl. Phys. B 273*, 413 (1986); C. G. Ćallan, D. Friedan, E. Martinec, and M. Perry, *Nucl. Phys. B 262*, 593 (1985).
5. W. Fischler and L. Susskind, *Phys. Lett. B 171*, 383 (1986); *Phys. Lett. B 173*, 262 (1986); C. Lovelace, *Nucl. Phys. B 273*, 413 (1986).
6. L. Dixon, J. Harvey, C. Vafa, and E. Witten, *Nucl. Phys. B 274*, 285 (1986).
7. E. Witten, *Phys. Lett. 149B*, 351 (1984).
8. M. Dine, N. Seiberg, and E. Witten, *Nucl. Phys. B 289*, 589 (1987).
9. L. Dixon, D. Friedan, E. Martinec, and S. Shenker, *Nucl. Phys. B 282*, 13 (1987).
10. D. Friedan, E. Martinec, and S. Shenker, *Nucl. Phys. B 271*, 93 (1986).
11. V. Kaplunovsky, unpublished.
12. S. Weinberg, in *Proceedings of the Oregon Meeting* (Eugene, Oregon, August, 1985) (R. C. Hwa, ed.), World Scientific, Singapore, 1986, p. 850; N. Seiberg, *Phys. Lett. 187B*, 56 (1987); A. Sen, SLAC-PUB-4383 (1987).

Chapter 7

Differential Equations in Moduli Space

Tohru Eguchi and Hirosi Ooguri

It is well known that conformal field theories are governed by their tight algebraic structures. Central charge of a conformal theory and dimension of its fields are dictated by the representation theory of Virasoro algebra [1, 2]. Furthermore, irreducibility of a representation, decoupling of null states, leads to differential equations for correlation functions [1]. These equations have been used to determine operator-product expansion coefficients [3, 4].

In this note we point out that, when conformal field theory is considered on a Riemann surface, irreducibility of a theory leads to differential equations in moduli space for its partition function. In the case of a torus we derive an ordinary differential equation in variable τ of order $m(m-1)/2$ [m is related to the central charge of degenerate representation of Virasoro algebra $c = 1 - 6/m(m+1)$] for conformal character functions. These differential equations follow from the existence of a null state at degree $m(m-1)$ in the module of highest-weight state $h = 0$ corresponding to the identity operator.

Let us first recapitulate basic steps in deriving differential equations on correlation functions. In the case of a complex plane, conformal Ward

TOHRU EGUCHI AND HIROSI OOGURI • Department of Physics, University of Tokyo, Tokyo, Japan 113.

identity is given by [1]

$$\langle T(z)\phi_1(w_1)\phi_2(w_2)\cdots\rangle = \sum_{i=1}\left[\frac{h_i}{(z-w_i)^2}+\frac{1}{z-w_i}\partial_{w_i}\right]\langle\phi_1(w_1)\phi_2(w_2)\cdots\rangle \tag{1}$$

Here ϕ_i's are conformal fields and h_i's are their conformal dimensions. $T(z)$ is the holomorphic part of the energy-momentum tensor. For the sake of illustration let us now consider the case of the Ising model which corresponds to the $m = 3$, $c = 1/2$ representation of Virasoro algebra. It is known in this case that there exists, for instance, a degeneracy at the second grade in the module of highest-weight state $h = 1/2$

$$[L_{-2} - \tfrac{3}{4}(L_{-1})^2]|h = 1/2\rangle = 0 \tag{2}$$

and this state must decouple from the rest of states in a unitary irreducible theory. Making use of the operator product expansion

$$T(z)\phi(w) \approx \left[\frac{h}{(z-w)^2}+\frac{1}{z-w}\partial_w\right]\phi(w)+\cdots$$

$$[T(z) = \sum_n L_n(w)(z-w)^{-n-2}] \tag{3}$$

and using the fact that $L_{-1}(w)\phi(w) = \partial_w\phi(w)$ we obtain

$$\frac{3}{4}\frac{\partial^2}{\partial w_1^2}\langle\phi_1(w_1)\phi_2(w_2)\cdots\rangle = \sum_{i=2}\left[\frac{h_i}{(w_1-w_i)^2}+\frac{1}{w_1-w_i}\partial_{w_i}\right] \times \langle\phi_1(w_1)\phi_2(w_2)\cdots\rangle \tag{4}$$

when ϕ_1 is a conformal field with $h = 1/2$. Equation (4) is a prototype of the differential equations in conformal field theory. To each null state in representation space corresponds a linear differential operator whose order is given by the grade of the null state.

In the case of a Riemann surface Ward identity acquires a new term, which describes the dependence on correlation functions on the moduli of Riemann surface. If we consider the case of a torus for simplicity, conformal Ward identity is given by [5]

$$\begin{aligned}\langle T(z)\phi_1(w_1)\phi_2(w_2)\cdots\rangle = \sum_i \{&h_i[\mathcal{P}(z-w_i)+2\eta_1] \\ &+[\zeta(z-w_i)+2\eta_1 w_i]\partial_{w_i}\}\langle\phi_1(w_1)\phi_2(w_2)\cdots\rangle \\ &+\frac{2\pi i}{Z(\tau)}\frac{\partial}{\partial\tau}\{Z(\tau)\langle\phi_1(w_1)\phi_2(w_2)\cdots\rangle\}\end{aligned} \tag{5}$$

Here $Z(\tau)$ is the partition function and $\mathcal{P}$ and ζ are Weierstrass $\mathcal{P}$ and ζ functions. $\eta_1 = \zeta(1/2) = 2\pi i \partial_\tau \log[1/\eta(\tau)]$, where $\eta(\tau)$ is the Dedekind η function. Local algebraic structure remains the same on the Riemann surface and (2) becomes an operator identity

$$\{L_{-2}(w) - \tfrac{3}{4}[L_{-1}(w)]^2\}\phi_{h=1/2}(w) = 0 \tag{6}$$

Then following the same steps leading to (4) we obtain a partial differential equation

$$\begin{aligned}\{\tfrac{3}{4}\partial^2_{w_1} - \eta_1 - 2\eta_1 w_1 \partial_{w_1}\}\langle\phi_1(w_1)\phi_2(w_2)\cdots\rangle &= \sum_{i=2} \{h_i[\mathcal{P}(w_1 - w_i) + 2\eta_1] + [\zeta(w_1 - w_i) + 2\eta_1 w_i]\partial_{w_i}\} \\ &\quad \times \langle\phi_1(w_1)\phi_2(w_2)\cdots\rangle + \frac{2\pi i}{Z(\tau)}\frac{\partial}{\partial\tau}\{Z(\tau)\langle\phi_1(w_1)\phi_2(w_2)\cdots\rangle\}\end{aligned} \tag{7}$$

Now let us discuss differential equations on conformal character functions. We stick to the case of the Ising model and consider null states generated out of identity operator $h = 0$. From the Kac formula [6] we know that there exists a null state at grades 1 and 6,

$$\begin{aligned}&L_{-1}(z)\mathbb{1} = 0 \\ &\{64[L_{-2}(z)]^3 + 93[L_{-3}(z)]^2 - 264L_{-4}(z)L_{-2}(z) - 108L_{-6}(z)\}\mathbb{1} \equiv \Phi(z) = 0\end{aligned} \tag{8}$$

A null state at grade 1 leads to a triviality; however, one at grade 6 gives nontrivial relations. If we consider the case of complex plane and compute, for instance, a three-point function involving $\Phi(z)$, using the Ward identity (1) we find

$$\langle\Phi(z)\phi(w_1)\phi(w_2)\rangle = h(h - 1/2)(h - 1/16)\frac{\text{const}}{(z - w_1)^6(z - w_2)^6(w_1 - w_2)^{2h-6}} \tag{9}$$

where ϕ is a conformal field with dimension h. Thus the null state $\Phi(z)$ behaves as a conformal field and has vanishing operator-product expansion coefficients with other fields with dimension $h = 0, 1/2, 1/16$. These are precisely the allowed values of highest weights in the Ising model.

Now let us go to the case of a torus and consider a one-point function of $\Phi(z)$. We obtain a differential equation on partition function from the relation

$$\text{tr}[\Phi(z)] = 0 \tag{10}$$

Here tr is taken over any of the highest-weight representations $V_{h=0}$, $V_{h=1/2}$, $V_{h=1/16}$ of the Ising model. Making use of the Ward identity

[5] and after some algebra we find

$$(2\pi i)^3 \frac{d^3}{d\tau^3} Z(\tau) + 12\eta_1(\tau)(2\pi i)^2 \frac{d^2}{d\tau^2} Z(\tau)$$

$$+ 2\pi i\left[-\frac{25}{64} g_2(\tau) + 24\eta_1(\tau)^2 \right] \frac{d}{d\tau} Z(\tau) - \frac{23}{4 \times 64} g_3(\tau) Z(\tau) = 0 \quad (11)$$

where g_2 and g_3 are coefficients in the expansions of the $\mathcal{P}$ function

$$\mathcal{P}(z) = \frac{1}{z^2} + \frac{1}{20} g_2 z^2 + \frac{1}{28} g_3 z^4 + \cdots \quad (12)$$

Equation (10) must be satisfied by each of the character functions $\chi_{h=0}, \chi_{h=1/2}, \chi_{h=1/16}$ of the representations $V_{h=0}, V_{h=1/2}, V_{h=1/16}$.

If we make a change of variable from τ to λ given by

$$\lambda(\tau) = \frac{e_2(\tau) - e_3(\tau)}{e_1(\tau) - e_3(\tau)} \quad (13)$$

where e_1, e_2, e_3 are related to g_2, g_3 as $g_2 = 2(e_1^2 + e_2^2 + e_3^2)$, $g_3 = 4e_1e_2e_3$ and obey $e_1 + e_2 + e_3 = 0$. λ is related to the parameters defining the torus as an elliptic curve, $y^2 = x(x - x_0)(x - x_1)$, as $\lambda = x_0/x_1$. Then (10) can be rewritten as

$$\left[\frac{d^3}{d\lambda^3} - \frac{2(2\lambda - 1)}{(1-\lambda)\lambda} \frac{d^2}{d\lambda^2} + \frac{391(\lambda^2 - \lambda) + 7}{8 \times 24(1-\lambda)^2\lambda^2} \frac{d}{d\lambda} - \frac{23}{24^3} \frac{(2-\lambda)(1+\lambda)(2\lambda - 1)}{(1-\lambda)^3\lambda^3}\right] Z(\lambda) = 0 \quad (14)$$

In this representation (14) is a differential equation of Fuchsian type; coefficient functions are rational functions of λ, and $\lambda = 0, 1, \infty$ are regular singular points. Solutions of (14) take particularly simple forms in terms of λ and in fact reproduce three character functions of the Ising model,

$$\begin{aligned}
\chi_{h=0} &= \lambda^{-1/24}(1-\lambda)^{-1/24} + \lambda^{-1/24}(1-\lambda)^{1/12} \\
&= \text{const} \times q^{-1/48} \frac{1}{2}\left[\prod_{n=1} (1 + q^{n-1/2}) + \prod_{n=1} (1 - q^{n-1/2})\right] \\
\chi_{h=1/2} &= \lambda^{-1/24}(1-\lambda)^{-1/24} - \lambda^{-1/24}(1-\lambda)^{1/12} \\
&= \text{const} \times q^{-1/48} \frac{1}{2}\left[\prod_{n=1} (1 + q^{n-1/2}) - \prod_{n=1} (1 - q^{n-1})\right] \\
\chi_{h=1/16} &= \lambda^{1/12}(1-\lambda)^{-1/24} \\
&= \text{const} \times q^{1/24} \prod_{n=1} (1 + q^n)
\end{aligned} \quad (15)$$

In general case null states occur at grade $m(m-1)$ and the order of differential equation becomes $m(m-1)/2$ since each power of L_{-2} is converted into a derivative $d/d\tau$ by Ward identity (note that L_{-1} does not occur since it annihilates the $h=0$ state). Then $m(m-1)/2$ independent solutions of the differential equation correspond to each of $\chi_{h_{r,s}}$ with $1 \leq s \leq r \leq m-1$.

In the case of a general Riemann surface, the partition function of a conformal theory will obey a set of partial differential equations in $3g-3$ modular parameters (g is the genus of the surface). The number of solutions to this system of equations must be finite. Thus they must constitute a holonomic set of partial differential equations. It is a challenging problem to elucidate their mathematical structure.

REFERENCES

1. A. A. Belavin, A. M. Polyakov, and A. B. Zamolodchikov, *Nucl. Phys. B 241*, 333 (1984).
2. D. Friedan, Z. Qiu, and S. Shenker, *Phys. Rev. Lett. 52*, 1575 (1984).
3. Vl. S. Dotsenko and V. Fatteev, *Nucl. Phys. B 240*, 312 (1984).
4. Y. Kitazawa *et al.*, University of Tokyo preprint UT 522 (1987).
5. T. Eguchi and H. Ooguri, *Nucl. Phys. B 282*, 308 (1987).
6. V. Kac, Proceedings of the International Congress of Mathematics, Helsinki, 1978; and *Lect. Notes Phys. 94*, 441 (1979).

Chapter 8

Consistent Quantum Mechanics of Chiral *p*-Forms

Marc Henneaux and Claudio Teitelboim

1. INTRODUCTION

p-form gauge potentials naturally arise in theories of fundamental extended objects. A *p*-form potential bears to a $(p - 1)$-dimensional object the same relation that the ordinary electromagnetic potential bears to a charged particle. It couples to the tangent of the object's history [1].

Of particular interest are *p*-forms whose field strength, a $(p + 1)$-form, is self-dual. Those fields, which will be called chiral *p*-forms in the sequel, play a central role in supergravity and in string theory [2].

There has been a difficulty so far in the analysis of the dynamics and the quantization of chiral *p*-forms, namely, the fact that the implementation of the chirality condition in the variational principle and in quantum mechanics is somewhat subtle and, if improperly handled, yields incorrect results.

In Ref. 3, a Lorentz and gauge invariant action that consistently incorporates the self-duality condition was given. This action can be directly used to pass to quantum mechanics. Furthermore, it leads to energy-momentum tensor components whose Poisson brackets obey the appropriate

---

MARC HENNEAUX • Faculté des Sciences, Université Libre de Bruxelles, B-1050 Brussels, Belgium and Centro de Estudios Científicos de Santiago, Santiago 9, Chile. CLAUDIO TEITELBOIM • Centro de Estudios Científicos de Santiago, Santiago 9, Chile and Center for Relativity, The University of Texas at Austin, Austin, Texas 78712.

"surface deformation algebra" [4, 5], enabling a classically consistent gravitational coupling.

It was this last property that was mainly used in Ref. 3 to establish the action for chiral p-forms. The purpose of this chapter is to present an alternative derivation based on more familiar methods, and to investigate in more detail the properties of the action of Ref. 3.

The alternative route for arriving at the action relies on an orthodox application of Dirac's method for constrained systems [6] to the action principle proposed in Ref. 7. In that action the self-duality condition is implemented by adding a term of the form $\lambda(F - {}^*F)^2$ to the usual action density which is proportional to the square F^2 of the field strength F.

The field λ is a Lagrange multiplier and carries no degree of freedom because of a new gauge invariance which the action acquires when the term $\lambda(F - {}^*F)^2$ is included. It can be eliminated classically. The elimination of λ can also be achieved quantum-mechanically since the new gauge invariance is anomaly free, as can straightforwardly be seen when the appropriate commutator dictated by the Dirac method is used.

Once λ is eliminated and the appropriate commutator defined, one gets the action that was presented in Ref. 3 and which permits direct passage to quantum mechanics. In that action, there is no remnant of the new gauge invariance acting on λ, because this gauge invariance leaves the chiral p-forms invariant.

The case $p = 0$ is of particular interest since it describes chiral bosons (left moving) in two space-time dimensions. These are basic elements of the heterotic string model [8, 9]. For this reason we will devote the next two sections to a careful analysis of the $p = 0$ case. The formalism is particularly transparent in this case.

For $p > 0$, the standard gauge invariance $A \to A + d\xi$ of p-form gauge fields also needs to be handled. The generators of this gauge invariance are mixed with the self-duality constraint. This requires a special treatment, which is given in the final section.

2. CHIRAL BOSONS—CLASSICAL ANALYSIS

2.1. Action

The field strength $F_\mu = \partial_\mu \varphi$ of a 0-form φ is self-dual if

$$ {}^*F = F \tag{1a} $$

i.e.,

$$ \frac{g_{\mu\nu}}{\sqrt{-g}} \varepsilon^{\nu\rho} \partial_\rho \varphi = \partial_\mu \varphi \tag{1b} $$

Here, $g_{\mu\nu}$ is the space-time metric. In flat space Minkowskian coordinates, (1b) reduces to (with $\varepsilon^{01} = -1$)

$$\dot{\varphi} = \varphi' \tag{1c}$$

which explicitly shows that φ is left moving. (We take $\eta_{00} = -\eta_{11} = -1$.)

The simplest way to incorporate the self-duality condition is to start from the action for an ordinary massless scalar field and add to it the term $\lambda(\dot{\varphi} - \varphi')^2$, where λ is a Lagrange multiplier. This yields the action of Ref. 7:

$$S[\varphi, \lambda] = \int dt\, dx[\tfrac{1}{2}(\dot{\varphi}^2 - \varphi'^2) + \lambda(\dot{\varphi} - \varphi')^2] \tag{2}$$

If one varies (2) with respect to λ, one gets

$$(\dot{\varphi} - \varphi')^2 = 0 \quad \Rightarrow \quad \dot{\varphi} = \varphi' \tag{3a}$$

Taking (3a) into account, one then finds that the φ-variational equation reads

$$\ddot{\varphi} = \varphi'' \tag{3b}$$

and clearly holds as a consequence of (3a). The self-duality condition (3a) thus completely contains the dynamics of a chiral boson. Furthermore, the multiplier λ is left completely arbitrary by the equations of motion.

2.2. Gauge Invariance

The multiplier λ is undetermined because (2) possesses a gauge invariance, given by

$$\delta\varphi = \varepsilon(\dot{\varphi} - \varphi') \tag{4a}$$

$$\delta\lambda = -\tfrac{1}{2}(\dot{\varepsilon} + \varepsilon') + \varepsilon(\dot{\lambda} - \lambda') - \lambda(\dot{\varepsilon} - \varepsilon') \tag{4b}$$

By choosing appropriately the gauge parameter ε, the multiplier λ can be made equal to any given function of t and x.

The gauge transformation (4) is somewhat analogous to a particular two-dimensional diffeomorphism, since (4a) coincides with

$$\delta\varphi = \mathcal{L}_\xi \varphi = \xi^\mu \varphi_{,\mu} \tag{4c}$$

where the vector field ξ^μ is given by $\xi^\mu = (\varepsilon, -\varepsilon)$. Furthermore, the transformation (4b) can be interpreted, in a suitable coordinate system explicitly given in Ref. 7, as the action of that same diffeomorphism on λ, provided that λ is interpreted as an appropriate component of the metric.

This similarity has led to the conjecture that (4) becomes anomalous quantum-mechanically, and that the quantization of chiral bosons is therefore generically inconsistent. We will show in this chapter that even though there is indeed a gravitational anomaly, the gauge transformation (4) is not anomalous when proper account is taken of the fact that the field φ carries only half the degree of freedom of a standard Klein–Gordon scalar field. Hence, the quantization of chiral bosons poses no problem in flat space. (The gravitational coupling is discussed in Sections 2.14 and 3.3.)

2.3. Counting the Number of Degrees of Freedom

The number of degrees of freedom is equal to half the number of gauge invariant initial data that must be specified in the Cauchy problem.

Because the multiplier λ is pure gauge, it should not be counted. For instance, one can eliminate it by choosing a gauge such that $\lambda = 0$. The residual gauge symmetries are then

$$\delta\lambda = 0 \quad \Rightarrow \quad \dot{\varepsilon} + \varepsilon' = 0 \quad \Rightarrow \quad \varepsilon = \varepsilon(t + x) \tag{5}$$

The number of degrees of freedom of the theory are thus all contained in φ. Since φ obeys a first-order differential equation, one should just specify φ at the initial time $t = 0$ to get a unique solution of the equations of motion. Besides, the gauge transformations (4)—and in particular, the residual gauge invariance (5)—have no action on φ when $\dot{\varphi} = \varphi'$, so that every solution of $\dot{\varphi} = \varphi'$ is physically distinct from every other. In that sense the gauge invariance (4) is not really present in the φ-sector, and as we will see, for that reason, φ cannot make (4) anomalous.

The fact that $\varphi(t = 0, x)$ is all there is to specify at the initial time indicates that φ plays simultaneously the role of a coordinate and a momentum, i.e., is somehow self-conjugate. This has the following consequences, which possess important implications quantum mechanically: (i) $[\varphi(t, x), \varphi(t, x')] \neq 0$: The fields φ at two different spacelike related points do not commute. (ii) In the path integral representation of the transition amplitude, one cannot give $\varphi(x)$ at the initial and final times t_i, t_f (contrary to what is done for a standard scalar field) since this would amount to specifying simultaneously the q's and the p's, which would contradict the uncertainty principle.

2.4. Hamiltonian Formulation

The safest way known to pass to quantum mechanics when a gauge invariance is present relies on the Hamiltonian formulation of the theory.

The canonically conjugate momenta derived from (2) read

$$\pi_\varphi = \dot{\varphi} + 2\lambda(\dot{\varphi} - \varphi') \tag{6}$$

$$\pi_\lambda = 0 \tag{7}$$

The Hamiltonian is equal to

$$H = \int (\dot{\varphi}\pi_\varphi + \dot{\lambda}\pi_\lambda - L)\, dx$$

$$= \int \left[\tfrac{1}{2}(\pi_\varphi^2 + \varphi'^2) - \frac{\lambda}{1+2\lambda}(\pi_\varphi - \varphi')^2\right] dx \tag{8}$$

The total Hamiltonian [6] is obtained by adding to (8) the primary constraint (7) multiplied by a Lagrange multiplier, which we denote by u,

$$H_T = \int \left[\tfrac{1}{2}(\pi_\varphi^2 + \varphi'^2) - \frac{\lambda}{1+2\lambda}(\pi_\varphi - \varphi')^2 + u\pi_\lambda\right] dx \tag{9}$$

The preservation in time of the primary constraint (7) implies

$$\dot{\pi}_\lambda = 0 \quad\Rightarrow\quad (\pi_\varphi - \varphi')^2 = 0 \tag{10a}$$

This equation is clearly equivalent to

$$\pi_\varphi = \varphi' = 0 \tag{10b}$$

Following Dirac [6], we will in the sequel replace (10a) by (10b). As analyzed in the next section, this is a mandatory step for correctly applying the standard Dirac analysis and, hence, properly building the quantum theory.

The constraint (10b) is preserved in time by the Hamiltonian (9), so that the consistency algorithm ends here. The theory is thus described by the Hamiltonian (9) and by the constraints (7), (10b).

2.5. First and Second Class Constraint Surfaces

2.5.1. Intrinsic Definition

The standard definition of first-class and second-class constraints is usually given in terms of the rank of the matrix of the brackets of the constraints,

$$C_{\alpha\beta} = [\chi_\alpha, \chi_\beta] \tag{11}$$

For instance, if $\det C_{\alpha\beta} \neq 0$ on the constraint surface, then the constraints $\chi_\alpha = 0$ are all second class, while if $C_{\alpha\beta} \approx 0$, the constraints are all first class.

This standard definition, however, is not invariant under arbitrary redefinitions of the constraints. This appears very strikingly in the second-class case. If one replaces $\chi_\alpha = 0$ by $\bar{\chi}_\alpha \equiv \chi_\alpha^2 = 0$ one gets $[\bar{\chi}_\alpha, \bar{\chi}_\beta] \approx 0$. It would seem that the same constraint surface can be viewed as either second class or first class.

In order to avoid this paradox, Dirac [6] imposed the condition that the Jacobian matrix of the differentials $d\chi_\alpha$ be of maximal rank. This condition can be rephrased as saying that the constraint functions χ_α can be taken, in the vicinity of $\chi_\alpha = 0$, as first coordinates of a new, regular coordinate system [10]. (This condition can be weakened to include the "reducible case," but this will not concern us here [11].)

When the constraints χ_α are chosen in that manner, the rank of the matrix $C_{\alpha\beta}$ is invariant under the allowed redefinitions of the constraints (which are those preserving the rank condition on $d\chi_\alpha$). Hence, the standard definition of first-class and second-class constraint surfaces is applicable only when the rank of $d\chi_\alpha$ is maximum.

When the condition on rank $(d\chi_\alpha)$ is not fulfilled, one cannot use the standard definition to determine whether the constraint surface is first class or second class. One must either replace $\chi_\alpha = 0$ by equivalent constraints $\chi'_A = 0$ that fulfill this condition, or one must use an alternative definition, invariant under arbitrary redefinitions of the constraint functions and equivalent to the standard one when the rank of $d\chi_\alpha$ is maximum. One such alternative definition is based on the properties of the two-form induced on the constraint surface by the phase space Poisson bracket.

It is well known that the Poisson bracket structure is equivalent to a rank-two contravariant antisymmetric tensor which is invertible and which, in turn, determines a two-form upon inversion. This two-form is closed by the Jacobi identity and is called the symplectic two-form. Given an arbitrary surface embedded in phase space, one can study the properties of the two-form induced on that surface by the symplectic two-form. If the induced two-form is regular, i.e., invertible, then the surface is second class. If, on the other hand, the induced two-form is noninvertible, there are first-class constraints as well. The number of independent first-class constraints is equal to the number of independent null eigenvectors of the induced two-form. If this number is maximum, namely, if it is equal to the dimension of phase space minus the dimension of the surface, the surface is first class. One also says that it is coisotropic, since the subspace orthogonal (in the symplectic product) to a first-class surface is tangent to that surface.

These definitions are equivalent to the standard ones when one represents the constraint surface by equations $\chi_\alpha = 0$ such that rank $d\chi_\alpha$ is maximum. However, they possess the advantage of being intrinsic, since they clearly do not depend on how one represents the constraint surface. For this reason they will be used here.

2.5.2. Quantization

In the case of a second-class constraint surface, one can define an induced Poisson bracket structure since the induced symplectic form is invertible. Hence the constraint surface possesses the ordinary structure of a standard phase space. According to Ref. 6, it is this smaller phase space that must be quantized. Since both the surface and its induced bracket are intrinsically defined, the quantum theory is formally independent of any particular representation of the constraint surface.

For the purpose of explicitly evaluating the induced bracket structure, it is convenient, however, to replace the constraints by ones that obey the maximum rank condition. After that replacement (and only then!), the bracket is given by

$$[A, B]^*[A, B] - [A, \chi_\alpha][\chi_\alpha, \chi_\beta]^{-1}[\chi_\beta, B]$$

an expression known as the "Dirac bracket."

When there are first class constraints present, there is no induced Poisson bracket structure on the constraint surface. However, one can show that the null directions of the induced two-form are surface-forming, i.e., can be integrated to yield well defined "orbits" on the constraint surface.

If it is physically legitimate to identify all points lying on a given orbit, then, one must pass to the quotient space of the orbits, known as the reduced phase space. The reduced phase space possesses a well defined Poisson bracket structure and can thus be a starting point for quantization. The quantum theory based on the reduced phase space turns out to be equivalent, at least in simple cases, to the quantum theory in which the full phase space is quantized and the constraints are imposed as conditions on the physical states.

If it is not legitimate to identify all points lying on the same orbits, then, it is not clear how to even start constructing the quantum theory. This is so because one does not know what the commutation relations should be, since there is no obvious Poisson bracket on the surface of the classical physical states.

If one takes the Hamiltonian as the starting point one can always consistently postulate that all points on a given orbit are to be identified. However, if one starts from a Lagrangian, the Hamiltonian dynamics thus obtained may cease to be equivalent to the original Lagrangian one. The statement that the Hamiltonian and Lagrangian dynamics are equivalent after identifying points on the same orbit is known as the "Dirac conjecture." There are counterexamples to it, although none of them is of known physical interest. These counterexamples typically occur when the secondary first-class constraints arise "squared" in the constraint algorithm. In that instance, these first-class constraints restrict the dynamics, but fail to generate gauge transformations. In the case of chiral bosons, one would get a counterexample to the Dirac conjecture, and one would not know how to pass to the quantum theory, if there were first-class constraints present among the secondary constraints (10b), $\pi_\varphi - \varphi' = 0$. These constraints are, however, all second class, provided one adopts suitable conditions at the spatial boundaries. We will therefore adopt in the sequel spatial boundary conditions which make all the constraints $\pi_\varphi - \varphi' = 0$ second class. The precise form of these boundary conditions will be given in Sections 2.10 and 2.11.

2.6. Dirac Bracket—Elimination of Second-Class Constraints

The constraint $\pi_\lambda = 0$ is clearly first class. The bracket of the constraints $\pi_\varphi - \varphi'$ is given by

$$[(\pi_\varphi - \varphi')(x), (\pi_\varphi - \varphi')(x')] = -2\delta'(x - x') \tag{12}$$

where the Poisson brackets are evaluated at equal times. The spatial boundary conditions alluded to above must thus be such that δ' is invertible, since only in this case will all the constraints $\pi_\varphi - \varphi'$ be second class.

Straightforward calculation yields for the Dirac bracket

$$[\varphi(x), \pi_\varphi(x')]^* = \tfrac{1}{2}\delta(x - x') \tag{13a}$$

or

$$[\varphi(x), \varphi'(x')]^* = \tfrac{1}{2}\delta(x - x') \tag{13b}$$

With the boundary conditions given below, equation (13b) is equivalent to

$$[\varphi(x), \varphi(x')] = -\tfrac{1}{4}\varepsilon(x - x') \tag{13c}$$

with $\varepsilon(x - x') = -1$ for $x < x'$ and $+1$ for $x > x'$.

An immediate consequence of (13c) is that at spacelike separations the fields φ fail to commute and cannot be diagonalized simultaneously.

We will from now on reduce the dynamics to the constraint surface $\pi - \varphi' = 0$, as one should since these constraints are second class. This will be done by eliminating π from the theory and using only the Dirac bracket (13), which permits indeed consistently setting π equal to φ' throughout. Since no confusion should arise, we will also drop the * on the Dirac bracket.

When π is set equal to φ', the action becomes

$$S[\varphi, \lambda, \pi_\lambda, u] = \int dt\, dx(\dot{\varphi}\varphi' + \dot{\lambda}\pi_\lambda - \varphi'^2 - u\pi_\lambda) \tag{14a}$$

and the canonical and total Hamiltonians are, respectively,

$$H = \int \varphi'^2\, dx, \qquad H_T = H + \int u\pi_\lambda\, dx \tag{14b}$$

The action (14a) is invariant under

$$\delta\varphi = 0 \tag{15a}$$

$$\delta\lambda = \varepsilon, \qquad \delta\pi_\lambda = 0, \qquad \delta u = \dot{\varepsilon} \tag{15b}$$

These gauge transformations are the Hamiltonian expression of the Lagrangian gauge transformations (4) in the phase space $(\varphi, \lambda, \pi_\lambda)$, which one arrives at upon elimination of the second-class constraints $\pi - \varphi' = 0$. We thus clearly see that once these constraints are properly incorporated, the gauge transformations do not act on the 0-form φ.

The canonical generator of (15) on the canonical variables $(\varphi, \lambda, \pi_\lambda)$ is given by

$$G[\varepsilon] = \int dx\, \epsilon \pi_\lambda \tag{16}$$

and is thus just a combination of the first-class constraint (7).

The pair (λ, π_λ) is pure gauge and can be dropped from the theory. When one sets $\pi_\lambda = 0$ in the action (14a), one finds that it reduces to

$$S[\varphi] = \int dt\, dx\, (\dot{\varphi}\varphi' - \varphi'^2) \tag{17}$$

which is the action proposed in Ref. 12 for describing a chiral boson. This action appears therefore to be the action resulting from (2) by application of the Dirac formalism for constrained systems. The fact that the action of Ref. 7 reproduces the action (17) when appropriately handled has been previously pointed out in Ref. 13.

By using the field strength $\varphi' \equiv \chi$ as new variable, one can rewrite the action (17) in the nonlocal form of Ref. 12.

2.7. Lorentz Invariance

The action (17) is invariant under Poincaré transformations. These transformations are obtained by evaluating the Dirac bracket of the scalar field φ with the standard Poincaré generators

$$P^0 = \tfrac{1}{2} \int dx\, (\pi^2 + \varphi'^2) = \int dx\, \varphi'^2 \tag{18a}$$

$$P^1 = \int dx\, \pi\varphi' = \int dx\, \varphi'^2 = P^0 \tag{18b}$$

and

$$M_{01} = \int dx\, (t + x)\varphi'^2 \tag{18c}$$

One finds

$$\delta\varphi = \varphi' \tag{19a}$$

for translations, and

$$\delta\varphi = (t + x)\varphi' \tag{19b}$$

for boosts.

Actually, the Dirac and Poisson brackets of any function with one of the Poincaré generators are equal, because the Poisson bracket of the Poincaré generators with the constraints $\pi - \varphi'$ all vanish. This is a consequence of the Poincaré invariance of the self-duality condition. It results from this observation that the Poincaré generators close in the Dirac bracket as they do in the Poisson bracket, i.e., according to the Poincaré algebra.

Of particular importance is the fact that the transformations (19) are linear in the fields. This feature, however, is generically destroyed by the interactions (see Section 2.13).

2.8. Comparison with Ordinary Klein–Gordon Field

The fact that a chiral scalar field does not commute with itself when evaluated at two different points even when these are spacelike related can be understood from a different point of view.

Consider an ordinary Klein–Gordon field ψ with conjugate momentum p,

$$S[\psi, p] = \int dt\, dx[\dot{\psi}p - \tfrac{1}{2}(p^2 + \psi'^2)] \tag{20}$$

The field equations imply $\ddot{\psi} = \psi$, so that ψ is a sum of a left-moving (chiral) and a right-moving (antichiral) scalar field.

It is possible to pass from ψ, p to new variables φ^R and φ^L which are the right- and left-moving components of ψ. This is done by defining

$$\varphi^L(x) = \frac{1}{2}\left[\psi(x) + \int_{-\infty}^{x} p(y)\, dy - \tfrac{1}{2}\int_{-\infty}^{\infty} p(y)\, dy\right] \tag{21a}$$

and

$$\varphi^R(x) = \frac{1}{2}\left[\psi(x) - \int_{-\infty}^{x} p(y)\, dy + \tfrac{1}{2}\int_{-\infty}^{\infty} p(y)\, dy\right] \tag{21b}$$

The field equations for ψ, p ($\dot{\psi} = p, \dot{p} = \psi''$) imply that φ^L and φ^R obey the appropriate equations for left- and right-moving bosons ($\dot{\varphi}^L = \varphi'^L, \dot{\varphi}^R = -\varphi'^R$).

We assume that $\psi, p \to 0$ as $x \to \pm\infty$ and that $\int_{-\infty}^{\infty} p(y)\, dy$ is finite. The left- and right-moving components of ψ are defined up to an arbitrary constant, since a constant is simultaneously right and left moving. That constant is taken in (21) such that φ^L and ψ^R commute [see (23c) below]. One then finds $\varphi^R(-\infty) = -\varphi^L(+\infty) = -\varphi^L(-\infty) = +\varphi^L(+\infty)$.

Now, a crucial property of the change of variables (21) is that φ^L and φ^R depend on both the scalar field ψ and its conjugate p. Furthermore, the transformation (21) is nonlocal in space. It thus comes as no surprise that left- and right-moving bosons obey, at equal times, the nonvanishing and nonlocal Poisson bracket relations found previously. From (21) and the equal time Poisson brackets

$$[\psi(x'), \psi(y')] = 0 = [p(x'), p(y')] \tag{22a}$$

$$[\psi(x'), p(y')] = \delta(x' - y') \tag{22b}$$

one finds

$$[\varphi^L(x'), \varphi'^L(y')] = \tfrac{1}{2}\delta(x' - y') \tag{23a}$$

$$[\varphi^R(x'), \varphi'^R(y')] = -\tfrac{1}{2}\delta(x' - y') \tag{23b}$$

$$[\varphi^L(x'), \varphi^R(y')] = 0 \tag{23c}$$

Relation (23a) coincides with (13b) above.

The relations (21) can be inverted to yield ψ and p as functions of φ^L, φ^R. One gets

$$\psi = \varphi^L + \varphi^R \tag{24a}$$

$$p = \varphi'^L - \varphi^R \tag{24b}$$

If one inserts (24) in the action (20), one finds

$$S = S^L[\varphi^L] + S^R[\varphi^R] + I(t_1, t_z) \tag{25a}$$

with

$$S^L = \int_{t_1}^{t_2} dt\, dx\, (\dot{\varphi}^L \varphi'^L - \varphi'^{L2}) \tag{25b}$$

$$S^R = \int_{t_1}^{t_2} dt\, dx\, (-\dot{\varphi}^R \varphi'^R - \varphi'^{R2}) \tag{25c}$$

The action (25b) is again the action for a left-moving boson that we found previously.

The last term $I(t_1, t_2)$ is a surface term at t_1 and t_2,

$$I(t_1, t_2) = \frac{1}{2}\left[\int_{-\infty}^{\infty} (\varphi'^L \varphi^R - \varphi'^R \varphi^L)\, dx\right]_{t_1}^{t_2} \tag{25d}$$

and its role will be discussed in more detail below. There are no analogous terms at spatial infinity, $x \to \pm\infty$, because of the inclusion of the appropriate constant in the definition (21) of the right- and left-moving components of ψ. (It is because of the absence of this term that the kinetic piece in the action does not mix φ^R and φ^L, and thus, $[\varphi^R, \varphi^L] = 0$.) The fields φ^L and φ^R are decoupled in the action, except in the boundary term at t_1 and t_2.

2.9. Boundary Conditions at t_1 and t_2

If one varies the action (17) for the left-moving boson φ, one finds, keeping all the terms,

$$\delta S[\varphi] = -2\int dt\,dx\,(\dot{\varphi}-\varphi')'\delta\varphi + \left[\int dx\,\delta\varphi\,\varphi'\right]_{t_1}^{t_2} + \left[\int dt\,\delta\varphi(\dot{\varphi}-2\varphi')\right]_{x_1}^{x_2} \tag{26}$$

where x_2 and x_1 are the spatial end points. If S is to be an extremum for the classical history, i.e., if $\delta S = 0$, then, each term on the right-hand side of (26) should be zero.

The vanishing of the "volume" term [first term on the right-hand side of (26)] leads to the field equations,

$$(\dot{\varphi}-\varphi')' = 0 \tag{27}$$

As we will show later, these equations imply $\dot{\varphi}-\varphi' = 0$.

The vanishing of the second term in (26), which is a surface term at the time boundaries, dictates what should be fixed at t_1 and t_2 in the variational principle. The condition under which this term is zero is analyzed in this section. The vanishing of the third term is discussed afterwards.

As we pointed out already, one cannot specify φ independently at t_1 and t_2, since φ obeys a first-order equation in time. Hence, one cannot assume that $\delta\varphi$ in (26) vanishes at both t_1 and t_2.

What are the appropriate boundary conditions at the time boundaries for a self-conjugate field, which obeys a first-order equation? Clearly, one should have only one datum per space point. Furthermore, the boundary conditions should involve t_1 and t_2 in a somewhat symmetric manner, since the surface term in the action should vanish at both t_1 and t_2. For instance, $\varphi(t_1, x) =$ given function of x would not do, since one would not find that the surface term vanishes at t_2. Furthermore, the term at t_2 is not equal to the variation of something, so it cannot be absorbed in a redefinition of the action.

Following what has proved successful for self-conjugate fermions [14], which possess analogous properties, we tentatively impose the conditions

$$\tfrac{1}{2}[\varphi(t_2, x) + \varphi(t_1, x)] = \xi(x) \tag{28a}$$

so that

$$\delta\varphi(t_2, x) = -\delta\varphi(t_1, x) \tag{28b}$$

in the variational principle. We will see that these conditions are classically permissible, in the sense that given $\xi(x)$ there is only one classical history that fulfills (28a), except maybe for unfortunate choices of $t_2 - t_1$ (see Section 2.11).

One finds that the surface term at t_1 and t_2 becomes

$$\int dx\, \delta\varphi(t_2, x)[\varphi'(t_2, x) + \varphi'(t_1, x)]$$

$$= \delta \int dx\, \varphi(t_2, x)[\varphi'(t_2, x) + \varphi'(t_1, x)]$$

$$= \tfrac{1}{2}\delta \int dx\, [\varphi(t_2, x) - \varphi(t_1, x)][\varphi'(t_2, x) + \varphi'(t_1, x)]$$

Therefore, if one improves the action as

$$S[\varphi] \to S^{\text{improved}} = S[\varphi] + J(t_1, t_2) \tag{29a}$$

with

$$J(t_1, t_2) = \tfrac{1}{2} \int dx\, [\varphi(t_1, x) - \varphi(t_2, x)][\varphi'(t_2, x) + \varphi'(t_1, x)] \tag{29b}$$

a modification that has no effect in the equations of motion, there is no surface term at the time boundaries in $\delta S^{\text{improved}}$.

The improved action (29) will be used from now on to describe a chiral boson. It turns out that the term $J(t_1, t_2)$ is extremely important in the path integral, and that if it is not included, one does not get the correct evolution operator.

In order to establish this result, (in Section 3.2) we will need to compare (29) with the Klein-Gordon action.

In the Klein-Gordon variational principle one usually assumes that the Klein-Gordon field ψ, which obeys a second-order equation, is given at both t_1 and t_2. It is, however, possible, and useful for our purposes, to consider a different action principle, in which ψ and its momentum p are treated in a more symmetric fashion. In this action principle one holds fixed the sums

$$\tfrac{1}{2}[\psi(t_2, x) + \psi(t_1, x)] = \Psi \tag{30a}$$

and

$$\tfrac{1}{2}[p(t_2, x) + p(t_1, x)] = P(x) \tag{30b}$$

Then, the action is an extremum on the classical history only if one includes [15] an appropriate term at t_1 and t_2, given by

$$S^{\text{K.G.}} \to \tilde{S}^{\text{K.G.}} = \int [p\dot{\psi} - \tfrac{1}{2}(p^2 + \psi'^2)]\, dt\, dx$$

$$- \tfrac{1}{2} \int dx\, [p(t_2, x) + p(t_1, x)][\psi(t_2, x) - \psi(t_1, x)] \tag{30c}$$

Now, the boundary conditions (30) are clearly equivalent to boundary conditions of the form (28) for the right- and left-moving components of ψ. Moreover, if one expresses the action (30c) in terms of φ^R and φ^L, using (25) and

$$\varphi^{R}_{L}(t, +\infty) = -\varphi^{R}_{L}(t, -\infty)$$

one obtains

$$\tilde{S}^{\text{K.G.}} = S^{\text{improved}}[\varphi^R] + S^{\text{improved}}[\varphi^L] \tag{30d}$$

i.e., the left- and right-moving components are now completely decoupled, even in the surface terms at the initial and final times. This will turn out to be crucial in the method we will adopt for establishing the correct path integral.

2.10. Boundary Conditions at Spatial Infinity (Real Line)

It remains to discuss the third term in the variation (26) of the action, which is the surface term at spatial infinity.

In view of our discussion of the Klein–Gordon field and of its right and left moving components, it appears natural to require that the chiral field obeys the conditions

$$\lim_{x\to\infty} \varphi(x) = a \qquad \lim_{x\to-\infty} \varphi(x) = -a \tag{31a}$$

and

$$\lim_{x\to\infty} \varphi'(x) = 0 = \lim_{x\to-\infty} \varphi'(x) \tag{31b}$$

at spatial infinity.

That these conditions are legitimate can be seen as follows:

(i) They make all the constraints $\pi - \varphi' = 0$ second class, as they should. Indeed, the equation

$$\left[\int \lambda(x)(\pi(x) - \varphi'(x))\, dx, \pi(x') - \varphi'(x')\right] \tag{32a}$$

implies

$$\lambda' = 0 \tag{32b}$$

and thus

$$\lambda = 0 \tag{32c}$$

since λ should belong to the same functional class as φ. The only constant contained in that class is zero by virtue of (3.1). There is therefore no combination of the constraints $\pi - \varphi'$ that is first class.

(ii) As a result, the action obtained by elimination of π is equivalent to the original action. And one finds, indeed, that the variational equation

(27) $(\dot{\varphi} - \varphi')' = 0$ implies by integration $\dot{\varphi} - \varphi' = C$, where the constant C must vanish since $C = \dot{\varphi}(+\infty) = -\dot{\varphi}(-\infty) = -C[\varphi'(\pm\infty) = 0]$. Without the elimination of the constant zero mode of φ, neither (i) nor (ii) would hold, and the action (27) would not yield equations equivalent to $\dot{\varphi} - \varphi' = 0$.

With the boundary conditions (31), the surface terms at $x = \pm\infty$ in (26) cancel, so that the improved action (29) is fully satisfactory as it stands.

We also note that the classical history is uniquely determined by the conditions (28a) at the time boundaries. This is because φ is a function of $t - x$ by the field equations, so that (28a) becomes

$$\tfrac{1}{2}[\varphi(x) + \varphi(x + \Delta t)] = \xi(x) \tag{33}$$

with $\varphi(x) \equiv \varphi(t_1, x)$ and $\Delta t = t_2 - t_1$. From (33), we infer

$$\varphi(x) = \sum_{n=1}^{N} [\xi(x - n\,\Delta t) + \xi(x + n\,\Delta t - \Delta t)](-1)^{n+1} + \tfrac{1}{2}(-1)^N[\varphi(x + N\,\Delta t) + \varphi(x - N\,\Delta t)] \tag{34a}$$

When $N \to \infty$, the second term in the right-hand side of (34a) goes to zero since $\varphi(+\infty) = -\varphi(-\infty)$. Hence, (34a) leads to

$$\varphi(x) = \sum_{n=1}^{\infty} [\xi(x - n\,\Delta t) + \xi(x + n\,\Delta t - \Delta t)](-1)^{n+1} \tag{34b}$$

The function $\xi(x)$ is such that $\xi(x - n\,\Delta t) + \xi(x + n\,\Delta t - \Delta t) \to 0$ as $n \to \infty$ because it clearly obeys boundary conditions at $x \to \pm\infty$ of the type (31). Furthermore, it must be such that the sum in (34b) converges. The equation (34b) shows that $\varphi(x)$ is completely determined by $\xi(x)$.

Conversely from the expression (34b), one easily checks that (33) holds. Indeed, $\varphi(x) = \sum_{1\le n\le N} [\xi(x - n\,\Delta t) + \xi(x + n\,\Delta t - \Delta t)](-1)^{n+1} + a_N$ and $\varphi(x + \Delta t) = \sum_{1\le n\le N} [\xi(x + \Delta t - n\,\Delta t] + \xi(x + n\,\Delta t)](-1)^{n+1} + b_N$, where, for large enough N, a_N and b_N are arbitrarily small. Thus, $\varphi(x) + \varphi(x + \Delta t) - 2\xi(x) = a_N + b_N + c_N$, where $c_N = \xi(x - N\,\Delta t)(-1)^{N+1} + \xi(x + N\,\Delta t)(-1)^{N+1}$. The right-hand side of this equation is arbitrarily small for $N \to \infty$ and must be independent of N. Therefore, it must be equal to zero.

As a final comment, we note that there may exist other acceptable boundary conditions as $x \to \pm\infty$. The criterion that these should fulfill is that they eliminate the constant mode from φ, so that $\lambda' = 0$ implies $\lambda = 0$ as in (32). We will not investigate this question further here.

2.11. Case of a Circle

We now assume that the spatial sections are compact, i.e., are circles. There is then no spatial surface term to worry about. We take those circles to be the interval $[0, \pi]$, with 0 and π identified, since we want to compare our results with the heterotic string theory, for which these are the standard conventions. For the same reason, we insert $1/\pi$ in front of the action (2) and we call the space variable σ instead of x.

It is useful to separate from $\varphi(\sigma)$ and $\pi(\sigma)$ their zero-mode component. Hence, we write

$$\varphi(\sigma) = \varphi_0 + \bar{\varphi}(\sigma) \tag{35a}$$

$$\pi(\sigma) = (1/\pi)\pi_0 + \bar{\pi}(\sigma) \tag{35b}$$

with

$$\int_0^\pi \varphi(\sigma)\, d\sigma = \pi\varphi_0, \qquad \int_0^\pi \pi(\sigma)\, d\sigma = \pi_0 \tag{35c}$$

and Poisson brackets

$$[\varphi_0, \pi_0] = 1, \qquad [\bar{\varphi}(\sigma), \bar{\pi}(\sigma')] = \delta(\sigma, \sigma') - 1/\pi \tag{35d}$$

(the other brackets vanish).

In terms of φ_0, π_0, $\bar{\varphi}$, and $\bar{\pi}$, the secondary constraints $\pi - \varphi' = 0$ read

$$\pi_0 = 0 \tag{36a}$$

$$\bar{\pi} - (1/\pi)\bar{\varphi}' = 0 \tag{36b}$$

and one immediately faces the difficulty that only the constraints (36b) are second class. The zero mode constraint (36a) is clearly first class, but it does not appear to correspond to any gauge invariance of the theory, since one cannot shift φ by an arbitrary function of time.

In order to avoid this problem—which would not occur if we had not included a zero mode component in φ and π—we will provisionally drop the first-class constraint (36a) and keep only the second-class constraints (36b). This amounts to replacing $\lambda(\dot{\varphi} - \varphi')^2$ by $\lambda(\dot{\bar{\varphi}} - \bar{\varphi}')^2$ in the Lagrangian action (2). In the next section, we will complete this step by enlarging the functional space to which $\varphi(\sigma)$ belongs in such a way that the left-moving condition on the zero mode can be reinserted.

The second-class constraints can be used to eliminate $\bar{\pi}(\sigma)$, as we did above. The final canonical action is then

$$S[\varphi_0, \pi_0, \bar{\varphi}(\sigma)] = \int dt \left[\dot{\varphi}_0\pi_0 - \tfrac{1}{2}\pi_0^2 + \frac{1}{\pi}\int d\sigma\, (\bar{\varphi}'\dot{\bar{\varphi}} - \bar{\varphi}'^2)\right]$$
$$+ \text{surface terms at } t_1 \text{ and } t_2 \tag{37}$$

The field $\bar{\varphi}(\sigma)$ can be expanded in Fourier modes,

$$\bar{\varphi}(\sigma) = \frac{i}{2}\sum_{n>0}\frac{1}{\sqrt{n}}(a_n\, e^{-2in\sigma} - a_n^*\, e^{2in\sigma}) \tag{38a}$$

and one finds

$$[a_n, a_n^*] = -i\delta_{n,n} \tag{38b}$$

The equations of motion imply

$$a_n(t) = a_n^0\, e^{-2in(t-t_0)} \tag{39a}$$

and one sees that a_n^0—and thus also $a_n(t)$—can be expressed in terms of $2\xi_n = a_n(t_0) + a_n(t_1)$ as

$$a_n^0 = \frac{2\xi_n}{1 + e^{-2in(t_1-t_0)}} \tag{39b}$$

The denominator in (39b) is different from zero, except for unfortunate choices of $t_1 - t_0$ for which $2(t_1 - t_0)$ is a rational number p/q times π, with p odd. As to the boundary conditions on the zero modes at t_0 and t_1, they should be of the standard type [e.g., $\varphi_0(t_0)$ and $\varphi_0(t_1)$ given, or $\pi_0(t_0)$ and $\pi_0(t_1)$], since the zero modes just form an ordinary canonical pair.

2.12. Winding Number—Heterotic String

When the field φ itself takes values on a circle, one can allow for topological solitons, described by a linear term in σ,

$$\varphi = \varphi_0 + L\sigma + \frac{i}{2}\sum_{n>0}\frac{1}{\sqrt{n}}(a_n\, e^{2in\sigma} - a_n^*\, e^{-2in\sigma}) \tag{40a}$$

L should be quantized as

$$L = 2Rn \tag{40b}$$

where n is an integer, called the winding number. As σ is increased by π, $\sigma \to \sigma + \pi$, φ becomes $\varphi + 2n\pi R$, which is identified with φ.

When the solitonic term is included in (40), one finds that the action (37) is modified as

$$S[\varphi_0, \pi_0, \bar{\varphi}(\sigma)] = \int dt\left[\dot{\varphi}_0\pi_0 - \tfrac{1}{2}\pi_0^2 - \tfrac{1}{2}L^2 + \frac{1}{\pi}\int_0^\pi d\sigma\,(\bar{\varphi}, \dot{\bar{\varphi}} - \bar{\varphi}'^2)\right] \tag{41}$$

where $\bar{\varphi}$ is still given by (38), i.e., contains only the oscillator variables. The energy and momentum are given by

$$H = \int T_{00}\, d\sigma = \tfrac{1}{2}\pi_0^2 + \tfrac{1}{2}L^2 + 2N \tag{42a}$$

$$P = \int T_{01}\, d\sigma = \pi_0 L + 2N \tag{42b}$$

with

$$N = \sum_{n>0} n a_n^* a_n \tag{42c}$$

Since L is a topological quantity, it would appear that it should be treated as a pure number, as was done in (41). However, it turns out that in the quantum theory, one can have transitions between the different topological sectors as a result of interactions. It is thus inconsistent to consider a single topological sector.

The transitions between the various L's can be described by means of an operator M which does not commute with L. It is convenient to have a classical analog of this operator, so we introduce a new variable M, with bracket

$$[L, M] = 1 \tag{43}$$

M can be taken to live on a circle of radius $1/2R$, so that (40b) is also implied by the quantization of the pair (L, M). Of course, $\dot{L} = 0$ in the free theory, so M should not appear in the free Hamiltonian.

The introduction of the pair L, M (with a kinetic term $M\dot{L}$ in the action) enables one to impose now the left-moving condition on the zero mode as well. Since $(\varphi_0 + L\sigma)^{\cdot} = \pi_0$ and $(\varphi_0 + L\sigma)' = L$, this condition is equivalent to

$$\pi_0 - L = 0 \tag{44a}$$

The condition (44a) is still first class by itself, but can be completed by another condition so that the full system is second class. This additional condition relates M to φ_0, so that M is not an independent variable after the constraints are eliminated, and becomes part of the already existent variables. We follow here an approach inspired by Ref. 8. Other equivalent treatments, which do not introduce M and do not impose (44b), may also be possible.

If one demands that the condition relating M to φ_0 be preserved by time and space translations, i.e., commute with H and P, one is led to the only possibility

$$M + \varphi_0 = 0 \tag{44b}$$

The elimination of L and M by means of the second-class constraints (44a) and (44b) yields the canonical action

$$S[\varphi_0, \pi_0, \bar{\varphi}(\sigma)] = \int dt \left[2\dot{\varphi}_0\pi_0 - \pi_0^2 + \frac{1}{\pi}\int_0^{\pi} d\sigma\, (\bar{\varphi}'\dot{\bar{\varphi}} - \bar{\varphi}'^2) \right] \tag{45a}$$

and the energy and momentum

$$H = \pi_0^2 + 2N \tag{45b}$$

$$P = \pi_0^2 + 2N = H \tag{45c}$$

The only bracket that is modified is $[\varphi_0, \pi_0]$; it becomes

$$[\varphi_0, \pi_0] = 1/2 \tag{45d}$$

It should be stressed that (45a) does not follow from the Siegel Lagrangian [7] alone. However, the modification concerns only the zero mode sector and simply consists in a proper incorporation of the winding number, by adding an extra variable (M) and an extra constraint [(44b)].

The allowed eigenvalues of the momentum π_0 are quantized in the quantum theory, as

$$\pi_0 = m/2R \tag{46}$$

so that the wave functions $\exp[i(2\pi_0)\varphi_0]$ are unchanged as φ_0 is shifted to $\varphi_0 + 2\pi R$ (recall that it is $2\pi_0$ that is conjugate to φ_0). This condition is consistent with (40b) and (44a) provided the radius R is restricted appropriately.

The heterotic construction is obtained by generalizing the above analysis to the case of 16 chiral bosons lying on a self-dual, integral, and even lattice [8].

2.13. Coupling of Chiral Bosons to a Complex Scalar Field

As discussed in Refs. 16 and 7, a chiral p-form can be coupled to a complex antisymmetric tensor of rank $p/2$.

For $p = 0$, the appropriate action is [7]

$$S[\varphi, B, \bar{B}, \lambda] = \int dt\, dx\, [-\tfrac{1}{2}F_\mu F^\mu - \partial_\mu \bar{B}\partial^\mu B - i\varphi\varepsilon^{\mu\nu}\partial_\mu \bar{B}\partial_\nu B + \lambda F_-^2] \tag{47}$$

where the field strength F_μ of φ is modified by the interaction and reads

$$F_\mu = \partial_\mu \varphi + \tfrac{1}{2}i\bar{B}\partial_\mu B - \tfrac{1}{2}i\partial_\mu \bar{B}B \tag{48}$$

The canonical analysis of the action (47) goes along the same lines as in the free case. In particular, one finds that the primary first-class constraint $\pi_\lambda = 0$ leads to the secondary second-class constraint

$$\pi_\varphi - F_1 = 0 \tag{49}$$

which is the canonical transcription of the self-duality condition $F_- = 0$. The determinant of the second-class constraints is again given by $\delta'(\sigma - \sigma')$ and thus can be inverted.

The major new feature of the interacting models is that the expression for π_φ in terms of the fields is no longer linear, but rather, involves the quadratic term $\bar{B}B' - \bar{B}'B$. As a result, the bracket of the momentum conjugate to B with φ contains $\bar{B}$, and furthermore, the Lorentz generators are no longer quadratic. This suggests difficulties in the quantum theory (realization of the Dirac bracket, Lorentz anomaly) which, however, will not be studied here.

2.14. Gravitational Coupling

The energy momentum tensor components of a chiral boson are given by

$$\mathcal{H} \equiv gT_{00} = \varphi'^2 \tag{50a}$$

$$\mathcal{H}_1 \equiv gT_{01} = \varphi'^2 \tag{50b}$$

$$gT_{11} = \varphi'^2 \tag{50c}$$

The trace is zero, as dictated by Weyl invariance. In addition, the left-moving condition implies that there is only one independent component of $T_{\alpha\beta}$, since T_{--} also vanishes identically,

$$T_{++} \equiv \tfrac{1}{2}(\mathcal{H} + \mathcal{H}_1) = \varphi'^2, \qquad T_{--} \equiv \tfrac{1}{2}(\mathcal{H} - \mathcal{H}_1) = 0 \tag{50d}$$

The energy-momentum tensor components obey the appropriate surface-deformation algebra [4, 5],

$$[\varphi'^2(\sigma), \varphi'^2(\sigma')] = [\varphi'^2(\sigma) + \varphi'^2(\sigma')]\delta'(\sigma, \sigma') \tag{51}$$

and accordingly, the system can be consistently coupled to gravity at the classical level.

If one parametrizes the two-dimensional metric as

$$g_{\alpha\beta} = g_{11}\begin{pmatrix} -N^2 + (N^1)^2 & N^1 \\ N^1 & 1 \end{pmatrix} \tag{52}$$

one finds that the action describing the propagation of a chiral boson in a given gravitational field reads

$$S[\varphi] = \int (\dot{\varphi}\varphi' - N\mathcal{H} - N^1\mathcal{H}_1)\, dt\, dx \tag{53}$$

The action for the propagation of d chiral bosons φ^A ($A = 0, \ldots, d-1$) is simply obtained by adding the actions for each individual chiral boson,

$$S[\varphi^A] = \int [\dot{\varphi}^A\varphi'^B - (N + N^1)\varphi'^A\varphi'^B]\eta_{AB}\, dt\, d\sigma \tag{54}$$

where η_{AB} is the metric in the internal space of the φ^A's.

One can also treat the gravitational field as a dynamical variable. The action (54) need not be supplemented by a gravitational action, since the Einstein action is a topological invariant in two dimensions. The equations obtained by varying (54) with respect to $g_{\alpha\beta}$ are

$$T_{\alpha\beta} = 0 \quad \Leftrightarrow \quad \varphi'^A\varphi'^B\eta_{AB} = 0 \tag{55}$$

In order to get a nonempty theory, the metric η_{AB} in internal space cannot be of definite sign.

As a final point, we note that there is no relation between $\varphi^A_{,\alpha}\varphi^B_{,\beta}\eta_{AB}$ and the metric $g_{\alpha\beta}$, since one finds, upon use of the field equations, that $\varphi^A_{,\alpha}\varphi^B_{,\beta}\eta_{AB}$ identically vanishes. This is in sharp contrast to what happens when the scalar fields contain both chiralities, where $g_{\alpha\beta}$ and $\varphi^A_{,\alpha}\varphi^B_{,\beta}\eta_{AB}$ are conformally related.

3. CHIRAL BOSONS—QUANTUM THEORY

3.1. Quantization in Minkowski Space

The quantization of a free chiral boson in Minkowski space is straightforward. The second-class constraints are eliminated before one goes to the quantum theory, and the Dirac brackets are turned into i-times commutators. There is thus no question as to whether one can impose $\pi - \varphi' = 0$ quantum-mechanically.

The only question that one may ask is whether the multiplier λ introduced classically can still be eliminated quantum-mechanically, i.e., whether the corresponding gauge invariance does not become anomalous in the space where $\pi - \varphi'$ is identified as the zero operator. The answer is clearly that there is no problem, since the generator (16) of that gauge invariance is linear in the momentum π_λ, which guarantees $[G[\varepsilon], G[\eta]] = 0$ even at the quantum level.

Finally, we note that the Lorentz generators close without anomaly, and that the spectrum of the theory just describes left-moving scalar particles.

3.2. Path Integral Quantization in Minkowski Space

The path integral quantization is also straightforward. The symbol $U(\xi, t_2 - t_1)$ of the evolution operator in the Weyl representation is given by the path integral

$$U[\xi; t_2 - t_1] = \int \mathcal{D}\varphi \exp iS^{\text{improved}} \tag{56}$$

where S^{improved} is the full action (29a) containing the appropriate endpoint terms at t_1 and t_2. The measure in (56) includes the determinant of the matrix of the Dirac bracket, which is just a c-number. The paths over which one sums in (56) are those that obey (28a).

That (56) should be correct is a consequence of the general analysis of [15], which implies that for an ordinary Klein–Gordon field, which obeys standard (q, p) commutation relations, the symbol $U[\psi, p; t_2 - t_1]$, with the

Weyl correspondence rule, is given by

$$U[\psi, p; t_2 - t_1] = \int \mathcal{D}\psi\mathcal{D}p \exp i\tilde{S}^{\text{K.G.}} \tag{57}$$

where $\tilde{S}^{\text{K.G.}}$ is the action (30c) with the additional appropriate end point term included.

The symbol (57) is also equal to

$$\begin{aligned} U[\psi, p; t_2 - t_1] &= \int \mathcal{D}\varphi^L\mathcal{D}\varphi^R \exp i(S^{\text{improved}}[\varphi^R] + S^{\text{improved}}[\varphi^L]) \\ &= \int \mathcal{D}\varphi^L \exp iS^{\text{improved}}[\varphi^L] \int \mathcal{D}\varphi^R \exp iS^{\text{improved}}[\varphi^R] \end{aligned} \tag{58}$$

as it merely follows from the change of integration variables $\psi, p \to \varphi^L, \varphi^R$. This change of variables is linear and one can independently Weyl-order the right- and left-moving parts of U. Therefore one has $U[\psi, p; t_2 - t_1] = U[\varphi^R; t_2 - t_1]U[\varphi^L; t_2 - t_1]$ with

$$U[\varphi^R; t_2 - t_1] = \int \mathcal{D}\varphi^R \exp iS^{\text{improved}}[\varphi^R] \tag{59a}$$

$$U[\varphi^L; t_2 - t_1] = \int \mathcal{D}\varphi^L \exp iS^{\text{improved}}[\varphi^L] \tag{59b}$$

Thus, the results of Ref. 15 also apply to a self-conjugate chiral field, as expressed by equation (56) above.

The integral (56) is Gaussian and easy to evaluate by standard methods. One finds that at the extremum, $\dot{\varphi} - \varphi'$, the "volume piece" of the action is zero, so that S^{improved} reduces to the end-point term at t_1 and t_2.

If one decomposes $\xi(\sigma)$ in terms of Fourier modes $a(k)$, normalized so that

$$[a(k), a^*(k)] = \delta(k - k') \tag{60}$$

one finds that $U[a, a^*; t_2 - t_1]$ is given by

$$\begin{aligned} U[a, a^*; t_2 - t_1] &= N \exp\left[-2i \int_0^\infty dk\, a^*(k)a(k) \frac{\sin k(t_2 - t_1)}{1 + \cos k(t_2 - t_1)}\right] \\ &= N \exp\left\{-2i \int_0^\infty dk\, a^*(k)a(k) \frac{\sin[k(t_2 - t_1)/2]}{\cos[k(t_2 - t_1)/2]}\right\} \end{aligned} \tag{61}$$

Here, N is the determinant of the quadratic part of the action and depends only on $t_2 - t_1$.

Let us stress again that (61) is not the kernel of the evolution operator in the holomorphic representation, but rather, it is its Weyl symbol. In order to get the kernel in the holomorphic representation, one should give different boundary conditions at t_2 and t_1, and add to the action the surface term appropriate to those boundary conditions [17].

It is easy to check that the expression (61) obeys the appropriate folding rule of Weyl symbols (given for instance in Ref. 15). Furthermore, when $t_2 - t_1$ is small, it reduces to

$$U[a, a^*; t_2 - t_1] \approx 1 - iH(t_2 - t_1) \tag{62a}$$

$$H = \int_0^\infty dk\, k a^*(k) a(k) \tag{62b}$$

as it should.

Also to be pointed out is the fact that the Weyl symbol of the evolution operator contains an infinite phase because the Weyl and normal orderings of the Hamiltonian differ from each other by the infinite zero point energy.

Finally, even in the presence of a finite number of oscillators, the Weyl symbol is singular when $t_2 - t_1$ is equal to an odd integer times half a period. For instance, in the case of a single harmonic oscillator of unit mass and frequency ω, (61) becomes

$$U[p, q; t_2 - t_1] = \frac{1}{\cos(\omega/2)(t_2 - t_1)} \exp[-2iH(p, q) \tan(\omega/2)(t_2 - t_1)] \tag{63a}$$

where the determinant N of the quadratic piece has now been explicitly written down

$$N = \frac{1}{\cos(\omega/2)(t_2 - t_1)} \tag{63b}$$

and where H is given by

$$H = \tfrac{1}{2}(p^2 + \omega^2 q^2) \tag{63c}$$

The reason why (63a) is singular when $t_2 - t_1 = (2n+1)(\pi/\omega)$ is that the Weyl reordering of the well-defined operator $\exp[i\hat{H}(t_2 - t_1)]$ involves an infinite number of rearrangements which yield, for those values of $t_2 - t_1$, a singular expression. From the point of view of the path integral, this fictitious singularity appears because the boundary conditions at t_1 and t_2 fail to determine a unique classical history.

3.3. Gravitational Anomaly

A condition traditionally viewed as necessary for a quantum-mechanically consistent gravitational coupling is that there should be no gravitational anomaly. By gravitational anomaly one means a modification of the algebra of surface deformations, which is the Hamiltonian translation of the algebra of diffeomorphisms.

We adopt the point of view of Faddeev where the anomaly appears in the algebra of the gauge generators [18], and we do not shift the gravitational anomaly into the Weyl anomaly. Thus the quantum theory remains manifestly Weyl invariant, but may not be invariant under diffeomorphisms. For a discussion of these matters, see Ref. 19, pp. 144–148. Also, we will work for definiteness on the circle.

The quantum algebra of the energy-momentum tensor component $T_{++}(\sigma)$ has already been computed within the context of string theory. The Fourier modes of $T_{++}(\sigma)$ are indeed just the standard left-moving Virasoro generators, which obey, in quantum mechanics

$$[L_m, L_n] = (m-n)L_{m+n} + \frac{d}{12}(m^3 - n)\delta_{m,-n} \tag{64}$$

where d is the number of chiral fields. There is thus an anomaly in (64), which makes the direct coupling of chiral bosons to gravity *a priori* inconsistent.

The right-moving components $\tilde{L}_m$ identically vanish, and thus, clearly remain anomaly-free.

One may write

$$[\tilde{L}_m, \tilde{L}_n] = \tilde{C}^r_{mn}\tilde{L}_r \tag{65a}$$

with any $\tilde{C}^r_{mn}$. In particular one may take

$$[\tilde{L}_m, \tilde{L}_n] = (m-n)\tilde{L}_{m+n} \tag{65b}$$

It is well known that the "physical" fields are not the only ones that contribute to the anomaly when proper account is taken of the (classical) gauge invariance in the quantization of a system with a gauge symmetry. There is also a ghost contribution, given by $(26/12)(m^3 - m)\delta_{m,n}$ in this case. The action (54) is indeed invariant under diffeomorphisms when the metric is treated as dynamical, and one thus needs to include the diffeomorphisms's ghosts.

So, the total left-moving Virasoro generators, including the ghost contribution, which is obtained by taking the brackets of the ghost momenta with the BRST charge, satisfy the algebra

$$[L^T_m, L^T_n] = (m-n)L_{m+n} + \frac{d-26}{12}(m^3 - m)\delta_{m,-n} \tag{66a}$$

$$L^T_m = L_m + L^{\text{ghost}}_m \tag{66b}$$

whereas the total right-moving Virasoro generators, with the choice (65b), would appear to yield

$$[\tilde{L}^T_m, \tilde{L}^T_n] = (m-n)\tilde{L}^T_{m+n} - (13/6)(m^3 - m)\delta_{m,-n} \qquad \text{(incorrect)} \tag{67a}$$

$$\tilde{L}^T_m = \tilde{L}_m + L^{\text{ghost}}_m \tag{67b}$$

For $d = 26$, the gravitational anomaly in the left-moving sector cancels. This is equivalent to the nilpotency of the corresponding BRST charge.

On the other hand, it would appear that there is a gravitational anomaly, due to the ghosts, in the right-moving sector, which is, however, classically trivial ($\tilde{L}_m = 0$).

The answer to this paradox is that the standard ghost spectrum in the right-moving sector is not correct when $\tilde{L}_m$ vanishes identically. One needs to add extra ghosts, of the commuting type (ghosts of ghosts), which take into account the fact that the constraints $\tilde{L}_m$ are not independent.

Indeed, there are relations among the $\tilde{L}_m$ of the form $Z^n_m \tilde{L}_n \equiv 0$, with $Z_m = \delta^n_m$. (The Z^m_n are in turn mutually independent so there are not "tertiary" ghosts.)

Following the general methods for handling ghosts of ghosts [20, 10, 11] one finds a BRST charge $\tilde{\Omega}$ for the right-moving sector given by

$$\tilde{\Omega} = -\tfrac{1}{2}\tilde{\mathcal{P}}_m \tilde{C}^m_{nr} \eta^n \eta^r + \tilde{\mathcal{P}}_r \tilde{\sigma}^r - \tilde{\pi}_m \tilde{C}^m_{nr} \tilde{\eta}^r \tilde{\sigma}^n \tag{68}$$

with $(\tilde{\eta}, \tilde{\mathcal{P}})$ and $(\tilde{\sigma}, \tilde{\pi})$ being, respectively, fermionic and bosonic canonically conjugate ghost pairs. One easily checks that $\tilde{\Omega}$ is classically nilpotent.

The total right-moving generators can be read from $\tilde{\Omega}$ as a Poisson bracket $\tilde{L}^T_m = [\tilde{\Omega}, -\tilde{\mathcal{P}}_m]$, which gives

$$\tilde{L}^T_m = -\tilde{\mathcal{P}}_n \tilde{C}^n_{mr} \eta^r + \tilde{\pi}_n \tilde{C}^n_{mr} \tilde{\sigma}^r \tag{69}$$

One sees that the $\tilde{L}^T_m$ consist of two contributions of exactly the same form in the commuting and anticommuting ghost sectors, each of which is in the adjoint representation of the gauge group. The anomalies of these two contributions cancel each other exactly. If, for definiteness, one chooses the $\tilde{C}^m_{nr}$ according to (65b) one finds that the commuting ghosts add an extra term $+(26/12)(m^3 - m)\delta_{m,n}$ to the right-hand side of (67a), which leaves no net anomaly in the right-moving sector,

$$[\tilde{L}^T_m, \tilde{L}^T_n] = (m - n)\tilde{L}^T_{m+n} \qquad \text{(correct)} \tag{70}$$

The theory (with ghosts included) is thus free from gravitational anomalies for $d = 26$.

4. CHIRAL p-FORMS

4.1. Chirality Condition

In order for chiral p-forms to exist it is necessary that F and $^*F$ should have the same number of components. It is also necessary that the square of the operation of taking the dual should give $+1$. These two demands restrict the space-time dimension to be equal to 2 modulo 4. Thus we will deal with a form $A = (p!)^{-1}A_{\mu_1\cdots\mu_p}\, dx^{\mu_1} \wedge \cdots \wedge dx^{\mu_p}$ and a field strength $F = dA = [(p+1)!]^{-1}F_{\mu_1\cdots\mu_{p+1}}\, dx^{\mu_1} \wedge \cdots \wedge dx^{\mu_{p+1}}$ in a space-time of dimension $d = 2p+2$ where p is even.

Since there is no difficulty in treating the classical coupling to gravity, we will assume from the outset that we are in a curved space-time. We will need to split both the field strength and the space-time metric into time and space. Thus we write

$$g_{\mu\nu} = \begin{bmatrix} -(N^\perp)^2 + g_{ij}N^iN^j & g_{ij}N^j \\ g_{ij}N^j & g_{ij} \end{bmatrix} \tag{71}$$

$$g^{\mu\nu} = \begin{bmatrix} -(N^\perp)^{-2} & N^i/(N^\perp)^2 \\ N^i/(N^\perp)^2 & g_{ij} - \dfrac{N^iN^j}{(N^\perp)^2} \end{bmatrix} \tag{72}$$

Here g^{ij} is the inverse of the spatial metric g_{ij} and one has $\det(g_{\mu\nu}) = -(N^\perp)^2 g$ with $g = \det(g_{ij})$.

An arbitrary vector (or tensor) may also be decomposed into components normal and perpendicular to the $t = \text{const}$ surfaces, as $V^\mu = V^\perp n^\mu + V^i\partial/\partial x^i$. The components of the normal are $n_\mu = (-N^\perp, 0)$, $n^\mu = (N^\perp)^{-1}(1, -N^i)$.

The field strength is split into "electric" and "magnetic" densities

$$\begin{aligned} \mathscr{E}^{i_1\cdots i_p} &= g^{1/2}F^{i_1\cdots i_{p\perp}} \\ &= (N^\perp)^{-1}g^{1/2}g^{i_1j_1}\cdots g^{i_pj_p}(F_{0j_1\cdots j_p} - N^kF_{kj_1\cdots j_p}) \end{aligned} \tag{73a}$$

$$\mathscr{B}^{i_1\cdots i_p} = \frac{1}{(p+1)!}\,\varepsilon^{i_1\cdots i_{2p+1}}F_{i_{p+1}\cdots i_{2p+1}} \tag{73b}$$

It follows from (73b) that $\mathscr{B}^{i_1\cdots i_p}$ is identically transverse, $\partial_{i_1}\mathscr{B}^{i_1\cdots i_p} = 0$.

The self-duality condition $F = {}^*F$ is then

$$\mathscr{E}^{i_1\cdots i_p} - \mathscr{B}^{i_1\cdots i_p} = 0 \tag{74}$$

(we take $\varepsilon_{01\cdots 2_{p+1}} = +1$) and contains the equation $d^*F = 0$ since dF identically vanishes.

A covariant action which enforces the self-duality condition (74) has been proposed in Ref. 7 and is obtained by adding to the usual action density, which is proportional to F^2, a term of the form

$$\Lambda^{\alpha\beta}F_\alpha^{(-)\gamma_1\cdots\gamma_p}F^{(-)}_{\beta\gamma_1\cdots\gamma_p} \tag{75a}$$

with

$$F^{(-)} = F - {}^*F \tag{75b}$$

Varying the action with respect to the Lagrange multiplier $\Lambda^{\alpha\beta}$, one obtains the equation

$$F_{\alpha}^{(-)\gamma_1\cdots\gamma_p} F^{(-)}_{\beta\gamma_1\cdots\gamma_p} = 0 \tag{75c}$$

whose 0–0 component reads

$$F_0^{(-)c_1 - c_p} F_{0c_1\cdots c_p} = 0 \tag{75d}$$

Since the spatial metric is positive definite, (75d) implies

$$F^{(-)}_{0c_1\cdots c_p} \tag{75e}$$

and therefore, the anti-self-dual part $F^{(-)}$ of F vanishes (the spatial components of $F^{(-)}$ are related to its temporal ones by $*F^{(-)} = -F^{(-)}$).

The action with (75a) included is just the generalization to chiral p-forms of the action (2) for a chiral boson. It possesses an extra gauge invariance—besides the standard gauge invariance of p-forms—which permits one to gauge away $\Lambda^{\alpha\beta}$. This extra gauge invariance is analogous to (4) and has no true action on the chiral p-forms.

Because chiral p-forms obey first-order differential equations, they are canonically self-conjugate. Failure to appropriately take this fact into account may lead to inconsistencies, and our first task, for this reason, is to derive the canonical formalism for chiral p-forms.

4.2. Canonical Formalism

Since the steps that lead to the canonical formalism are similar to those followed in the case of chiral bosons, we will simply sketch here the salient new features.

The dynamical variables are the canonically conjugate pairs $(A_{i_1\cdots i_p}, \pi^{i_1\cdots i_p})$ formed by the spatial components of the p-form potentials and their momenta.

The momenta $\pi^{i_1\cdots i_p}$ are subject to the standard first-class constraints associated with the gauge invariance

$$A_{i_1\cdots i_p} \to A_{i_1\cdots i_p} + \partial_{[i_1}\Lambda_{i_2\cdots i_{p-1}]} \tag{76a}$$

which read

$$\partial_{i_1}\pi^{i_1\cdots i_p} = 0 \tag{76b}$$

In addition, the canonical variables are also constrained by the chirality condition (74), which becomes, in terms of the A's and the π's,

$$\pi^{i_1\cdots i_p} - \mathcal{B}^{i_1\cdots i_p} = 0 \tag{77}$$

The major new element compared to chiral bosons is that (77) is no longer pure second class. Because of the transversality of $\mathscr{B}^{i_1\cdots i_p}$, one finds instead that (77) implies (76b). This shows also that all the constraints are actually contained in the self-duality condition.

Because the only gauge invariance of the theory is given by (76a), there should be no other first-class constraint in (77) besides (76b). Under appropriate conditions to be given below, this is indeed so, and the first-order action is thus

$$S = \int [\pi^{j_1\cdots j_p}\dot{A}_{j_1\cdots j_p} - \mathscr{H} - \lambda_{i_1\cdots i_p}(\pi - \mathscr{B})^{i_1\cdots i_p}]\, dx\, dt \tag{78}$$

where $\lambda_{i_1\cdots i_p}$ are Lagrange multipliers enforcing (77), and where $\mathscr{H}$ is given by

$$\mathscr{H} = N^{\perp}\mathscr{H}_{\perp} + N^{i}\mathscr{H}_{i} \tag{79a}$$

$$\mathscr{H}_{\perp} = g^{-1/2}\mathscr{B}^2 \tag{79b}$$

$$\mathscr{H}_{k} = {}^{*}\mathscr{B}_{kk_1\cdots k_p}\mathscr{B}^{k_1\cdots k_p} \tag{79c}$$

with $\mathscr{B}^2 = g^{i_1 j_1}\cdots g^{i_p j_p}\mathscr{B}_{i_1\cdots i_p}\mathscr{B}_{j_1\cdots j_p}$.

The energy and momentum densities $\mathscr{H}_{\perp}$ and $\mathscr{H}_i$ are really defined as functions of the canonical variables up to combinations of the constraints, but the form (79b) and (79c) turns out to be the most convenient.

4.3. Conditions on *A*

If there were extra first-class constraints besides (76b), the action (78) would yield equations of motion containing more arbitrary functions than those implied by the gauge invariance (76a), and hence, it would not be equivalent to the original Lagrangian action. Therefore, one way to arrive at the appropriate boundary conditions to be imposed on A (if any) is to make sure that when those conditions hold, extremizing (78) leads to the correct equations.

Now, as the Lagrange multiplier method indicates, the variational principle based on (78) is equivalent to the variational principle in which the constraints are solved (since the constraints multiplying $\lambda_{i_1\cdots i_p}$ fulfill the maximal rank condition). If one eliminates $\pi^{j_1\cdots j_p}$ from (78), one gets the gauge invariant action

$$S = \int dt\, dx\, [Ng^{-1/2}(\mathscr{E}\cdot\mathscr{B} - \mathscr{B}^2)] \tag{80}$$

which, although no longer canonical, is still in first-order form.

The electric density $\mathscr{E}^{i_1\cdots i_p}$ depends on $A_{0i_1\cdots i_p}$. However, this dependence drops out from (80), as it should since $A_{0i_1\cdots i_p}$ does not appear in

(78). This is because $\mathcal{B}$ is identically transverse. Furthermore, for the same reason, the longitudinal component of $A_{i_1\cdots i_p}$ does not contribute to (80) either. Thus the action (80) is really a functional of $\mathcal{B}^{i_1\cdots i_p}$ only. If one extremizes (80) with respect to $A_{i_1\cdots i_p}$ one obtains

$$\partial_0 \mathcal{B}^{i_1\cdots i_p} = \frac{1}{p!}\partial_i \varepsilon^{ii_1\cdots i_p k_1\cdots k_p}[N^{\perp}g^{-1/2}\mathcal{B}_{k_1\cdots k_p} + N^k \, {}^*\mathcal{B}_{kk_1\cdots k_p}] \tag{81}$$

Equation (81) has the same content as (74). Indeed, (74) does not restrict $A_{0i_1\cdots i_{p-1}}$ at all and hence leaves the evolution of the longitudinal part of $A_{i_1\cdots i_p}$ arbitrary. Therefore, the spatial curl of (74) which is given by (81), contains the same information as (74) itself, provided that the topology of the spatial sections is such that any closed p-form is also exact. If this is not so, one should restrict A in such a way that $\partial_0 A - {}^*dA$ belongs to the trivial cohomological class when it is closed [i.e., $d(\partial_0 A - {}^*dA) = 0 \Rightarrow \partial_0 A - {}^*dA = d\Lambda$]. Note that in the $p = 0$ case, it is the second possibility that arises (the closed 0-forms are the constants, and are not exact), and thus appropriate behavior at the spatial boundary had to be demanded.

4.4. Alternative Lagrangian Action

The first-order action (80) is the analog of the first-order action (17) proposed in Ref. 12 in the case of chiral bosons. It appears therefore as a generalization of that action for arbitrary chiral p-forms. It is not only gauge invariant, but also Lorentz invariant (see below) and can be used as an (equivalent) alternative starting point of the theory. [The canonical reformulation of (80) straightforwardly leads back to (78).]

Just as in the case of chiral bosons, the action (80) should be supplemented by an end-point term at the time boundaries. The form of this term is similar to the one found previously, and for this reason, it will not be reproduced here.

4.5. Dirac Bracket

We now come back to the Hamiltonian formulation of the theory. As we have seen, the canonical variables are the conjugate pairs $(A_{i_1\cdots i_p}, \pi^{i_1\cdots i_p})$, which are subject to both first-class and second-class constraints.

In order to develop the quantum theory, it is necessary to eliminate the second-class constraints and to work with the corresponding Dirac bracket. As a preliminary, it is therefore necessary to disentangle in (77) the first-class constraints from the second-class ones.

One possibility for making the split of the constraints is to decompose the forms into longitudinal and transverse components. This is unfortunately awkward, and furthermore, it is nonlocal. One can partly bypass this difficulty (which we have not been able to solve in a fully satisfactory manner) by evaluating the Dirac bracket for a complete set of gauge invariant observables only.

The reason why it is simpler to consider only gauge invariant objects is that their Dirac brackets do not depend on the particular split of the constraints that has been made. Moreover, they still yield a complete description of the physics.

A particular complete set of gauge invariant observables is given by the magnetic densities $\mathscr{B}^{i_1\cdots i_p}$ (obeying to $\mathscr{B}^{i_1\cdots i_p}_{\ ,i_1} = 0$). Their Dirac brackets are equal to

$$[\mathscr{B}^{i_1\cdots i_p}(x), \mathscr{B}^{j_1\cdots j_p}(x')]^* = \frac{1}{2p!}\varepsilon^{i_1\cdots i_p j_1\cdots j_p j}\delta_{,j}(x, x') \tag{82}$$

Note that these brackets are metric independent (the δ-function is defined as a density without use of the metric) and consistent with the transverse character of $\mathscr{B}$. They clearly indicate that chiral p-forms are self-conjugate.

4.6. Surface-Deformation Algebra—Lorentz Invariance

With (82), one can evaluate the brackets of the surface deformation generators (energy and momentum densities) (79). One gets the standard algebra [4, 5]

$$[\mathscr{H}_\perp(x), \mathscr{H}_\perp(x')]^* = (g^{ij}\mathscr{H}_j(x) + g^{ij}(x')\mathscr{H}_j(x'))\delta_{,j}(x, x') \tag{83a}$$

$$[\mathscr{H}_i(x), \mathscr{H}_\perp(x')]^* + 2\left[\frac{\delta\mathscr{H}_\perp(x')}{\delta g_{kj}(x)} g_{ik}(x)\right]_{,j} = \mathscr{H}_\perp(x)\delta_{,i}(x, x') \tag{83b}$$

$$[\mathscr{H}_i(x), \mathscr{H}_j(x')]^* = \mathscr{H}_i(x')\delta_{,j}(x, x') + \mathscr{H}_j(x)\delta_{,i}(x, x') \tag{83c}$$

which guarantees that the evolution from a given initial spacelike surface to a given final one is independent of the sequence of intermediate surfaces employed to calculate the evolution. The second term on the left-hand side of (83b) is present because $\mathscr{H}_\perp$ is explicitly dependent on the metric. This term is necessary to yield the full Lie derivative of $\mathscr{H}_\perp$. No such term is present in the other bracket relation (83a) involving $\mathscr{H}_\perp$, because of locality of $\mathscr{H}_\perp$ in g_{ij} [5]. This path independence property expressed by (83) is equivalent to general covariance. It follows therefore that the action (80) is invariant under changes of the space-time coordinates provided $g_{\mu\nu}$ is transformed in the usual way and provided $\mathscr{B}^{i_1\cdots i_k}$ is transformed appropriately.

The transformation law for $\mathscr{B}$ is obtained from the generators (79) themselves. Under an infinitesimal space-time reparametrization $\xi^\mu = \xi^\perp n^\mu + \xi^i\, \partial/\partial x^i$ one finds

$$\delta\mathscr{B}^{i_1\cdots i_p} = \frac{1}{p!}\partial_i\varepsilon^{ii_1\cdots i_p k_1\cdots k_p}[\xi^\perp g^{-1/2}\mathscr{B}_{k_1\cdots k_p} + N^{k*}\mathscr{B}_{kk_1\cdots k_p}] \tag{84}$$

This equation has of course the same form as the evolution equation (81). This is so because for a generally covariant system the time evolution may be regarded as the unfolding in time of a space-time reparametrization.

Equation (84) gives the transformation law of a chiral p-form under a change of the space-time coordinates. It may be obtained from the ordinary transformation law $\delta A_{\mu_1 \cdots \mu_p} = F_{\mu_1 \cdots \mu_p \mu} \xi^\mu$ by replacing $\mathscr{E}$ by $\mathscr{B}$ in it. However, it is to be emphasized that (84) leaves (80) invariant without using the equations of motion. What happens of course is that $\delta A = F\xi$ does not leave the action invariant, which just means that it is not the correct transformation law.

When ξ^μ is taken to be a Killing vector of Minkowski space and $g_{\mu\nu}$ is set equal to $\eta_{\mu\nu}$ equations, (84) define the action of the Poincaré group on a chiral p-form. For example, for boost one may take $\xi^\perp = \varepsilon_i^\perp x^i$, $\xi^i = \varepsilon^{\perp i} t$, whereas for a rotation $\xi^\perp = 0$, $\xi^i = \varepsilon^i_j x^j$. The Poincaré algebra closes without use of the equations of motion, and furthermore is linearly realized on the fields, as can be seen from (84). In that sense, one may say that Lorentz invariance is manifest.

4.7. Spectrum (Flat Space)

The facts that the Lorentz transformations act linearly on the magnetic densities, and that the brackets (82) are c-numbers, imply that the flat space quantum theory is Poincaré invariant (no danger of anomalies). An explicit representation of (82) can be easily obtained by Fourier analyzing $\mathscr{B}^{i_1 \cdots i_p}$.

Now, when $d = 4q + 2$ (q integer, $p = 2q$) the little group for a particle of zero mass and fixed momentum is effectively $O(4q)$. The representation in terms of antisymmetric tensors of rank $2q = p$ then breaks into two irreducible components given by self-dual and anti-self-dual tensors in the $4q$ transverse dimensions. These two representations correspond to dual and anti-self-dual p-forms. The states in the quantum theory based on the action (80) are just those of the self-dual representation.

Note that if we were to consider n independent p-forms (n independent "chiral bosons" if $p = 0$) the spectrum would just be n times over that of one form alone. There appears to be nothing that selects a particular "critical" value for the number of fields n.

4.8. Path Integral

Lastly, we indicate how to go over to quantum mechanics through the path integral. To do so safely, it is best to use the Hamiltonian form.

The first-class constraints (76b) may be brought into the action with a Lagrange multiplier $A_{0 i_1 \cdots i_{p-1}}$. This reproduces the Hamiltonian form of the standard action $[-2(p+1)!]^{-1} \int F^2$.

One must also incorporate the piece of (77) that is not longitudinal. That piece is second class and is what brings in the chirality condition into the theory.

Now, the second-class constraints appear in the integration measure through the square root of the determinant (the Pfaffian) of the matrix of their Poisson brackets. To evaluate that matrix one needs to bring in a metric to define what one means by transverse. That metric may most naturally be taken to be the actual metric g_{ij}, but an auxiliary metric such as δ_{ij} could also be used. Different choices of the metric correspond to redefining the second-class constraints by adding a term proportional to the first-class piece (76b). That change would not alter the brackets of the second-class constraints because the bracket of (76b) with (77) vanishes identically. Therefore the Pfaffian in question is independent of the metric and may be absorbed into an overall normalization of the path integral.

Alternatively one may consider the matrix of the brackets of all the constraints (77) which is proportional to $\varepsilon^{i_1\cdots i_p j_1\cdots j_p s}\delta_{,s}(x, x')$, and express the Pfaffian as a functional integral over real anticommuting form fields $\chi_{i_1\cdots i_p}$ defined on a $(2p+1)$-dimensional spatial section. The corresponding action is then of the Chern–Simons form $\int \chi \wedge d\chi$. This action possesses a gauge invariance $\chi \to \chi + d\xi$ where ξ is a $(p-1)$ form. That invariance is present because of the presence of a first-class constraint among (77). To evaluate the functional integral one must impose a gauge condition, which is what corresponds to splitting the transverse component of (77) in a definite way. However, if the quantization of the Chern–Simons term does not introduce anomalies the functional integral should be independent of the splitting.

The action to be path integrated is therefore

$$S = \int [\pi^{i_1\cdots i_p}\dot{A}_{i_1\cdots i_p} - \mathscr{H} - \lambda_{i_1\cdots i_p}(\pi - \mathscr{B})^{i_1\cdots i_p}]\, dx\, dt \tag{85}$$

i.e., just (78). The path integral is to be taken over the fields A, π, and λ. The action (85) is invariant under the gauge transformation

$$A_{i_1\cdots i_p} \to A_{i_1\cdots i_p} + (d\xi)_{i_1\cdots i_p} \tag{86a}$$

$$\lambda_{i_1\cdots i_p} \to \lambda_{i_1\cdots i_p} - (d\dot{\xi})_{i_1\cdots i_p} \tag{86b}$$

This gauge invariance must be handled according to the usual Faddeev–Popov or BRST methods, including ghosts of ghosts [11, 20, 10]. It will not be discussed here since the fact that our forms are chiral introduces no new features.

The action is also Weyl invariant, i.e., invariant under $A_{\mu_1\cdots\mu_p} \to A_{\mu_1\cdots\mu_p}$, $g_{\mu\nu} \to e^{\lambda(x,t)}g_{\mu\nu}$. This invariance would have to be reexamined quantum mechanically, but such considerations are beyond the formal remarks given here. The Weyl invariance implies that the energy-momentum tensor is traceless. This is indeed so. One has $g^{1/2}T_{\perp}^{\perp} = \mathscr{H}_{\perp}$, $g^{1/2}T_{i}^{\perp} = \mathscr{H}_{i}$, with $\mathscr{H}_{\perp}$, $\mathscr{H}_{i}$ given by (79), while $gT_{j}^{i} = 2p\mathscr{B}^{ii_1\cdots i_{p-1}}\mathscr{B}_{ji_1\cdots i_{p-1}} - \delta_{j}^{i}\mathscr{B}^{2}$.

ACKNOWLEDGMENTS. This work was supported by the National Science Foundation under grant No. 8600584 to The University of Texas at Austin, by organized research funds of The University of Texas, by a grant of the Tinker Foundation to the Centro de Estudios Científicos de Santiago, and by a NATO Grant for International Collaboration in Research. One of us (M.H.) is Chercheur qualifié au Fonds National de la Recherche Scientifique (Belgium).

REFERENCES

1. M. Kalb and P. Ramond, *Phys. Rev. D 9*, 2273 (1974); P. G. O. Freund and R. I. Nepomechie, *Nucl. Phys. B 199*, 482 (1982); C. Teitelboim, *Phys. Lett. 167B*, 63, 69 (1986); see also M. Henneaux and C. Teitelboim, *Found. Phys. 16*, 593 (1986).
2. W. Nahm, *Nucl. Phys. B 135*, 149 (1978); M. B. Green and J. H. Schwarz, *Phys. Lett. 109B*, 444 (1982); N. Marcus and J. H. Schwarz, *Phys. Lett. 115B*, 111 (1982).
3. M. Henneaux and C. Teitelboim, Dynamics of chiral (self-dual) *p*-forms, *Phys. Lett. 206B*, 650 (1988).
4. P. A. M. Dirac, *Lectures on Quantum Mechanics*, Academic Press, New York, 1964; J. Schwinger, *Phys. Rev. 127*, 324 (1962); C. Teitelboim, *Ann. Phys. (N.Y.) 79*, 542 (1973).
5. C. Teitelboim, The Hamiltonian structure of space-time, in *General Relativity and Gravitation: One Hundred Years After the Birth of Albert Einstein* (A. Held, ed.), Plenum Press, New York, 1980.
6. P. A. M. Dirac, *Can. J. Math. 2*, 129 (1950).
7. W. Siegel, *Nucl. Phys. B 238*, 307 (1984).
8. D. J. Gross, J. A. Harvey, E. J. Martinec, and R. Rohm, *Phys. Rev. Lett. 54*, 502 (1985); *Nucl. Phys. B 256*, 253 (1985).
9. L. Brink and M. Henneaux, *Principles of String Theory*, Chap. 5, Plenum Press, New York, 1988.
10. M. Henneaux, *Phys. Rep. 126*, 1 (1985).
11. I. A. Batalin and E. S. Fradkin, *Phys. Lett. 122B*, 157 (1983); J. Fisch, M. Henneaux, J. Stasheff, and C. Teitelboim, Existence, uniqueness and cohomology of the classical BRST charge with ghosts of ghosts, *Commun. Math. Phys. 120*, 379 (1989).
12. R. Floreanini and R. Jackiw, *Phys. Rev. Lett. 59*, 1873 (1987).
13. M. Bernstein and J. Sonnenschein, *Phys. Rev. Lett. 69*, 1772 (1988). Treatments of the action proposed in Ref. 7 which do not use the Dirac method lead to inconsistent quantization. See C. Imbimbo and A. Schwimmer, *Phys. Lett. 193B*, 455 (1987); J. Labastida and M. Pernici, *Nucl. Phys. B297*, 557 (1988); L. Mezincescu and R. I. Nepomechie, *Phys. Rev. D 37*, 3067 (1988).
14. C. A. P. Galvão and C. Teitelboim, *J. Math. Phys. 21*, 1863 (1980).
15. M. Henneaux and C. Teitelboim, *Ann. Phys. (N.Y.) 143*, 127 (1982).
16. J. H. Schwarz and P. C. West, *Phys. Lett. 126B*, 301 (1983); J. H. Schwarz, *Nucl. Phys. B 226*, 269 (1983).
17. L. D. Faddeev and A. A. Slavnov, *Gauge Fields: Introduction to Quantum Theory*, Benjamin, Reading, 1980.

18. L. D. Faddeev, *Phys. Lett. 145B*, 81 (1984).
19. M. B. Green, J. H. Schwarz, and Edward Witten, *Superstring Theory*, Vol. 1, Cambridge University Press, Cambridge, 1987.
20. I. A. Batalin and E. S. Fradkin, *Riv. Nuovo Cimento 9*(10), 1–48 (1986).

Chapter 9

First and Second Quantized Point Particles of Any Spin

Marc Henneaux and Claudio Teitelboim

1. INTRODUCTION

1.1. Spin-Zero Particle—Reparametrization Invariance

String theory has both brought new interest and shed new light, into the interplay between "first" and "second quantized" theories of many identical systems. In particular the interplay between the, technically very different, gauge invariances of both levels of the theory has been extensively discussed.

This chapter is devoted to analyzing these issues in systems simpler than the string, but still possessing many of the fascinating features of the latter. These systems provide a consistent framework for discussing in a unified manner free particles of arbitrary spin (the problem of interactions is still largely unstudied).

The treatment herein reviews and generalizes our previous work on particles of spin zero and one half [1].

MARC HENNEAUX • Faculté des Sciences, Université Libre de Bruxelles, B-1050 Brussels, Belgium and Centro de Estudios Científicos de Santiago, Santiago 9, Chile. CLAUDIO TEITELBOIM • Centro de Estudios Científicos de Santiago, Santiago 9, Chile and Center for Relativity, The University of Texas at Austin, Austin, Texas 78712.

If one desires a description of the dynamics of a relativistic particle that is both complete and also manifestly Lorentz invariant, one is led to incorporate reparametrization invariance along the world line. We will regard this gauge symmetry as an essential element.

In the spin-0 case, the canonical action reads

$$S[x^\mu, p_\mu, N] = \int d\tau\,(p_\mu \dot{x}^\mu - N\mathcal{H}) \tag{1a}$$

$$\mathcal{H} = \tfrac{1}{2}(p^2 + m^2) \tag{1b}$$

and is invariant under the transformations generated by $\mathcal{H}$,

$$\delta x^\mu = \varepsilon p^\mu \tag{2a}$$

$$\delta p^\mu = 0 \tag{2b}$$

$$\delta N = \dot{\varepsilon} \tag{2c}$$

The transformations (2a)-(2c) are just the canonical transcriptions of the standard reparametrizations along the world line (for the explicit correspondence see Ref. 1), and one can thus say that the spinless point particle offers the simplest example of a diffeomorphism-invariant theory.

In view of this invariance, it is sometimes useful to reformulate the point particle in the tensor language appropriate to generally covariant models. In that language, the x^μ's are viewed as scalar fields in (1 + 0)-dimensions, whereas the Lagrange multiplier N turns out to be the component of the einbein describing one-dimensional gravity (which is trivial). The p^2-term in $\mathcal{H}$ is the unique component of the energy-momentum tensor of the scalar fields in one dimension, m^2 appears as a cosmological constant, and the equation

$$p^2 + m^2 = 0 \tag{3}$$

enforced by the multiplier N, is just the Einstein equation $G_{00} + \lambda g_{00} = T_{00}$, since G_{00} identically vanishes.

The second-order action corresponding to (1a) is obtained by eliminating the momenta p^μ from (1a) and reads

$$S[x^\mu, N] = -\tfrac{1}{2}\int d\tau\,\sqrt{-g}\,(g^{\alpha\beta}\,\partial_\alpha x^\mu\,\partial_\beta x_\mu + m^2) \tag{4}$$

($\alpha, \beta = 0$; $g_{00} = -N^2$). If one eliminates N as well, one recovers the action $S = -m\int ds$, where s is the length of the world line.

1.2. Spin-1/2 Particle—Square Root of Mass Shell Condition

In order to describe particles of spin 1/2, one extends the Abelian algebra of the gauge generator $\mathcal{H}$

$$[\mathcal{H}, \mathcal{H}] = 0 \tag{5}$$

by adding new gauge generators and new variables describing internal degrees of freedom. The appropriate extension for the case at hand is given by a single fermionic generator $\mathcal{S}$, which is the square root of $\mathcal{H}$ [1, 2]:

$$[\mathcal{S}, \mathcal{S}] = -i\mathcal{H} \tag{6a}$$

$$[\mathcal{S}, \mathcal{H}] = 0 \tag{6b}$$

Here, [,] stands for the graded Poisson bracket. The fermionic generator $\mathcal{S}$ explicitly reads [3, 4]

$$\mathcal{S} = \theta^{\mu} p_{\mu} + m\theta^5 \tag{6c}$$

Upon quantization, the new anticommuting variables θ and θ^5 become γ-matrices and the system describes a Dirac particle.

The fermionic function $\mathcal{S}$ generates a local gauge symmetry and, as such, is constrained to vanish:

$$\mathcal{S} \approx 0 \tag{7}$$

When applied to physical states, this constraint yields the Dirac equation.

The gauge transformation generated by $\mathcal{S}$ is a local supersymmetry, since it squares to a diffeomorphism. The Dirac electron is thus $N = 1$ locally supersymmetric along the world line and is the simplest system that exhibits (local) supersymmetry.

One can again use a language adapted to supersymmetry invariance. The fields (x^{μ}, θ^{μ}) form an $N = 1$ matter supermultiplet. The equation $\mathcal{S} \approx 0$ results from varying the "gravitino" field ψ_{α}—which is the superpartner of the metric $g_{\alpha\beta}$ and which possesses no dynamics in one dimension—and expresses (when $m = 0$) that the "supercurrent" should be zero. Accordingly, the above system describes $N = 1$ supergravity coupled to an $N = 1$ scalar multiplet. The action can be written in a manifestly supersymmetric form [4]. When $m \neq 0$, the field θ^5 appears as an extra Goldstone spin-1/2 field, related to the cosmological constant [5].

As shown in Ref. 1, straightforward application to the above model of canonical path integral methods for constrained systems (see, for example, Refs. 6 and 7) yields a super-proper-time representation of the symbol of the Feynman propagator for a Dirac electron. The symbol is a classical function, and, in its functional representation, the integration over the "super-proper-times"—which has a definite measure in the canonical approach—plays a role analogous to the integral over moduli in string theory. For this reason the super-proper-times are also called (super)moduli.

1.3. Higher Spins—Other Extensions of the Mass Shell Constraint Algebra

The above incorporation of spin 1/2 by enlarging the algebra of the mass shell constraint suggests considering further extensions of $[\mathcal{H}, \mathcal{H}] = 0$.

We will study in this chapter extensions that constitute a direct generalization of (6), namely,

$$[\mathcal{S}_i, \mathcal{S}_j] = -ik_{ij}\mathcal{H}, \qquad i = 1, \ldots, N \tag{8a}$$

$$[\mathcal{S}_i, \mathcal{H}] = 0 \tag{8b}$$

where k_{ij} are constants with the symmetry

$$k_{ij} = -(-1)^{\varepsilon_i\varepsilon_j} k_{ji} \tag{8c}$$

Here, ε_i is the Grassmann parity of the gauge generators $\mathcal{S}_i$.

We will show that one can easily write down explicit realizations of the extended algebras (8), by adding extra internal degrees of freedom. The corresponding models contain particles of spin $\geq 1/2$. Actually, all spins can be generated in this way. We will also see that the spin content of the models is not irreducible in general. Some models even contain an infinite tower of different spins. If desired, one can truncate the theory to a definite irreducible representation of the Lorentz group by adding extra constraints. This is consistent within the free theory; but we will not do so here, because there are indications that consistent interacting theories of higher spins require an infinite number of them [8]. Hence, the reducibility of the multiplets present in the models discussed here may be a virtue rather than a shortcoming.

1.4. Extended Supersymmetry along the World Line

Some particular extensions of $[\mathcal{H}, \mathcal{H}] = 0$ are of special interest. If all the square roots $\mathcal{S}_i$ are fermionic, k_{ij} in (8a) is symmetric and may therefore be assumed to be diagonal. If $k_{ij} = \delta_{ij}$, the model possesses a further manifest global $SO(N)$ invariance and all the square roots are on an equal footing. The algebra (8) of the gauge generators is then recognized as the algebra of N-extended supersymmetries in one dimension.

The models that realize (8) for $k_{ij} = \delta_{ij}$ are straightforward extensions of the $N = 1$ model and are obtained by adding extra anticommuting degrees of freedom. They can be viewed as supersymmetric σ models in one dimension with a target manifold (the space of the x^μ) which is Minkowskian.

As we have seen, $N = 1$ corresponds to a spin-1/2 particle. The case $N = 2$ turns out to describe a system of p-form gauge fields while for $N \geq 3$, fields of mixed symmetries arise.

Incidentally, this reformulation of some higher spin models as extended supersymmetric nonlinear σ-models in one dimension sheds new light on the problem of consistent gravitational coupling of fields with spin ≥ 2. Indeed, the space-time metric appears in this view as the metric of the target manifold, and it is known that when the number of supersymmetries increases, one gets strong constraints on the target manifold metric [9]. These constraints arise only for $N > 2$ in one dimension, since there is no problem in coupling a system of p-forms ($N = 2$ case) to an arbitrary gravitational background.

1.5. Contraction of Virasoro Algebra

Another extension of interest is given by an infinite number of purely bosonic square roots L_m (with m an arbitrary positive or negative integer), obeying

$$[L_m, L_n] = -2im\delta_{n,-m}L_0 \tag{9a}$$

$$L_0 = \mathcal{H}, \qquad L_{-m} = L_m^* \tag{9b}$$

If one starts from the classical Virasoro algebra of string theory (see, e.g., Ref. 10)

$$[\tilde{L}_m, \tilde{L}_n] = i(n-m)\tilde{L}_{n+m} \tag{9c}$$

and redefines

$$L_0 = \frac{1}{\alpha'}\tilde{L}_0, \qquad L_m = \frac{1}{\sqrt{\alpha'}}\tilde{L}_m \tag{9d}$$

one finds, in the limit $\alpha' \to \infty$ (with L_0, L_m kept finite) that L_0 and L_m obey the algebra (9). Hence, the infinite bosonic extension (9) of the mass shell constraint turns out to be a contraction of the Virasoro algebra.* Similar considerations apply to the super-Virasoro algebra.

The system (9) possesses a global $SO(\infty)$ invariance, which can be made explicit by redefining $L_m \to \sqrt{|m|}\,L_m$, since the structure constants are then seen to be an invariant tensor (δ_{mn}) of the rotation group (here L_m and L_m^* are regarded as components of two different vectors). The limit $\alpha' \to \infty$, in which the Planck mass is set to zero [11], turns out to describe an infinite number of massless gauge fields, of increasing spins. It is somewhat analogous to the "strong coupling limit" of gravity [1, 12]. The string model, with α' finite, is related to the model (9) by a low-energy symmetry breaking of the $SO(\infty)$ invariance which lifts the mass degeneracy, since now there is no rescaling that would bring the structure constants into rotationally invariant tensors.

* We will not investigate here whether there is a two-dimensional geometry associated with the algebra (9), in the same way that the two-dimensional conformal geometry is associated with the Virasoro algebra.

It appears from these examples that the extensions (8) of the mass shell constraint algebra open the door to interesting structures.

1.6. Space-Time Supersymmetry

The models considered in this chapter possess manifest Lorentz invariance. We will not try to incorporate space-time supersymmetry in them, although this leads to interesting possibilities ("superparticle" [13], "spinning superparticle" [14]). The reason why we will not discuss here space-time supersymmetric models is that no covariant quantum formulation of the single-particle system which permits a direct passage to a satisfactory second-quantized formalism exists so far. Hence, our considerations will shed some light on the bosonic and "old" fermionic string field theories, but not on the "new" superstring formalism of Green and Schwarz [15].

1.7. Contents of the Chapter

The chapter is organized as follows. In the next section, we discuss in more detail the possible extensions of the mass shell algebra. We show that in the pure fermionic case, the extension (8) covers actually the most general nontrivial situation. In the bosonic case, there exist many other extensions, but we restrict ourselves to those that have the same form as in the fermionic case.

We then turn to explicit realizations of the extended algebra (8). The models that we consider contain "internal" extra harmonic oscillator coordinates besides the space-time variables x^μ, p_μ.

The analysis of the spectrum is given next, first in the light cone gauge, and then along Dirac and BRST lines. A "no ghost" theorem is proven which establishes complete equivalence between these approaches.

The second quantized, non-gauge-fixed theory is then considered along the lines of string field theory. We find that the BRST symmetry associated with one-dimensional world-line gauge invariances generates space-time gauge symmetries at the second quantized level, as in the case of the string. The following points are stressed: (i) there is no need to restrict the ghost number of the second quantized field in the gauge invariant approach; (ii) it is necessary to impose an analog of the string G-parity truncation in the case of fermionic extensions; (iii) because of the first quantized BRST decoupling theorem, one can also go to the light-cone gauge in the second quantized models.

Although we treat only the free theory in Minkowski space, a few remarks are given at the end concerning possible interactions.

Some of the ideas presented here have already been put forward independently by other authors [16], along different lines. Our approach, based on an extension of the diffeomorphism gauge algebra, and the results on the BRST cohomology, appear, however, to be new.

2. MORE ON THE EXTENSIONS OF THE MASS SHELL ALGEBRA

2.1. Fermionic Case: Algebra

When the square roots $\mathcal{S}_i$ are fermionic, it is easy to work out the most general algebraic extension of the mass shell algebra:

$$[\mathcal{H}, \mathcal{H}] = 0 \tag{10a}$$

By "algebraic extension," we mean that the $\mathcal{S}_i$ form, together with $\mathcal{H}$, a graded algebra with true structure constants, which do not involve the dynamical variables.

It may be of interest to consider more general extensions, in which $\mathcal{S}_i$ and $\mathcal{H}$ do not form a true algebra and refer explicitly to a definite realization with structure functions involving the fields. In any case the restriction to the algebraic case turns out to be sufficient for dealing with free theories.

The graded commutator of two $\mathcal{S}_i$'s must close on $\mathcal{H}$ and $\mathcal{S}_k$,

$$[\mathcal{S}_i, \mathcal{S}_j] = B_{ij}\mathcal{H} + C_{ij}^k\mathcal{S}_k$$

Since $[\mathcal{S}_i, \mathcal{S}_j]$ is bosonic, C_{ij}^k must be anticommuting. However, the only anticommuting constant is zero, so that C_{ij}^k must vanish. This leads to

$$[\mathcal{S}_i, \mathcal{S}_j] = B_{ij}\mathcal{H} \tag{10b}$$

The matrix B_{ij} is symmetric and can be diagonalized. Its diagonal elements can be normalized to $\pm i, 0$ (i arises because we take $\mathcal{S}_i$ to be real, so that $[\mathcal{S}_i, \mathcal{S}_j]$ is pure imaginary). We will restrict ourselves to the case where the eigenvalues are either 0 or $-i$. This makes iB_{ij} positive semidefinite and leads—as will be seen below (Sections 4 and 5)—to a physical subspace without negative norms. The presumption is strong, although we have not done a full analysis, that a negative sign in iB_{ij} would lead to negative norms for physical states and thus to inconsistency.

The same argument leading to (10b) yields

$$[\mathcal{S}_i, \mathcal{H}] = a_i^j\mathcal{S}_j \tag{10c}$$

with a_i^j constant and obeying (from the Jacobi identity applied to three $\mathcal{S}$'s),

$$a_i^m B_{jk} + a_j^m B_{ki} + a_k^m B_{ij} = 0 \tag{10d}$$

Now, if at least one diagonal element of B_{ij} is different from zero, say B_{11}, then one finds from (10d) that $a_i^m = 0$. Indeed, (10d) with $i = j = k = 1$

implies $a_1^m = 0$, while if $i = j = 1$, $k \neq 1$, one gets $a_k^m = 0$. In that case, the general extension of (10a) reads

$$[\mathcal{H}, \mathcal{H}] = 0 \tag{11a}$$

$$[\mathcal{S}_i, \mathcal{S}_j] = -ik_{ij}\mathcal{H} \tag{11b}$$

$$[\mathcal{S}_i, \mathcal{H}] = 0 \tag{11c}$$

$$k_{ij} = \text{diag}(1, 0) \tag{11d}$$

If B_{ij} vanishes, (10d) is automatically fulfilled, there is no condition on a_i^m, and one has

$$[\mathcal{H}, \mathcal{H}] = 0 \tag{12a}$$

$$[\mathcal{S}_i, \mathcal{S}_j] = 0 \tag{12b}$$

$$[\mathcal{S}_i, \mathcal{H}] = a_i^j \mathcal{S}_j \tag{12c}$$

This second possibility is not of direct interest to us, for it does not correspond to a square root of $\mathcal{H}$. The new generators do not reproduce $\mathcal{H}$ upon anticommutation, and the connection between $\mathcal{S}_i$ and $\mathcal{H}$ is not as tight as in (11). The algebra (12) possesses actually a semidirect product structure. We will exclude the possibility (12) from now on. Actually we will not only assume that $B_{ij} \neq 0$ but will allow no vanishing eigenvalue. There is no real loss of generality in this, since the generators $\mathcal{S}_i$ associated with the eigenvalue zero form a direct product with the other generators, and can be studied separately, if desired.

2.2. Alternative Form of the Algebra

Having disposed of the eigenvalue zero we are left with

$$k_{ij} = \delta_{ij} \tag{13}$$

The algebra possesses then a manifest global $O(N)$ symmetry, which rotates the square roots among themselves,

$$[\mathcal{H}, \mathcal{H}] = 0 \tag{14a}$$

$$[\mathcal{S}_i, \mathcal{S}_j] = -i\delta_{ij}\mathcal{H} \tag{14b}$$

$$[\mathcal{S}_i, \mathcal{H}] = 0 \tag{14c}$$

$$\mathcal{H}^* = \mathcal{H}, \qquad \mathcal{S}_i^* = \mathcal{S}_i \tag{14d}$$

$$\mathcal{S}_i \rightarrow \mathcal{S}'_i = R_i^j \mathcal{S}_j \tag{14e}$$

For the explicit construction of the models, as well as for straightforward comparison with the bosonic case, it is convenient to redefine the generators in a way that breaks this manifest $O(N)$ invariance. This is done by combining the generators $\mathcal{S}_i$ $(i = 1, \ldots, N)$ in pairs.

If N is even, $N = 2n$, we define

$$\mathcal{S}_1^A = \mathcal{S}_{2A-1} + i\mathcal{S}_{2A} \tag{15a}$$

$$\mathcal{S}_2^A = \mathcal{S}_{2A-1} - i\mathcal{S}_{2A} \tag{15b}$$

$(A = 1, \ldots, n)$. One finds

$$[\mathcal{S}_a^A, \mathcal{S}_b^B] = -2i\delta^{AB} d_{ab}\mathcal{H} \tag{15c}$$

$$\mathcal{S}_a^{A*} = d_{ab}\mathcal{S}_b^A, \qquad a = 1, 2 \tag{15d}$$

with

$$d_{ab} = \begin{pmatrix} 0 & 1 \\ 1 & 0 \end{pmatrix} \tag{15e}$$

If N is odd, $N = 2n + 1$, we perform the same redefinitions for the first $2n$ generators, and set in addition

$$\mathcal{S} \equiv \mathcal{S}_{2n+1} \tag{16a}$$

$$\mathcal{S}^* = \mathcal{S} \tag{16b}$$

so that

$$[\mathcal{S}, \mathcal{S}] = -i\mathcal{H} \tag{16c}$$

$$[\mathcal{S}, \mathcal{S}_a^A] = 0 \tag{16d}$$

The gauge algebras (15) and (16) are the algebras of $2n$ or $(2n + 1)$-extended supersymmetry, respectively. In the form (15) and (16), only a global $O(n)$ invariance is manifest.

2.3. Bosonic Case

As mentioned in the Introduction, in the bosonic case, there are many more possible extensions of the mass shell condition than in the fermionic one. Indeed, from just the point of view of the algebra—but without inquiring about specific realizations—any algebra can be regarded as an extension of the Abelian subalgebra which consists of any one of its generators. We will not consider here the general case, but will restrict ourselves to the same algebra as in the fermionic case, namely,

$$[\mathcal{H}, \mathcal{H}] = 0 \tag{17a}$$

$$[\mathcal{S}_i, \mathcal{S}_j] = k_{ij}\mathcal{H} \tag{17b}$$

$$[\mathcal{S}_i, \mathcal{H}] = 0 \tag{17c}$$

$$k_{ij} = -k_{ji} \tag{17d}$$

The form (17b) may be justified by demanding that for a proper "square root," the commutator of two $\mathcal{S}$'s should include only $\mathcal{H}$, but not the $\mathcal{S}$'s themselves—a possibility automatically excluded in the fermionic case, since it would need an anticommuting structure constant. However, this would still leave open a possible extra term on the right-hand side of (17c).

To avoid a direct product structure, we assume, as before, that $\det k_{ij} \neq 0$. This implies that the number N of square roots is even, i.e., $N = 2n$ (the determinant of an odd-dimensional skew matrix is zero). By redefinitions similar to (15), one can thus rewrite (17) as

$$[\mathcal{H}, \mathcal{H}] = 0 \tag{18a}$$

$$[\mathcal{S}^A_a, \mathcal{S}^B_b] = -2i\delta^{AB}\varepsilon_{ab}\mathcal{H} \tag{18b}$$

$$[\mathcal{S}^A_a, \mathcal{H}] = 0 \tag{18c}$$

$$A = 1, \ldots, n; \qquad a = 1, 2 \tag{18d}$$

$$\varepsilon_{ab} = -\varepsilon_{ba}, \qquad \varepsilon_{12} = 1 \tag{18e}$$

$$\mathcal{H}^* = \mathcal{H}, \qquad \mathcal{S}^{A^*}_a = d_{ab}\mathcal{S}^A_b \tag{18f}$$

2.4. Mixed Case

It results from our previous discussion that the basic building blocks of the relevant extensions can be of three different types:

1. A single real fermionic square root.
2. A pair of two complex conjugate fermionic square roots, as in (15c) and (15d).
3. A pair of two complex conjugate bosonic square roots, as in (18).

The general systems studied here are a combination of these.

3. EXPLICIT REALIZATIONS

3.1. $N = 1$ Supersymmetry along the World Line

We first treat the massless case. The massive case will be considered in Section 3.4.

The simplest system is a single real fermionic square root which corresponds to $N = 1$ supersymmetry along the world line. It can be realized by introducing d real fermionic variables θ^μ obeying

$$[\theta^\mu, \theta^\nu] = -\tfrac{i}{2}\eta^{\mu\nu} \tag{19a}$$

Here, d is the space-time dimension, and $\eta_{\mu\nu} = (-, +, \ldots, +)$. The square root $\mathcal{S}$ is given by

$$\mathcal{S} = \theta^{\mu} p_{\mu} \tag{19b}$$

and is real,

$$\mathcal{S}^* = \mathcal{S} \tag{19c}$$

3.2. $N = 2$ Supersymmetry Along the World Line

The next case corresponds to $N = 2$ supersymmetry and contains a pair of complex conjugate fermionic square roots.

It can be obtained by introducing two sets of real fermionic variables $\theta_1^{\mu}, \theta_2^{\mu}$,

$$[\theta_a^{\mu}, \theta_b^{\nu}] = -\frac{i}{2}\delta_{ab}\eta^{\mu\nu} \tag{20a}$$

or, what is the same, a set of Fermi oscillators,

$$[a^{\mu}, a^{\nu *}] = -i\eta^{\mu\nu} \tag{20b}$$

$$[a^{\mu}, a^{\nu}] = [a^{\mu *}, a^{\nu *}] = 0 \tag{20c}$$

$$a^{\mu} = \theta_1^{\mu} + i\theta_2^{\mu}, \qquad a^{\mu *} = \theta_1^{\mu} - i\theta_2^{\mu} \tag{20d}$$

In analogy with (19b), the complex generator $\mathcal{S}$ is given by

$$\mathcal{S} = p_{\mu} a^{\mu} \tag{20e}$$

and obeys

$$[\mathcal{S}, \mathcal{S}^*] = -2i\mathcal{H} \tag{20f}$$

$$[\mathcal{S}, \mathcal{S}] = [\mathcal{S}^*, \mathcal{S}^*] = 0 \tag{20g}$$

3.3. Bosonic Square Roots

The fermionic formulas also apply to the bosonic case. If $b^{\mu}, b^{\mu *}$ are bosonic harmonic oscillator variables,

$$[b^{\mu}, b^{\nu *}] = -i\eta^{\mu\nu} \tag{21a}$$

one defines

$$\mathcal{S} = p_{\mu} b^{\mu} \tag{21b}$$

and finds

$$[\mathcal{S}, \mathcal{S}^*] = -2i\mathcal{H} \tag{21c}$$

$$[\mathcal{S}, \mathcal{S}] = [\mathcal{S}^*, \mathcal{S}^*] = 0 \tag{21d}$$

as above.

In analogy with the string case, we will call the space-time pairs (x^μ, p_μ) and the real fermionic variables θ^μ associated with the singled-out real fermionic constraint (16a) (if any) the "zero mode variables." At the same time, the constraints $p^2 = 0$ and $p \cdot \theta = 0$ will be referred to as the "zero mode constraints."

3.4. Mass

From the point of view of gravity along the world line, the square of the rest mass corresponds to a cosmological constant. Furthermore, the mass can be easily incorporated "à la Kaluza-Klein," by considering a massless model in $d + 1$ dimensions, and by restricting the last component of the momentum to take the definite value m. This is consistent because p^μ has vanishing brackets with all constraints. The massive models contain therefore $d + 1$ additional (pairs of) internal degrees of freedom, instead of just d. Because their analysis is carried out along the same lines as in the massless case, we will set the mass equal to zero from now on.

3.5. Gauge Invariance

The constraints

$$\mathscr{H} \approx 0, \qquad \mathscr{S}_i \approx 0 \tag{22}$$

are first-class and generate gauge invariance. Displacements generated by $\mathscr{H}$ are given by

$$\begin{aligned} \delta x^\mu &= [x^\mu, \eta(\tau)\mathscr{H}] \\ &= \eta p^\mu \end{aligned} \tag{23a}$$

$$\delta p^\mu = 0, \qquad \delta\theta^\mu = \delta a^\mu = \delta b^\mu = \delta a^{\mu *} = \delta b^{\mu *} = 0 \tag{23b}$$

The new gauge transformations are generated by the square roots $\mathscr{S}_i$ and explicitly read

$$\delta x^\mu = [x^\mu, i\varepsilon\mathscr{S}] = i\varepsilon\theta^\mu \tag{24a}$$

$$\delta p_\mu = 0 \tag{24b}$$

$$\begin{aligned} \delta\theta^\mu &= [\theta^\mu, i\varepsilon\mathscr{S}] \\ &= -\tfrac{1}{2}\varepsilon p^\mu \end{aligned} \tag{24c}$$

$(N = 1)$, or

$$\delta x^\mu = [x^\mu, i(\varepsilon^*\mathscr{S} + \varepsilon\mathscr{S}^*)] = i(\varepsilon^* a^\mu + \varepsilon a^{\mu *}) \tag{25a}$$

$$\delta p_\mu = 0 \tag{25b}$$

$$\delta a^\mu = -\varepsilon p^\mu, \qquad \delta a^{\mu *} = -\varepsilon^* p^\mu \tag{25c}$$

($N = 2$ fermionic), or

$$\delta x^\mu = [x^\mu, \varepsilon^* \mathcal{S} + \varepsilon \mathcal{S}^*] = \varepsilon^* b^\mu + \varepsilon b^{\mu *} \tag{26a}$$

$$\delta p_\mu = 0 \tag{26b}$$

$$\delta b^\mu = -i\varepsilon p^\mu, \qquad \delta b^{\mu *} = i\varepsilon^* p^\mu \tag{26c}$$

($N = 2$ bosonic).

In (24), ε is an arbitrary time-dependent fermionic, real function, while in (25) and (26), ε is an arbitrary fermionic or bosonic complex function.

The important property of the new gauge symmetries is that if one repeats them twice, one gets a time reparametrization (23). This is a characteristic feature of local supersymmetry, but it also holds for the bosonic square roots.

3.6. Lorentz Invariance

The models considered are Lorentz invariant. The Poincaré generators are

$$P^\mu = p^\mu \tag{27a}$$

$$M^{\mu\nu} = \tfrac{1}{2}(p^\mu x^\nu - p^\nu x^\mu) + i\theta^\mu \theta^\nu \tag{27b}$$

($N = 1$),

$$M^{\mu\nu} = \tfrac{1}{2}(p^\mu x^\nu - p^\nu x^\mu) + \frac{i}{2}(a^{*\mu} a^\nu - a^{*\nu} a^\mu) \tag{27c}$$

($N = 2$ fermionic), or

$$M^{\mu\nu} = \tfrac{1}{2}(p^\mu x^\nu - p^\nu x^\mu) + \frac{i}{2}(b^{*\mu} b^\nu - b^{*\nu} b^\mu) \tag{27d}$$

($N = 2$ bosonic).

The new variables transform as vectors under Lorentz transformations. The constraints $\mathcal{H}$ and $\mathcal{S}_i$ are Lorentz scalars.

Finally, we note that the Lorentz-invariant "occupation numbers"

$$N_A = a_A^{*\mu} a_{A\mu} \tag{28}$$

(one for each type of oscillator) also define global symmetries since they (weakly) commute with the constraints.

4. LIGHT-CONE GAUGE QUANTIZATION

The quickest way to work out the physical spectrum is to impose the light-cone gauge.

Because the gauge symmetries are noninternal, the problem of the imposition of a good gauge condition in the path integral is not straightforward: there may exist "moduli" [1]. It appears, however, that these subtleties are not relevant when discussing the spectrum. For this reason, we will not address them in this chapter. We will in the sequel develop the theory for an arbitrary number of oscillators. In order to simplify the notations, we will drop the indices which label the oscillators and the square roots, so that, for instance, a^μ stands for all the fermionic oscillators a^μ_A.

4.1. Light-Cone Gauge Conditions

The light-cone gauge is a good gauge condition for motions with $p^+ \neq 0$, and we therefore assume $p^+ \neq 0$ throughout. Here, $p^+ = (p^0 + p^{d-1})/2, p^- = (p^0 - p^{d-1})/2 = -p_+$.

By an appropriate reparametrization, one can set x^+ equal to an arbitrary function of τ. In particular, one can take

$$x^+ = p^+\tau \tag{29}$$

This condition completely fixes the parametrizations, because $0 = \delta x^+ = \eta p^+$ implies $\eta = 0\,(p^+ \neq 0)$.

Similarly, one can use the gauge freedom generated by the square roots $\mathcal{S}_i$ to set the + component of all the internal degrees of freedom θ^μ, a^μ, or b^μ equal to zero,

$$\theta^+ = 0 \tag{30a}$$

$$a^+ = 0, \qquad a^{*+} = 0 \tag{30b}$$

$$b^+ = 0, \qquad b^{*+} = 0 \tag{30c}$$

Again, these conditions completely freeze the new gauge invariances (24), (25), or (26) because $p^+ \neq 0$.

4.2. Light-Cone Gauge Action

The light-cone gauge action is obtained by solving for the gauge conditions and the constraints inside the canonical action

$$S[x^\mu, p_\mu, \theta^\mu, a^\mu, a^{*\mu}, b^\mu, b^{*\mu}, N, M^i]$$
$$= \int d\tau(\dot{x}^\mu p_\mu - i\dot{\theta}^\mu\theta_\mu + ia^{*\mu}\dot{a}_\mu + ib^{*\mu}\dot{b}_\mu - N\mathcal{H} - M^i\mathcal{S}_i) \tag{31}$$

Here, M^i are the Lagrange multipliers for the constraints $\mathcal{S}_i \approx 0$. The action (31) should be supplemented by appropriate end-point terms at τ_1 and τ_2 (Ref. 17; and Ref. 3 third reference), but we will not write them explicitly here.

One finds

$$S[x^-, p^+, x^i, p_i, \theta^i, a^i, a^{*i}, b^i, b^{*i}]$$
$$= \int d\tau\,(\dot{x}^i p_i - \dot{x}^- p^+ - i\dot{\theta}^i \theta_i + ia^{*i}\dot{a}_i + ib^{*i}\dot{b}_i - H - \dot{p}^+ p^- \tau) \tag{32}$$

where i is now a $(d-2)$-dimensional transverse index, while p^- and the light-cone gauge Hamiltonian H are given by

$$p^- = \frac{1}{2p^+} H \tag{33a}$$

$$H = \tfrac{1}{2}\sum_i (p^i)^2 \tag{33b}$$

By the redefinition

$$x^- \to u^- = x^- - p^- \tau$$

and the elimination of a total time derivative in the action, one can get rid of the last term in (32) [10], and thereby obtain a Lagrangian that does not explicitly depend on time,

$$S[u^-, p^+, x^i, p_i, \theta^i, a^i, a^{*i}, b^i, b^{*i}]$$
$$= \int d\tau\,(\dot{x}^i p_i - \dot{u}^- p^+ - i\dot{\theta}^i \theta^i + ia^{*i}\dot{a}^i + ib^{*i}\dot{b}^i - H) \tag{34}$$

The Dirac brackets can be read off from the kinetic term of (34) and are

$$[x^i, p_j]^* = \delta^i_j \tag{35a}$$

$$[u^-, p^+]^* = -1 \tag{35b}$$

$$[\theta^i, \theta^j]^* = -\frac{i}{2}\delta^{ij} \tag{35c}$$

$$[a^i, a^{*j}]^* = -i\delta^{ij} \tag{35d}$$

$$[b^i, b^{*j}]^* = -i\delta^{ij} \tag{35e}$$

All other basic brackets vanish.

4.3. Light-Cone Gauge Lorentz Generators

The light-cone gauge Lorentz generators are simply obtained by eliminating $x^+, p^-, \theta^\pm, a^\pm, a^{*\pm}, b^\pm$, and $b^{*\pm}$ from (27), using the constraints and the gauge conditions. They now act through the Dirac bracket.

Because θ^-, a^-, or b^- are linear in the independent internal variables,

$$\theta^- = \frac{1}{p^+} p_i \theta^i \tag{36a}$$

$$a^- = \frac{1}{p^+} p_i a^i \tag{36b}$$

$$b^- = \frac{1}{p^+} p_i b^i \tag{36c}$$

the Lorentz generators are still quadratic in the oscillators.

4.4. Spectrum

The physical space must yield a representation of the independent degrees of freedom. These obey commutation-anticommutation relations, which follow from the Dirac bracket according to the quantization rule

$$\text{graded commutator} = i \times (\text{Dirac bracket}) \tag{37}$$

Hence, one gets from (35)

$$[x^i, p_j] = i\delta^i_j, \qquad [u^-, p^+] = -i \tag{38a}$$

$$[\theta^i, \theta^j] = \tfrac{1}{2}\delta^{ij} \tag{38b}$$

$$[a^i, a^{*j}] = \delta^{ij} \tag{38c}$$

$$[b^i, b^{*j}] = \delta^{ij} \tag{38d}$$

where [,] now denotes the graded commutator, i.e., it is the commutator unless the two arguments are odd in the classical theory, in which case it is the anticommutator.

Let us first consider the case when there are no θ^i. Then, the appropriate Hilbert space is given by the direct product of the space of functions of x^i and u^- and the Fock space generated from the vacuum $|0\rangle$ by the creation operators a^{*i} or b^{*i}.

$$a^i|0\rangle = 0, \qquad b^i|0\rangle = 0 \tag{39}$$

In the fermionic case, the states are antisymmetric in the $SO(d-2)$ indices carried by the oscillators, while they are symmetric in the bosonic case.

When the θ^i are present, one needs in addition to represent the anticommutation relations

$$\theta^i\theta^j + \theta^j\theta^i = \tfrac{1}{2}\delta^{ij} \tag{40}$$

which yields a Clifford algebra in $d-2$ dimensions. The corresponding representation space is $2^{d-2/2}$-dimensional (assuming d to be even).* The vacuum (39) is therefore degenerate and carries a $(d-2)$-dimensional spinorial index.

A general physical state can thus be characterized by its momentum (p^i, p^+), by its oscillator occupation numbers and, in the presence of the θ^μ's, by its spin in θ^i-space. Note that for $p^+ > 0$, one gets particles, while for $p^+ < 0$, one has antiparticles since $p^+ < 0$ corresponds to $p^0 < 0$. Of course, these states should be identified if the particles are their own antiparticles.

The Hilbert space is manifestly positive definite, since the oscillators obey the standard creation and destruction operator algebra, whereas the γ-matrices of (40) are Hermitian in the inner product u^+u. Note that if one had a minus sign in the right-hand side of (38b) or (38c), the fermionic variables would create negative norm states. This is what forces a definite sign for k_{ij} in the algebra (11b) of the (real) fermionic generators for the models at hand.

The above construction of the Hilbert space closely parallels what is done in string theory [10].

4.5. Lorentz Transformation Properties of the States

Because the light-cone gauge Lorentz generators are still bilinear in the oscillators, there is no ordering problem and no Lorentz anomaly in the quantum mechanics. The states therefore yield representations of the Poincaré group.

The mass shell condition $p^2 \approx 0$ implies that the states are all massless. The relevant little group which completes the characterization of the representation is thus effectively $SO(d-2)$.

How the states transform under $SO(d-2)$ is easily determined from the little group generators M^{ij} (with $p^i = 0$). The oscillators carry a $SO(d-2)$ vector index, so that the transformation properties of an arbitrary state are given by (symmetrized or antisymmetrized) tensor products of the vector representation times the representation of the vacuum in the case of a single set of oscillators. To get representations with tensors of mixed symmetry [18], one simply needs to introduce many different oscillators.

If the number of square roots of the mass-shell condition is even, the ground state is a scalar and one generates only integer-spin states. With an

* The fermionic oscillators a^i, a^{*i} need to include "γ_{d+1}" in order to anticommute with θ^i. More explicitly, starting from the oscillators a, a^* acting on a Fock space and commuting with the θ's—which act on a different space—one fulfills $a\theta + \theta a = 0$, $a^*\theta + \theta a^* = 0$ by redefining $a \to \gamma_{d+1}a$ with $\gamma_{d+1} = \gamma_0 \cdots \gamma_{d-1}$.

odd number of square roots, the ground state is a spinor and one gets half-integer spin representations.

In particular, the $N = 2$ supersymmetric theory, with two fermionic square roots, contains one set of fermionic oscillators, which generate antisymmetrized states $a^{*[i_1} \cdots a^{*i_p]}|0\rangle$, $p = 0, \ldots, d-2$. This corresponds to a set of massless p-form gauge fields $(0 \leq p \leq d-2)$.

It is perhaps worthwhile to emphasize that the quantum mechanical models obtained by adding extra internal degrees of freedom yield a reducible representation of the Poincaré group. In the free theory, one may truncate the spectrum to a definite irreducible representation without running into inconsistencies. This is done by imposing extra conditions which select an irreducible subspace. But, as stated in the Introduction, the reducibility of the representation may be an advantage when it comes to discussing interacting models.

4.6. Absence of Critical Dimension

The massless models analyzed above are quantum-mechanically consistent and Lorentz invariant in any number of dimensions. This is also true for the massive models, with extra oscillators corresponding to one extra dimension, which can be treated along entirely similar lines.

The situation is in sharp contrast with what happens in string theory, where one finds critical dimensions. There is no problem with Lorentz invariance in the models studied here because one has always enough states to fill in representations of the little group. This property does not hold in the case of the string [10], where the states form manifest $SO(d-2)$ multiplets, which must, for the massive levels, combine to form representations of the larger little group $SO(d-1)$.

This indicates that the introduction of mass in the present massless models by breaking the rotational symmetry among the oscillators rather than by adding extra oscillators in the manner of Kaluza–Klein is likely to be a very subtle and interesting question.

5. DIRAC QUANTIZATION

5.1. Representation Space

Instead of eliminating redundant degrees of freedom by fixing the gauge, one can carry all the dynamical variables into the quantum theory. This approach maintains manifest Lorentz invariance. Because gauge degrees of freedom are now included, the space of states contains unphysical states. Physical states are selected by appropriately enforcing the constraints.

Since every dynamical variable is realized as an independent operator, the space of states is the direct product of the space of functions of x^μ, of the Fock space generated from the vacuum $|0\rangle$

$$a^\mu|0\rangle = 0, \qquad b^\mu|0\rangle = 0 \tag{41}$$

by the creation operators $a^{*\mu}$, $b^{*\mu}$, and of the representation space of the Clifford algebra

$$[\theta^\mu, \theta^\nu] = \tfrac{1}{2}\eta^{\mu\nu} \tag{42}$$

Relation (42) identifies the θ's as the Dirac matrices.

This space contains negative norm states, which arise because $\eta^{\mu\nu}$ in (42) is not positive definite and because the temporal destruction–creation operators do not obey the standard commutation relations. Instead, one has

$$a^0 a^{*0} + a^{*0} a^0 = -1, \qquad b^0 b^{*0} - b^{*0} b^0 = -1 \tag{43}$$

with a minus sign on the right-hand side.

5.2. Physical States

Straightforward application of the Dirac method would say that the physical states should be annihilated by all the constraints, namely,

$$p^2|\psi\rangle = 0 \tag{44a}$$

$$p \cdot \theta|\psi\rangle = 0 \tag{44b}$$

$$p \cdot a|\psi\rangle = p \cdot b|\psi\rangle = 0 \tag{44c}$$

$$p \cdot a^*|\psi\rangle = p \cdot b^*|\psi\rangle = 0 \tag{44d}$$

These conditions are, however, much too strong, and even inconsistent if there is at least one bosonic oscillator. This is because there is no Fock space state annihilated by the creation operator $p \cdot b^*$, except the uninteresting zero state itself.

Hence, even though there is no anomaly in the gauge algebra [the quantum gauge operators still close according to (11) and (17), without central charge], one is forced to weaken the conditions (44) simply because one has chosen a Fock representation space. The weakened conditions are obtained by dropping the creation part (44d) of the constraints, and read

$$p^2|\psi\rangle = 0, \qquad p \cdot \theta|\psi\rangle = 0 \tag{45a}$$

$$p \cdot a|\psi\rangle = 0, \qquad p \cdot b|\psi\rangle = 0 \tag{45b}$$

Now, the replacement of (44) by (45) is an important conceptual step. Indeed, it is not a priori clear that by imposing only half of the constraints one is still guaranteeing full gauge invariance of the physical states. It is true that the constraints are fulfilled in the mean, $\langle\psi_1|$ constraints $|\psi_2\rangle = 0$ if $|\psi_1\rangle$ and $|\psi_2\rangle$ are physical, but, as examples taken from string theory below the critical dimension indicate [10], this may not be enough to ensure decoupling of the "longitudinal modes" from the physical spectrum, i.e., to enforce full gauge invariance.

What is needed is that not all the solutions of the weaker conditions (45) be physically relevant. It turns out that this is what happens for the models at hand: the states created from the vacuum by $p \cdot a^*$ or $p \cdot b^*$ are physical, i.e., obey (45), but decouple from all other physical states including themselves ("null spurious states"). For this reason, they can—and should—be factored out. Once this is done, one gets a physical space with only "transverse states," and the second half of the gauge invariance is recovered.

One thus sees that gauge invariance is enforced in two steps in the Fock representation. First, one imposes half of the constraints, $p \cdot a|\psi\rangle = 0$ and $p \cdot b|\psi\rangle = 0$. Second, one removes the physical states created by the other half of the constraints $p \cdot a^*$ and $p \cdot b^*$. This is possible because those states decouple, as we explicitly show in the next section.

5.3. No Negative Norm States Theorem

In order to prove that the unwanted states created by $p \cdot a^*$ and $p \cdot b^*$ drop out from physical amplitudes, we will closely parallel the steps followed in string theory [10].

Since p^μ commutes with the constraints, it can be diagonalized, and one can then work with states of definite momentum. For these to be physical, the momentum should obey the mass shell condition $p^2 = 0$, which will therefore be assumed from now on.

Similarly, we will assume that the other zero mode constraint $p \cdot \theta|\psi\rangle = 0$ is fulfilled. Acting with the oscillators on a state that obeys $p \cdot \theta|\psi\rangle = 0$, one still gets a state that obeys that condition, since the oscillators commute or anticommute with θ^μ. It is therefore consistent to freely act with the oscillators and, at the same time, to assume $p \cdot \theta|\psi\rangle = 0$ throughout.

We introduce a null vector k^μ such that

$$k_\mu p^\mu = -1, \qquad k^2 = 0 \tag{46}$$

[We take $p^\mu \neq 0$ so that (46) possesses a solution.]

The transverse states $|T\rangle$ are defined by

$$k \cdot a|T\rangle = 0, \qquad p \cdot a|T\rangle = 0 \tag{47a}$$

$$k \cdot b|T\rangle = 0, \qquad p \cdot b|T\rangle = 0 \tag{47b}$$

These states are clearly physical. Furthermore, the subspace which they span has a positive definite inner product. Indeed, the general solution of

(47a) and (47b) is a linear combination of states of the form

$$(a_i^*)^{n_i}(b_j^*)^{n_j}|0\rangle \tag{47c}$$

where

$$a_i^* = e_i^\mu a_\mu^*, \qquad b_j^* = e_j^\mu b_\mu^* \tag{47d}$$

$$e_i \cdot k = e_i \cdot p = 0 \tag{47e}$$

The vectors e_j^μ are spacelike and can be chosen to be orthonormal,

$$e_i \cdot e_j = \delta_{ij} \tag{47f}$$

so that the transverse oscillators a_i^* and b_i^* obey the standard commutation relations and accordingly create positive norm states only.

Theorem 1. The most general physical state is given by

$$|\psi\rangle = |T\rangle + |\text{ns}\rangle \tag{48}$$

where $|T\rangle$ is a transverse state, and where $|\text{ns}\rangle$ is a "null spurious state," i.e., a physical state orthogonal to all physical states (including itself).

Proof. The states $(k \cdot a^*)^m(p \cdot a^*)^n(k \cdot b^*)^r(p \cdot b^*)^t|T\rangle$, where $|T\rangle$ ranges over the transverse states, span the full space of states, because $k \cdot a^*, p \cdot a^*, e_i \cdot a^*, k \cdot b^*, p \cdot b^*$, and $e_i \cdot b^*$ form a basis of creation operators.

Now, one has

$$[k \cdot a, p \cdot a^*] = -1 = [p \cdot a, k \cdot a^*] \tag{49a}$$

$$[k \cdot b, p \cdot b^*] = -1 = [p \cdot b, k \cdot b^*] \tag{49b}$$

$$[p \cdot a, p \cdot a^*] = [p \cdot b, p \cdot b^*] = 0 \tag{49c}$$

Therefore, a linear combination of states of the form $(k \cdot a^*)^m(p \cdot a^*)^n(k \cdot b^*)^r(p \cdot b^*)^t|T\rangle$, where $|T\rangle$ is any transverse state, is physical if and only if it does not involve the oscillators $k \cdot a^*$ or $k \cdot b^*$ ($m = r = 0$), i.e., if it consists only of terms

$$(p \cdot a^*)^n(p \cdot b^*)^t|T\rangle \tag{50}$$

Next set

$$|\psi\rangle = |T_0\rangle + |\text{ns}\rangle \tag{51a}$$

where $|T_0\rangle$ is the transverse state appearing in $|\psi\rangle$ for $n = t = 0$, and $|\text{ns}\rangle$ contains all the other terms with n or $t \neq 0$. The state $|\text{ns}\rangle$ contains at least one oscillator $p \cdot a^*$ or $p \cdot b^*$ and accordingly, is not transverse. In addition, its scalar product with any physical state $|\psi'\rangle$ is zero,

$$\langle \psi' | \text{ns} \rangle = 0 \tag{51b}$$

since $p \cdot a^*$ or $p \cdot b^*$ annihilates $\langle \psi'|$ by the physical state condition.

The decomposition (51a) of a general physical state yields therefore the decomposition claimed in the theorem, since $|T_0\rangle$ and $|\mathrm{ns}\rangle$ possess the required properties.

5.4. Scalar Products

The above theorem shows that the Dirac quantization is equivalent to the light-cone gauge one, since one gets exactly the same spectrum in both cases, once the null spurious states are discarded.

Now, the scalar product of a general state of the Dirac quantum space formally involves an integral over x^μ. It also involves the scalar product in the Clifford algebra representation space of the θ's, as well as the Fock space scalar product.

The Fock space scalar product is clearly well defined and positive definite in the physical subspace (45), since physical states contain only transverse oscillators.

This is not so for the part of the scalar product involving x^μ and θ^μ. Because physical states are on the mass shell, one finds that the integral over x^μ is infinite, even if one considers wave packets with different spatial momenta. This is because the "extra" integral over x^0 (or p^0) gives infinity. At the same time, the θ^μ component of the scalar product yields zero.

$$\begin{aligned}\langle\psi_1|\psi_2\rangle &= \langle\psi_1| - 2[k\cdot\theta, p\cdot\theta]|\psi_2\rangle \\ &= 0\end{aligned}$$

if both $|\psi_1\rangle$ and $|\psi_2\rangle$ are annihilated by the real zero mode constraint $p\cdot\theta$.

This means that the scalar product defined in the space of all physical and unphysical states cannot directly be used to determine physical amplitudes. Actually, things are even worse because physical states are not normalizable and thus, strictly speaking, they do not live in the space of states under consideration.

The same difficulties arise in the case of string models, where again, the oscillator modes allow for a definition of a scalar product that is sensible even for physical states, but where the zero modes (x^μ and θ^μ) lead to ill-defined expressions not amenable to direct interpretation.

One way to solve this problem, which perhaps is not the most elegant, is to enlarge the space of states that are allowed to begin with, so as to include states that obey the mass shell condition. For instance, one may simply impose an appropriate fall-off at spatial infinity only, without restriction at timelike infinity ($x^0 \to \pm\infty$).

In what concerns the zero modes, one does not define a scalar product to begin with. A scalar product is defined only for on-the-mass-shell states (obeying also the Dirac equation if there are zero mode fermionic variables). This is done by using the isomorphism with the states in the light-cone gauge, for which a well-defined scalar product exists. This procedure is analogous to the usual treatment of the Klein–Gordon particle where a scalar product is defined only for solutions of the equations of motion.

The scalar product defined in this manner for on-the-mass-shell states can be related to the original scalar product involving an integral over x^0 by a formal factorization of $\delta(0)$ [1]. When the anticommuting zero mode variables θ^μ are present, one should also factor out zero.

Although the above solution to the scalar product problem enables one to compute any physical scattering amplitude, it is not fully satisfactory from a conceptual point of view. Indeed, since no distinction is made a priori between pure gauge and "physical" operators in the "big" linear space in which these operators act, it would have been more in the line of the Dirac quantization to define a scalar product in the big space which would have kept all states (physical and unphysical) on the same footing. While the Fock representation with negative norms makes this possible for the oscillator variables, we have not found a way to implement this feature in the space of the zero modes.

6. BRST QUANTIZATION

6.1. BRST Charge-Ghost Number

By following the general BRST method, one finds that the BRST charge is given by

$$\Omega = \tfrac{1}{2}p^2\eta + (p\cdot\theta)q + \tilde{\Omega} - \tfrac{1}{2}\mathscr{P}q^2 - 2\bar{\mathscr{P}}M \tag{52a}$$

where we have set

$$\tilde{\Omega} = (a\cdot p)c^* + (a^*\cdot p)c + (b\cdot p)d^* + (b^*\cdot p)d \tag{52b}$$

$$M = c^*c + d^*d \tag{52c}$$

Here, $(\eta, \bar{\mathscr{P}})$ and $(q, \bar{\Pi})$ are the fermionic and bosonic ghost pairs associated with the mass shell condition and the fermionic zero mode constraint $\theta\cdot p = 0$, respectively. The remaining ghosts, associated with the oscillator constraints, are $(c, c^*, \bar{\pi}, \bar{\pi}^*)$ (bosonic) and $(d, d^*, \bar{\gamma}, \bar{\gamma}^*)$ (fermionic).

We have adopted the following conventions:

$$\eta^* = \eta, \qquad \bar{\mathscr{P}}^* = \bar{\mathscr{P}}, \qquad [\eta, \bar{\mathscr{P}}] = 1 \tag{53a}$$

$$q^* = q, \qquad \bar{\Pi}^* = \bar{\Pi}, \qquad [q, \bar{\Pi}] = i \tag{53b}$$

$$[c, \bar{\pi}^*] = 1, \qquad [\bar{\pi}, c^*] = 1 \tag{53c}$$

$$[d, \bar{\gamma}^*] = 1, \qquad [\bar{\gamma}, d^*] = 1 \tag{53d}$$

where the bracket stands again for the quantum-mechanical graded commutator.

Straightforward computations yield

$$2\Omega^2 = [\Omega, \Omega]$$

$$= -2p^2M + [\bar{\Omega}, \bar{\Omega}] \tag{54a}$$

and

$$[\bar{\Omega}, \bar{\Omega}] = 2p^2M \tag{54b}$$

so that Ω is quantum mechanically nilpotent,

$$\Omega^2 = 0 \tag{54c}$$

for any space-time dimension.

The separation (52a) of Ω into $\bar{\Omega}$ and pieces involving the zero mode ghosts will prove useful when we study the BRST cohomology.

The ghost number operator is given by

$$\mathcal{G} = \tfrac{1}{2}(\eta\bar{\mathcal{P}} - \bar{\mathcal{P}}\eta) + \frac{i}{2}(q\bar{\Pi} + \bar{\Pi}q)$$

$$+ c^*\bar{\pi} - \bar{\pi}^*c + d^*\bar{\gamma} - \bar{\gamma}^*d \tag{55a}$$

and is such that

$$[\mathcal{G}, A] = (gh\, A)A \tag{55b}$$

for any operator A of definite ghost number.

While the BRST charge is Hermitian, the ghost number operator is anti-Hermitian,

$$\Omega^* = \Omega \tag{56a}$$

$$\mathcal{G}^* = -\mathcal{G} \tag{56b}$$

Because the eigenvalues of $\mathcal{G}$ are real, this implies that the eigenstates of $\mathcal{G}$ with nonzero eigenvalue possess zero norm. Actually, the statements about hermiticity must be taken with a grain of salt since the norms may not be well defined (recall Section 5.4, and see discussion in Section 6.2 below).

6.2. Representation Space

Because the BRST formalism involves new variables, one must enlarge the Hilbert space to accommodate the (graded) commutator relations (53).

We will represent the zero mode commutation relations (53a) and (53b) in the space of polynomials in η and q, with

$$\bar{\mathscr{P}} = \frac{\partial}{\partial \eta}, \qquad \bar{\Pi} = \frac{1}{i}\frac{\partial}{\partial q} \tag{57}$$

Since $\eta^2 = 0$, the expansion of the states in powers of η terminates at first order. Such a feature does not hold for q, which is commuting, and any power q^n $(n \geq 0)$ can arise. We will, in the sequel, consider polynomials of arbitrary but finite degree in q.

We choose to work with polynomials in q because this allows one to interpret the ghosts as exterior 1-forms along the gauge orbits and because this leads to a sensible BRST cohomology. The monomials q^n are n-forms and, furthermore, they possess definite real ghost number n.

It is clear that monomials in q^n are not normalizable in the positive definite scalar product, which makes q and its momentum Hermitian operators. However, in the space of square integrable functions of q, one cannot find eigenvectors of the ghost number operators with real eigenvalues. Moreover, although we have not investigated the question in detail, it is not completely clear that the BRST cohomology in this space reproduces the expected results. It appears therefore more appropriate to consider instead the space of polynomials in q, as here.

Accordingly, the naive scalar product $\int dq f^*(q) g(q)$ cannot be used and must be modified. However, to our knowledge, no fully satisfactory answer exists in the "big" space of all the variables. But again, just as in the Dirac method of quantization, one can bypass this difficulty by defining a scalar product only after the cohomology has been computed (see next section). This scalar product is defined for physical states only and does not treat all operators of the BRST formalism (physical and unphysical) on an equal footing.

The representation of the remaining commutation relations (53c) and (53d) will be achieved by assuming that the vacuum is annihilated by c, $\bar{\pi}$, d, and $\bar{\gamma}$,

$$c|0\rangle = \bar{\pi}|0\rangle = d|0\rangle = \bar{\gamma}|0\rangle = 0 \tag{58}$$

and by regarding c^*, $\bar{\pi}^*$, d^*, and $\bar{\gamma}^*$ as creation operators. Because of the noncanonical form of the commutation relations, the subspace generated by the ghost creation operators contains negative norm states.

6.3. BRST Cohomology

Because the BRST charge is nilpotent, one can discuss its cohomology. Physical states in the BRST formalism are defined by

$$\Omega|\psi\rangle = 0 \tag{59a}$$

and two physical states which differ by a "BRST null" (or simply, "null") state, i.e., by a term of the form $\Omega|\chi\rangle$ should be identified,

$$|\psi\rangle \sim |\psi\rangle + \Omega|\chi\rangle \tag{59b}$$

[19, 7].

The BRST theory is satisfactory if the BRST cohomological classes defined by (59) coincide with the transverse states. We explicitly show here that this is indeed the case for the above models, and that (59) appropriately selects the physical subspace without needing any further condition. The demonstration is again modeled on what is done for strings [20, 21].

Our first step in computing the BRST cohomology consists in getting rid of the zero mode ghosts. This is done as follows.

Theorem 2. In any BRST cohomological class, one can find a representative that is annihilated by $\bar{\mathscr{P}}$, i.e., that does not involve the zero mode ghost η associated with the mass shell condition.

Proof. By making explicit the η dependence of $|\psi\rangle$,

$$|\psi\rangle = |a\rangle + |b\rangle\eta$$

one finds that $|b\rangle$ transforms as

$$|b\rangle \to |b\rangle + \Box|c\rangle$$

when one adds to $|\psi\rangle$ the exact state $\Omega|c\rangle$, with $|c\rangle$ independent of η. By choosing $|c\rangle$ such that $\Box|c\rangle = -|b\rangle$, one arrives at the desired result.

It should be pointed out here that the state $|c\rangle$ solution of $\Box|c\rangle = -|b\rangle$ may blow up as $x^0 \to \pm\infty$. This would occur when $|b\rangle$ is a solution of the equation $\Box|b\rangle = 0$ since then the equation $\Box|c\rangle = -|b\rangle$ describes an infinite number of forced harmonic oscillators at the resonance frequency. Hence, in order for the above considerations to make sense, the space of states should contain wave functions with no restriction at timelike infinity, and in particular, functions that are not necessarily square integrable. The necessity to allow a more flexible behavior of the states as $x^0 \to \pm\infty$ was already encountered in Section 5.3, where scalar product questions were analyzed.

If the space of states did not contain any of the solutions of $\Box|c\rangle = -|b\rangle$ when $|b\rangle$ is at the resonance frequency, then one could not remove $|b\rangle$. The state $|b\rangle\eta$ would then not be pure gauge and the BRST cohomology would give twice as many physical states as one would expect. This doubling of states has found so far no physical interpretation and hence, does not appear to be reasonable. To remove it, one should either impose a truncation (which is not implied by the formalism itself) or one should enlarge, as here, the space of states so that $|b\rangle\eta$ becomes pure gauge. This second possibility will receive another justification below, when we turn to the second quantization. As we will see, $|b\rangle$ contains no dynamical field and is indeed pure gauge in the field theory.

Once η is eliminated, it is easy to see that the physical states should be on the mass-shell, since $\Omega|\psi\rangle = 0$ implies, with $\bar{\mathcal{P}}|\psi\rangle = 0$,

$$\left.\begin{array}{l}\Omega|\psi\rangle = 0\\ \bar{\mathcal{P}}|\psi\rangle = 0\end{array}\right\} \Rightarrow \{(\theta\cdot p)q + \tilde{\Omega}\}|\psi\rangle = 0, \qquad \Box|\psi\rangle = 0 \tag{60}$$

If $|\psi\rangle$ possesses definite spatial momentum, then it must also possess definite p^0 by (60), and we can thus assume p^μ to be diagonal: physical states can be Fourier-transformed (while $|\chi\rangle$ above does not necessarily possess a well-defined Fourier transform in time).

Thus we work in a subspace where p^μ is diagonal, with $p^2 = 0$. In this subspace, $\tilde{\Omega}$ is nilpotent by (54b).

If the model describes half-integer spin particles, there is a zero mode fermionic constraint ($\theta\cdot p = 0$), and one must as a next step show that one can get rid of the corresponding (commuting) ghost as well. This is the content of the following

Theorem 3. In any BRST cohomological class, one can find a representative that does not depend on q.

Proof. The states can be expanded as

$$|\psi\rangle = \sum_{n=0}^{N} |a_n\rangle(q)^n \tag{61a}$$

with N arbitrary but finite by assumption. One has

$$\Omega|\psi\rangle = \sum_{n=0}^{N} \tilde{\Omega}|a_n\rangle(q)^n + \sum_{n=0}^{N} (\theta\cdot p)|a_n\rangle(q)^{n+1} \tag{61b}$$

From (61b), it follows that $|a_N\rangle$ is annihilated by $(\theta\cdot p)$. This implies

$$|a_N\rangle = (\theta\cdot p)|b_{N-1}\rangle \tag{61c}$$

with

$$|b_{N-1}\rangle = -2(k\cdot\theta)|a_N\rangle \tag{61d}$$

since

$$[\theta\cdot p, k\cdot\theta] = -1/2 \tag{61e}$$

Therefore, by adding the exact state $\Omega(-|b_{N-1}\rangle q^{N-1})$ to $|\psi\rangle$, one can eliminate the $|a_N\rangle q^N$ component of $|\psi\rangle$.

By going on in the same fashion for the powers of order $N-1$, $N-2,\ldots$, one can assume that $|\psi\rangle$ does not involve q,

$$|\psi\rangle = |a_0\rangle \tag{62a}$$

with

$$\theta \cdot p|a_0\rangle = \Box|a_0\rangle = 0, \qquad \bar{\Omega}|a_0\rangle = 0 \tag{62b}$$

by the BRST condition.

To conclude the computation of the BRST cohomology, one defines the counting operator $\tilde{N}$ by

$$\begin{aligned}\tilde{N} = {} & c^*\bar{\pi} + \bar{\pi}^*c + d^*\bar{\gamma} + \bar{\gamma}^*d - (k\cdot a)^*(p\cdot a)\\ & - (k\cdot b)^*(p\cdot b) - (p\cdot a)^*(k\cdot a) - (p\cdot b)^*(k\cdot a)\end{aligned} \tag{63a}$$

The operator $\tilde{N}$ counts the number of ghost modes, as well as the number of gauge modes created by the operators $k\cdot a^*$, $p\cdot a^*$, $k\cdot b^*$, or $p\cdot b^*$. Its eigenvalues are positive integers, and the transverse states are completely characterized by

$$\tilde{N}|T\rangle = 0 \tag{63b}$$

A central property of $\tilde{N}$ is that it is a BRST null operator,

$$\tilde{N} = [K, \tilde{\Omega}] \tag{64a}$$

with

$$K = -(k\cdot a^*\bar{\pi} + \bar{\pi}^*k\cdot a + k\cdot b^*\bar{\gamma} + \bar{\gamma}^*k\cdot b) \tag{64b}$$

From (64a), it follows that $[\tilde{N}, \tilde{\Omega}] = 0$, so that one can work out the cohomology of $\tilde{\Omega}$ at a fixed eigenvalue n of $\tilde{N}$. One then arrives at the following theorem.

Theorem 4. Physical states with $n \neq 0$ are exact.

Proof. If $\tilde{\Omega}|a_0\rangle = 0$ and $\tilde{N}|a_0\rangle = n|a_0\rangle$, $n \neq 0$, then, one finds from (64a)

$$\begin{aligned}|a_0\rangle &= \frac{1}{n}\tilde{N}|a_0\rangle\\ &= \frac{1}{n}(K\tilde{\Omega} + \tilde{\Omega}K)|a_0\rangle\\ &= \tilde{\Omega}|\chi\rangle \qquad (\tilde{\Omega}|a_0\rangle = 0)\end{aligned}$$

with $|\chi\rangle = (1/n)K|a_0\rangle$.

The operator K plays the role of a contracting homotopy, and the mechanism by which the ghost and gauge modes disappear is known as the "quartet mechanism" [19].

The results that we have established in this section can be collected in the following theorem.

Theorem 5. The most general BRST invariant state is given by

$$|\psi\rangle = |T\rangle|0\rangle + \Omega|\chi\rangle \tag{65}$$

where $|T\rangle$ is a transverse state and where $|0\rangle$ is the vacuum of the ghosts and of the oscillators $k \cdot a^*, p \cdot a^*, k \cdot b^*, p \cdot b^*$, with no η or q dependence.

We have therefore demonstrated that the BRST theory reproduces the results of other quantization methods. To arrive at this conclusion, there was no need to add an extra condition on the ghost number of the physical states: the BRST condition by itself, together with the factorization by the BRST exact states, was enough to establish (65) in the space of states considered here.

Another point worth mentioning is that the presence of null states in the physical subspace does not appear as an accident in the BRST formalism. It rests instead on a conceptually sounder basis since it simply follows from the nilpotency and hermiticity of the BRST charge. In addition, even though the conditions $a \cdot p|\psi\rangle = b \cdot p|\psi\rangle = 0$ are not directly imposed on the physical states, they emerge after one has suitably fixed the BRST gauge freedom of adding BRST exact states. In that sense, one can view the Dirac method as resulting from a partial gauge fixing of the BRST one, which still admits a nontrivial residual gauge invariance.

Finally, we note that by using the isomorphism between physical states and states of the light cone gauge, expressed in Theorem 5, one can define a scalar product in the space $\operatorname{Ker}\Omega/\operatorname{Im}\Omega$ of BRST cohomological classes. In what concerns the oscillator variables, the scalar product so defined is the restriction to physical states of the Fock space inner product, but this is not so for the zero modes, since for them there is no satisfactory scalar product in the big space to begin with. [The physical states possess an ill-defined norm for the naive scalar product in the big space involving an integration over x^0 and q (which yields infinity) and over θ^μ and the zero mode ghost η (which yields zero).]

6.4. Lagrange Multipliers

In order to discuss the path integral, it is necessary to include the Lagrange multipliers associated with the constraints as canonical variables. Hence, for each constraint, we introduce one canonical pair of new variables with same Grassmann parity, according to the scheme

$$p^2 = 0 \to (N, p_N) \tag{66a}$$

$$\theta \cdot p = 0 \to (M, p_M) \tag{66b}$$

$$p \cdot a = 0, p \cdot a^* = 0 \to (\lambda^*, p_\lambda), (\lambda, p_\lambda^*) \tag{66c}$$

$$p \cdot b = 0, p \cdot b^* = 0 \to (\mu^*, p_\mu), (\mu, p_\mu^*) \tag{66d}$$

The new variables are pure gauge, and thus do not change the physical content of the theory. This is because the new momenta vanish

$$p_N \approx 0, \quad p_M \approx 0, \quad p_\lambda \approx 0, \quad p^*_\lambda \approx 0, \quad p_\mu \approx 0, \quad p^*_\mu \approx 0 \quad \text{(67a)}$$

These new first class constraints contribute an additional piece to the BRST generator,

$$\Omega = \Omega^{\text{old}} + \Omega' \quad \text{(67b)}$$

with Ω^{old} given by the former expression (52) and Ω' equal to

$$\Omega' = p_N\mathscr{P} + p_M\Pi + \pi^* p_\lambda + p^*_\lambda \pi + \gamma^* p_\mu + p^*_\mu \gamma \quad \text{(67c)}$$

The ghost pairs $(\bar{\eta}, \mathscr{P})$, $(\bar{q}, \Pi)$, $(\bar{c}, \pi^*)$, $(\bar{c}^*, \pi)$, $(\bar{d}, \gamma^*)$, and $(\bar{d}^*, \gamma)$ are usually referred to as the antighost canonical pairs.

The simple bilinear form (67c) of the new term Ω' added to the original BRST operator is a consequence of the Abelian nature of the constraints (67a). One clearly gets

$$[\Omega', \Omega'] = 0, \qquad [\Omega', \Omega^{\text{old}}] = 0 \quad \text{(67d)}$$

and thus, nilpotency of the complete BRST charge (67b) still holds.

Because of the Abelian nature of Ω', the new variables can easily be shown to disappear from the BRST cohomology (see, e.g., Ref. 22). Therefore, the physical subspace associated with (67b) is still spanned by the (on-shell) transverse states.

7. SECOND QUANTIZED THEORY

7.1. Free Field Action

The above models can be in different spin states. Each of these states is characterized by a wave function $\phi_{\mu_1\cdots\mu_k}(x)$ of the space-time coordinates x^μ, which after the freedom $|\psi\rangle \to |\psi\rangle + \Omega|\chi\rangle$ has been suitable fixed, obeys the Klein–Gordon or Dirac equation

$$\Box\phi_{\mu_1\cdots\mu_k} = 0 \quad \text{or} \quad \gamma^{\mu_0}\partial_{\mu_0}\phi_{\mu_1\cdots\mu_k} = 0 \quad \text{(68a)}$$

as well as the transversality conditions

$$\partial^{\mu_1}\phi_{\mu_1\cdots\mu_k} = 0 \quad \text{(68b)}$$

expressing the Dirac constraint $a \cdot p|\psi\rangle = 0$ or $b \cdot p|\psi\rangle = 0$.

As we have shown, these equations result from a partial gauge fixing of the BRST formalism, and admit the residual gauge invariance $\phi_{\mu_1\cdots\mu_k} \to \phi_{\mu_1\cdots\mu_k} + \partial_{\mu_1}\Lambda_{\mu_2\cdots\mu_k}$ where $\Lambda_{\mu_2\cdots\mu_k}$ obeys (68a).

In order to develop a many-particle theory, one regards the wave function $\phi_{\mu_1\cdots\mu_k}$ as a field operator creating or destroying particles in the corresponding spin state. In the gauge (68b), the space-time development of the field operators is still given by (68a).

For the purpose of constructing the second quantized formalism, it appears necessary to derive the field equations (68a) from an action principle. This action principle should be local in space-time and Lorentz invariant.

Now, it is well known that the local formulation of gauge theories is most transparent when gauge invariance is fully maintained, so that one would like to find an action principle that leads to equations equivalent to (68), but without the gauge condition (68b).

It should be clear from our previous analysis that the way to achieve this goal is not to start from the Dirac quantum formalism, since this one appears to be already partially gauge fixed. Rather, one should adopt the first quantized BRST theory as a starting point, since in it, the transversality condition (68b) arises only after the freedom $|\psi\rangle \rightarrow |\psi\rangle + \Omega|\chi\rangle$ has been partly frozen.

So, our aim is to find an action principle that implies the equation

$$\Omega|\psi\rangle = 0 \tag{69a}$$

and which is gauge invariant under

$$|\psi\rangle \rightarrow |\psi\rangle + \Omega|\chi\rangle \tag{69b}$$

From the analysis of the BRST cohomology, it results that the true degrees of freedom are automatically those associated with the light cone gauge spectrum. The other components of $|\psi\rangle$ are either pure gauge or auxiliary.

Because the field equations (69a) are linear in $|\psi\rangle$ (free field theory), the action should be quadratic in $|\psi\rangle$. The simplest possibility is, in the case of a complex field

$$S = -(\psi, \Omega\psi) \tag{70}$$

where (,) is a nondegenerate bilinear form obeying the conditions

$$\text{(i)}\quad (\chi_1, \Omega\chi_2) = (\Omega\chi_1, \chi_2) \tag{71a}$$

$$\text{(ii)}\quad (\psi, \Omega\psi) = (\psi, \Omega\psi)^* \tag{71b}$$

The requirement (71a) ensures that one can consistently treat ψ and ψ^* as independent fields in the action principle. The condition (71b) guarantees that the action is real. The gauge invariance (69b) is manifestly enforced in (70) since Ω is nilpotent.

If the field ψ itself obeys some reality conditions, i.e., if ψ^* is completely determined by ψ as

$$\psi^* = R\psi, \qquad RR^* = I \tag{72a}$$

and if these reality conditions are such that

$$(\chi_1, \Omega\chi_2) = (\Omega\chi_2, \chi_1) \tag{72b}$$

for any pairs of fields that obey (72a), then the action (70) should be replaced by

$$S = -\tfrac{1}{2}(\psi, \Omega\psi) \tag{73}$$

In that case, ψ^* should not be regarded as an independent field in the variational principle, but rather, should be related to ψ by (72a).

7.2. Bilinear Form

The problem of formulating the second quantized variational principle is thus equivalent to the problem of finding an appropriate nondegenerate bilinear form obeying (71a) and (71b) and, in the real case, (72b) as well.

It is clear that the searched-for bilinear form is far from being uniquely determined by the above demands. For instance, if (χ_1, χ_2) is a solution of (71a) and (71b) such that $(\chi_1, \chi_2)^* = (\chi_2, \chi_1)$, then $(\chi_1, \chi_2)' = (\chi_1, A\chi_2)$ is also a solution if the operator A commutes with Ω and is Hermitian for $(\ , \ )$. To select the bilinear form, further considerations are thus needed.

The situation is exactly the same as in string theory, and the analysis leads to exactly the same conclusions: while a natural bilinear form can be chosen when the system describes integer spin particles (bosonic or Neveu-Schwarz strings) [23], the same choice meets difficulties if applied to models with half-integer spin particles (Ramond string). The origin of the problem can be traced to the zero mode fermionic constraint: one would like the action $(\psi, \Omega\psi)$ to yield the first-order kinetic term $\int a_0\theta \cdot pa_0 \, d^4x = \int a_0 \not{p} a_0 \, d^4x$ for the physical fermions ($\psi = a_0 + \Omega\chi$), but the straightforward attempts yield $\int a_0 \Box a_0 \, d^4x$ instead.

For this reason, we will from now on restrict our study to models with integer spin particles only, and will not discuss further the interesting suggestions that have been put forward to overcome the difficulty present in the models containing half-integer spin fields [24].

This means that we will assume the zero-mode fermionic constraint $\theta \cdot p \approx 0$ to be absent: the fermionic internal degrees of freedom come by pairs, and the constraints are

$$p^2 \approx 0, \qquad a \cdot p \approx 0, \qquad a^* \cdot p \approx 0, \qquad b \cdot p \approx 0, \qquad b^* \cdot p \approx 0 \tag{74}$$

with BRST charge

$$\Omega = \tfrac{1}{2}p^2\eta + \tilde{\Omega} - 2\bar{\mathcal{P}}M \tag{75a}$$

$$\tilde{\Omega} = (a \cdot p)c^* + (a^* \cdot p)c + (b^* \cdot p)d + d^*(p \cdot b) \tag{75b}$$

$$M = c^*c + d^*d \tag{75c}$$

The bilinear form which gives an acceptable second quantized action is just the first quantized scalar product

$$(\psi_1, \psi_2) = \langle\psi_1|\psi_2\rangle \tag{76a}$$

$$\langle\psi_1|\psi_2\rangle = \int d^4x\, d\eta (\psi_1(x, \eta)|\psi_2(x, \eta))_F \tag{76b}$$

where $(\psi_1(x, \eta)|\psi_2(x, \eta))_F$ denotes the Fock space inner product associated with the oscillators. This scalar product makes all real operators self-adjoint and therefore $\Omega = \Omega^+$, so that the conditions (71) clearly hold.

The objections raised against (76) in the discussion of the first quantized theory, namely, the fact that (76b) is ill defined and of the form $0 \cdot \delta(0)$ for physical states, are not applicable here. This is because we no longer interpret (76) as a probability amplitude. Furthermore, the action $S[\psi] = \langle\psi|\Omega|\psi\rangle$ is well defined when evaluated between two spacelike hypersurfaces (say, $x^0 = x_1^0$, and $x^0 = x_2^0$), provided the field ψ appropriately falls off at space-like infinity, a condition that does not rule out the solutions of the equations $\Omega\psi = 0$. The fact that the action may be infinite for an infinite time interval is no longer a problem.

7.3. Ghost Number −1/2 Gauge

With the definition (55a) of the ghost number operator, the only nontrivial cohomology of the BRST charge is at ghost number −1/2, as theorem (65) indicates. This means that in the expansion of $|\psi\rangle$ according to ghost number,

$$|\psi\rangle = \sum_{n=-\infty}^{+\infty} |\psi_{n-1/2}\rangle \tag{77}$$

the physical fields should be found at $n = 0$, while the fields contained in $|\psi_{n-1/2}\rangle$ with $n \neq 0$ do not carry physical information and are either auxiliary or pure gauge.

Now, let $|a\rangle$ and $|b\rangle$ be two states of respective ghost numbers g_a and g_b, with $g_a \neq 0$, $g_b \neq 0$. One has

$$\langle a|b\rangle = \langle a|\frac{1}{g_b}\mathcal{G}|b\rangle = -\frac{g_a}{g_b}\langle a|b\rangle \tag{78}$$

because g_a and g_b are real, but $\mathcal{G}$ is antihermitian. From (78), it follows that $\langle a|b\rangle$ vanishes, unless $g_a = -g_b$.

This implies that the action $\langle\psi|\Omega|\psi\rangle$ reads

$$S[\psi] = -\langle\psi|\Omega|\psi\rangle = \sum_n -\langle\psi_{n-1/2}|\Omega|\psi_{n-1/2}\rangle \tag{79}$$

The only term that involves the physical fields is that with $n = 0$,

$$-\langle\psi_{-1/2}|\Omega|\psi_{-1/2}\rangle$$

The auxiliary and pure gauge fields contained in $|\psi_{n-1/2}\rangle$, $n \neq 0$, do not couple to the physical fields and can therefore be consistently set equal to zero in the action ("before variation"). This defines the "ghost number $-1/2$ gauge,"

$$\mathscr{G}|\psi\rangle = -\tfrac{1}{2}|\psi\rangle \quad \Leftrightarrow \quad |\psi\rangle = |\psi_{-1/2}\rangle$$

$$|\psi_{n-1/2}\rangle = 0, \qquad n \neq 0 \tag{80}$$

which, as our analysis of the BRST cohomology has shown, can always be reached by the addition of an appropriate exact state.

In the gauge (80), the action reads

$$S[\psi] = -\langle\psi_{-1/2}|\Omega|\psi_{-1/2}\rangle \tag{81}$$

and the residual gauge invariance is given by

$$|\psi_{-1/2}\rangle \to |\psi_{-1/2}\rangle + \Omega|\varepsilon_{-3/2}\rangle \tag{82a}$$

$$\mathscr{G}|\varepsilon_{-3/2}\rangle = -\tfrac{3}{2}|\varepsilon_{-3/2}\rangle \tag{82b}$$

It thus appears that there is no need to restrict the ghost number of $|\psi\rangle$ in the gauge invariant formalism. This restriction arises only as a partial gauge choice. Since this gauge choice is particularly convenient, however, it will be assumed from now on.

A complete gauge choice, which removes the freedom (82a), would be obtained by imposing further that $|\psi_{-1/2}\rangle$ contains (light cone) transverse excitations only. This additional requirement is permissible as our analysis of the BRST cohomology indicates. However, this complete gauge fixing breaks manifest Lorentz invariance, and will therefore not be imposed in the sequel [only (80) will be assumed].

7.4. Component Expansion

In the holomorphic representation [17], the field $|\psi_{-1/2}\rangle$ and the gauge parameter $|\varepsilon_{-3/2}\rangle$ become functions of x, η, and of the oscillator variables $a^*, b^*, c^*, d^*, \bar{\pi}^*, \bar{\gamma}^*$:

$$|\psi_{-1/2}\rangle \to \psi(x, \eta, a^*, b^*, c^*, d^*, \bar{\pi}^*, \bar{\gamma}^*)|0\rangle \tag{83a}$$

$$|\varepsilon_{-3/2}\rangle \to \varepsilon(x, \eta, a^*, b^*, c^*, d^*, \bar{\pi}^*, \bar{\gamma}^*)|0\rangle \tag{83b}$$

In (83a) and (83b), $|0\rangle$ stands for the Fock vacuum annihilated by $\bar{\mathscr{P}}$ and p^μ. It carries ghost number $-1/2$, so that ψ and ε possess ghost numbers 0 and -1, respectively. In the sequel, we will often drop the symbol $|0\rangle$, and the functions ψ and ε should be thought of as multiplied by the Fock vacuum annihilated by $\bar{\mathscr{P}}$ and p_μ even if this is not explicitly indicated.

It is convenient to isolate the η dependence of the states as

$$\psi = A + \eta B \tag{84a}$$

$$\varepsilon = \lambda + \eta\mu \tag{84b}$$

with

$$gh\, A = 0, \qquad gh\, B = -1 \tag{84c}$$

$$gh\, \lambda = -1, \qquad gh\, \mu = -2 \tag{84d}$$

Here A, B, λ, and μ involve only the space-time coordinate x^μ and the oscillators.

Direct evaluation of the action yields

$$S[\psi] = \tfrac{1}{4}\int d^4x\,(A'(x)|\Box A(x))_F + \tfrac{1}{2}\int d^4x\,(A'(x)|\tilde{\Omega}B(x))_F$$
$$-\tfrac{1}{2}\int d^4x\,(B'(x)|\tilde{\Omega}A(x))_F + \int d^4x\,(B'(x)|MB(x))_F \tag{85a}$$

where we have set

$$A' = A_0 - A_1, \qquad B' = B_0 - B_1 \tag{85b}$$

with $\varepsilon(A_0) = \varepsilon(B_0) = 0$, $\varepsilon(A_1) = \varepsilon(B_1) = 1$, $A = A_0 + A_1$, $B = B_0 + B_1$. The flip of sign for the Grassmann parity +1 component of A and B in (85a) results from an anticommutation with η.

The action is invariant under the gauge transformations (82a), which read explicitly

$$A \to A + \tilde{\Omega}\lambda - 2M\mu \tag{86a}$$

$$B \to B - \tfrac{1}{2}\Box\lambda - \tilde{\Omega}\mu \tag{86b}$$

To further analyze the action (85a), we introduce the level operator N,

$$N = \tilde{N} + N_T \tag{87a}$$

$$= a^{*\mu}a_\mu + b^{*\mu}b_\mu + c^*\bar{\pi} + \bar{\pi}^*c + d^*\bar{\gamma} + \bar{\gamma}^*d \tag{87b}$$

which is BRST invariant,

$$[N, \Omega] = 0 \tag{87c}$$

The operator N counts the occupation number for *all* the modes (physical and unphysical). Contrary to $\tilde{N}$, which counts only unphysical modes, the level operator N is Lorentz invariant and not BRST exact. The eigenvalues of N are positive integers, and states with different occupation numbers are orthogonal.

Therefore, we expand A and B as

$$A = \sum_{n\geq 0} A_n, \qquad gh(A_n) = 0, \qquad NA_n = nA_n \tag{88a}$$

$$B = \sum_{n\geq 0} B_n, \qquad gh(B_n) = -1, \qquad NB_n = nB_n \tag{88b}$$

and find that the action becomes

$$S[\psi] = \sum_{n\geq 0} S_n$$

$$S_n = \tfrac{1}{4}\int d^4x\,(A'_n(x)|\Box A_n)_F + \tfrac{1}{2}\int d^4x\,(A'_n(x)|\tilde{\Omega}B_n(x))_F$$

$$-\tfrac{1}{2}\int d^4x\,(B'_n(x)|\tilde{\Omega}A_n(x))_F + \int d^4x\,(B'_n(x)|MB_n(x))_F \tag{89a}$$

The gauge transformations reads

$$A_n \to A_n + \tilde{\Omega}\lambda_n - 2M\mu_n \tag{89b}$$

$$B_n \to B_n - \tfrac{1}{2}\Box\lambda_n - \tilde{\Omega}\mu_n \tag{89c}$$

with

$$\lambda = \sum_{n\geq 0} \lambda_n, \qquad gh(\lambda_n) = -1, \qquad N\lambda_n = n\lambda_n \tag{89d}$$

$$\mu = \sum_{n\geq 0} \mu_n, \qquad gh(\mu_n) = -2, \qquad N\mu_n = n\mu_n \tag{89e}$$

7.5. *p*-Form Gauge Fields

The analysis of the field theoretical models at this point goes along the same lines independently of the number of oscillators. In order to unclutter the formulas, we will therefore treat only a specific example, which shows already all the features of the general case. That specific example is described by a single set of fermionic oscillators, i.e., it is the $N = 2$ supersymmetric model studied above. We chose the oscillators to be fermionic because an extra interesting feature arises: it is necessary to truncate the field theory to a definite parity sector. That truncation is the same as the G-parity truncation encountered in string models [10]. No similar restriction appears to be required in the pure bosonic case.

The physical requirement that makes the truncation necessary is positivity of the energy. As we have shown, the physical fields are the transverse components of $A_n(x)$. Therefore, the term in (89a) that determines the sign of the energy is

$$\frac{1}{4}\int d^4x\,(A_n'^T(x)|\Box A_n^T(x))_F$$

$$= -\frac{1}{4}\int d^4x\,(\partial^\mu A_n'^T(x)|\partial_\mu A_n^T(x))$$

$$= -\frac{(-1)^n}{4}\int d^4x\,(\partial^\mu A_n^T(x)|\partial_\mu A_n^T(x))_F \tag{90}$$

where we have used the fact that $A_n^T(x)$, which contains n fermionic oscillators $a_i^{T*}(x)$, possesses Grassmann parity $(-1)^n$. [It is assumed that the coefficients of the oscillators in the expansion of A_n are commuting functions of x^μ, since the fields describe integer-spin particles. Otherwise, one would violate the spin-statistics relation.]

It is clear from (90) that the sign of the kinetic term will be positive or negative according to whether n is even or odd. Accordingly, in order to get a definite sign, it is necessary to truncate the theory to a definite parity of the occupation number n. This guarantees at the same time that the field ψ possesses definite Grassmann parity.*

The fact that the field theory based on fermionic oscillators forces a truncation to a definite parity sector was recognized for the first time in the case of the fermionic string models [24].

Once appropriately truncated, the second quantized theory describes a collection of free massless p-form gauge fields, with p either even or odd. At each level there is a physical gauge in which the only remaining fields are those corresponding to the $\binom{p}{d-2}$ physical helicities.

To see explicitly how this happens, let us analyze in detail the second level of the theory. One gets from (88)

$$A_2 = A_{\lambda\mu}(x)a^{\lambda*}a^{\mu*} + A(x)c^*\bar{\pi}^* \tag{91a}$$

$$B_2 = iB_\lambda(x)a^{\lambda*}\bar{\pi}^* \tag{91b}$$

where we take $A_{\lambda\mu}$, A, and B_λ to be real (so that the action is real). The action at the second level reads

$$S = \frac{1}{2}\int d^4x\,A_{\lambda\mu}\Box A^{\lambda\mu} + \frac{1}{4}\int d^4x\,A\Box A$$
$$+ 2\int d^4x\,A^{\lambda\mu}\partial_\lambda B_\mu + \int d^4x\,A\partial^\mu B_\mu - \int d^4x\,B_\lambda B^\lambda \tag{92}$$

* In order to accommodate both parities, one may try to change the bilinear form (70), but this appears unnatural.

and is invariant under

$$A \to A + \partial^\lambda \varphi_\lambda - 4\Phi \tag{93a}$$

$$B_\lambda \to B_\lambda - \tfrac{1}{2}\Box \varphi_\lambda + 2\partial_\lambda \Phi \tag{93b}$$

$$A_{\lambda\mu} \to A_{\lambda\mu} + \tfrac{1}{2}(\partial_\lambda \varphi_\mu - \partial_\mu \varphi_\lambda) \tag{93c}$$

with

$$\lambda_2 = i\varphi_\lambda a^{\lambda *}\bar{\pi}^*, \qquad \mu_2 = \Phi \bar{\pi}^* \bar{\pi}^* \tag{93d}$$

By redefining Φ as $\Phi = \Phi' + \partial^\lambda \varphi_\lambda / 4$, one can rewrite (93a)-(93c) in the equivalent form

$$A \to A - 4\Phi' \tag{94a}$$

$$B_\lambda \to B_\lambda + \tfrac{1}{2}\partial^\rho(\partial_\lambda \varphi_\rho - \partial_\rho \varphi_\lambda) + 2\partial_\lambda \Phi' \tag{94b}$$

$$A_{\lambda\mu} \to A_{\lambda\mu} + \tfrac{1}{2}(\partial_\lambda \varphi_\mu - \partial_\mu \varphi_\lambda) \tag{94c}$$

This new form shows explicitly that A is pure gauge and can be set equal to zero by an appropriate choice of Φ'. The residual gauge invariance is characterized by $\Phi' = 0$, φ_λ arbitrary. The partially gauge fixed action is given by

$$S[A_{\lambda\mu}, B_\rho] = \tfrac{1}{2}\int d^4x\, A_{\lambda\mu}\Box A^{\lambda\mu} - 2\int d^4x\, \partial_\mu A^{\mu\lambda} B_\lambda - \int d^4x\, B_\lambda B^\lambda \tag{95}$$

The field B_λ is auxiliary and can be eliminated by means of its own equations of motion. This does not break the gauge invariance of the theory under arbitrary φ_λ in (94). After this is done, one finds the standard action

$$S[A_{\lambda\mu}] = -\tfrac{1}{6}\int d^4x\, G_{\lambda\mu\nu}G^{\lambda\mu\nu} \tag{96a}$$

$$G_{\lambda\mu\nu} = \partial_\lambda A_{\mu\nu} + \partial_\mu A_{\nu\lambda} + \partial_\nu A_{\lambda\mu} \tag{96b}$$

describing a physical 2-form $A_{\lambda\mu}$. This action is invariant under (94c), as it should be.

We thus see that the BRST second quantized formalism yields an appropriate local action possessing the required gauge invariances. What is true for the second level also holds at the other levels, as can straightforwardly be checked along the same lines. We can thus conclude that the BRST formalism gives reasonable answers, but in a more powerful way.

7.6. Remarks on Interactions

The problem of introducing interactions in the models constructed here is clearly an important and challenging one. We will not investigate it in this chapter, but rather, we will merely report here some possible lines of approach.

As a first step, one may try to couple the models to a given electromagnetic or gravitational background. This question is part of the difficult problem of defining consistent electromagnetic or gravitational couplings for higher spin fields. The interest of the approach based on the present models is that it sheds a new light on this question, since the analysis can be carried out in terms of the "first quantized" degrees of freedom, which are in finite number, rather than in terms of the second-quantized field components themselves.

A model interacting with a given background is consistent if, after the interactions have been incorporated, the algebra of the constraints $\mathcal{H}$ and $\mathcal{S}_i$ is still first class. In that case, one can construct again the BRST charge and the associated second quantized theory with gauge invariance $\psi \rightarrow \psi + \Omega\chi$.

In order to preserve the first-class property of the constraints $\mathcal{H}$ and $\mathcal{S}_i$, it may turn out to be necessary to add extra internal degrees of freedom [25, 26]. These new variables generate new fields at the second quantized level, which thus appear necessary for a consistent description of the coupling of higher spin fields to a given gravitational or electromagnetic background.

A more ambitious line of research is to investigate self-interacting models. Few encouraging results have been obtained so far. Nevertheless, this question should be studied further since it would enlighten the problem of interacting string field theory. It is hoped to return to it in the future.

Note added in proof. After this work was completed, we received two preprints [27, 28] in which the first quantized description of higher spin particles was discussed along the lines of the gauged $N = 2$ supersymmetric model of Ref. 29. The main difference between our approach and these very interesting works is that we do not gauge the global $O(N)$ symmetry mentioned in Section 2.2 by constraining the corresponding generator to vanish.

ACKNOWLEDGMENTS. This work has been partly supported by a grant from the Tinker Foundation to the Centro de Estudios Científicos de Santiago, by the NSF under grant No. PHY 8600384 to the University of Texas at Austin and by a NATO Collaboration research grant. One of us (MH) is Chercheur qualifié au Fonds National de la Recherche Scientifique (Belgium).

REFERENCES

1. C. Teitelboim, *Phys. Rev. D 25*, 3159 (1982); Proper time approach to quantized supergravity, in *Superspace and Supergravity* (S. W. Hawking and M. Roček, eds.), Cambridge University Press, Cambridge, 1981; M. Henneaux and C. Teitelboim, *Ann. Phys.* (*N.Y.*) *143*, 127 (1982).
2. C. Teitelboim, *Phys. Rev. Lett. 38*, 1106 (1977).
3. A. Barducci, R. Casalbuoni, and L. Lusanna, *Nuovo Cimento A 35*, 377 (1976); F. A. Berezin and M. S. Marinov, *Ann. Phys.* (*N.Y.*) *104*, 336 (1977); C. A. P. Galvão and C. Teitelboim, *J. Math. Phys. 21*, 1863 (1980).
4. L. Brink, P. Di Vecchia, and P. Howe, *Phys. Lett. 65B*, 471 (1976); *B118*, 76 (1977); S. Deser and B. Zumino, *Phys. Lett. 65B*, 369 (1976); L. Brink, S. Deser, P. Di Vecchia,

P. Howe, and B. Zumino, *Phys. Lett. 64B*, 435 (1976); see also F. Raundel, *Phys. Rev. D 21*, 2823 (1980).
5. S. Deser and B. Zumino, *Phys. Rev. Lett. 38*, 1433 (1977).
6. E. S. Fradkin and G. A. Vilkovisky, *Phys. Lett. 55B*, 224 (1975); I. A. Batalin and G. A. Vilkovisky, *ibid. 69B*, 309 (1977); E. S. Fradkin and T. E. Fradkina, *ibid. 72B*, 343 (1978).
7. M. Henneaux, *Phys. Rep. 126*, 1 (1985); see also Chap. 10 in *Quantum Mechanics of Fundamental Systems 1* (C. Teitelboim, ed.), Plenum Press, New York, 1988.
8. A. K. H. Bengtsson, *Phys. Rev. D 32*, 2031 (1985); F. A. Berends, G. H. J. Burgers, and H. van Dam, *Nucl. Phys. B 260*, 295 (1985); E. S. Fradkin and M. A. Vasiliev, *Nucl. Phys. B 291*, 141 (1987).
9. B. Zumino, *Phys. Lett. 87B*, 203 (1979).
10. M. B. Green, J. H. Schwarz, and E. Witten, *Superstring Theory*, Vol. 1, Cambridge University Press, Cambridge, 1987; L. Brink and M. Henneaux, *Principles of String Theory*, Plenum Press, New York, 1988.
11. D. J. Gross, *Phys. Rev. Lett. 60*, 1229 (1988).
12. M. Henneaux, M. Pilati, and C. Teitelboim, *Phys. Lett. 110B*, 123 (1982).
13. R. Casalbuoni, *Nuovo Cimento 35A*, 289 (1976); L. Brink and J. H. Schwarz, *Phys. Lett. 100B*, 310 (1981); D. V. Volkov and A. J. Pashnev, *Theor. Math. Phys. 44*, 321 (1980).
14. J. Kowalski-Glikman, J. W. van Holten, S. Aoyama, and J. Lukierski, *Phys. Lett. 201B*, 487 (1988).
15. M. B. Green and J. H. Schwarz, *Nucl. Phys. B181*, 502 (1981); *B198*, 252 (1982); *B198*, 441 (1982); *Phys. Lett. 136B*, 367 (1984).
16. A. K. H. Bengtsson, *Phys. Lett. B 182*, 321 (1986); *189B*, 337 (1987); I. Bengtsson, *Class. Quant. Grav. 3*, 1033 (1986); J. Gomis, M. Novell, and K. Rafanelli, *Phys. Rev. D 34*, 1072 (1986); S. Ouvry and J. Stern, *Phys. Lett. B 177*, 335 (1986); I. G. Koh and S. Ouvry, *ibid. 179*, 115 (1986); M. Bellon and S. Ouvry, *ibid. 187*, 93 (1987); R. Casalbuoni and D. Dominici, *ibid. 186*, 63 (1987); N. V. Krasnikov, *Phys. Lett. B195*, 377 (1987); J. M. F. Labastida, *Phys. Rev. Lett. 58*, 531 (1987); P. P. Srivastava and N. A Lemos *Phys. Rev. D 15*, 3568 (1977).
17. L. D. Faddeev and A. A. Slavnov, *Gauge Fields: Introduction to Quantum Theory*, Benjamin, Reading, 1980.
18. T. Curtright, *Phys. Lett. B 165*, 304 (1985).
19. T. Kugo and I. Ojima, *Suppl. Progr. Theor. Phys. 66*, (1979).
20. M. Kato and K. Ogawa, *Nucl. Phys. B 212*, 443 (1983); M. D. Freeman and D. Olive, *Phys. Lett. B175*, 151 (1986); I. B. Frenkel, H. Garland, and G. J. Zuckerman, *Proc. Natl. Acad. Sci. U.S.A. 83*, 8442 (1986); H. Neuberger, *Phys. Lett. B188*, 214 (1987).
21. M. Henneaux, *Phys. Lett. B177*, 35 (1986); *183*, 59 (1987).
22. M. Henneaux, *Found. Phys. 17*, 634 (1987).
23. A. Neveu, H. Nicolai, and P. West, *Phys. Lett. B 167*, 307 (1986); E. Witten, *Nucl. Phys. B 268*, 253 (1986); see also W. Siegel and B. Zwiebach, *ibid. 263*, 105 (1986); T. Banks and M. Peskin, *Nucl. Phys. B 264*, 513 (1986); K. Itoh, T. Kugo, H. Kunimoto, and H. Ooguri, *Progr. Theor. Phys. 75*, 162 (1986); J. G. Taylor, *Phys. Lett. B186*, 57 (1987).
24. E. Witten, *Nucl. Phys. B 276*, 291 (1986); T. Banks, M. E. Peskin, C. R. Preitschopf, D. Friedan, and E. Martinec, *Nucl. Phys. B 274*, 71 (1986); A. Le Clair and J. Distler, *Nucl. Phys. B 273*, 552 (1986); J. P. Yamron, *Phys. Lett. B 174*, 69 (1986); H. Ooguri, *Phys. Lett. B 172*, 204 (1986); H. Terao and S. Uehara, *Phys. Lett. B 173*, 134 (1986); Y. Kasama, A. Neveu, H. Nicolai, and P. West, *Nucl. Phys. B 278*, 833 (1986).
25. I. A. Batalin and E. S. Fradkin, *Nucl. Phys. B 279*, 514 (1987).
26. M. Henneaux and H. Waelbroeck, *Class. Quant. Grav.* (to be published).
27. P. Howe, S. Penati, M. Pernici, and P. Townsend, preprint IFUM 345/FT (August 1988).
28. R. Marnelius and U. Mårtensson, Göteborg preprint (November 1988).
29. L. Brink, P. Di Vecchia, and P. Howe, *Nucl. Phys. B 118*, 76 (1977).

Chapter 10

Strings in Space

Leonard Susskind, Marek Karliner, and Igor Klebanov

String theory was originally invented to describe hadrons.* Ultimately this idealized mathematical theory of hadrons failed, owing in part to the inability to couple strings to the external local fields, such as the electromagnetic field. The reason for this failure is the infinity of normal mode zero point fluctuations spreading the string over all space [2]. In this chapter we will examine in detail the spatial properties of fundamental strings. We will also speculate on how they compare with the strings of large-N_{color} gauge theory.† We will be particularly interested in the following characteristics of the ground state of the fundamental string:

1. What is the average size of the spatial region occupied by the string?
2. What is the average length of the string?
3. Is the string smooth on small scales or does it exhibit rough or fractal-like behavior?
4. How densely is space filled with string?

* For a review see [1]. † For a review see [3].

LEONARD SUSSKIND • Physics Department, Stanford University, Stanford, California 94305. MAREK KARLINER AND IGOR KLEBANOV • Stanford Linear Accelerator Center, Stanford University, Stanford, California 94305.

In order to answer these questions and to provide some intuition we have constructed a numerical method for generating "snapshots" of the ground state of the string. For that purpose we use the exact wave function of free string in the light cone gauge to generate a statistical ensemble of strings. In fact we find that the overwhelming majority of the ensemble have similar qualitative features. In the first part of the chapter we show the "snapshots" and discuss their important features. In particular we find that both the average length of string and the average size of the region occupied by the string are infinite. The relationship between the two divergences is such that string actually packs the space densely. We also find that the string is microscopically very smooth, with no tendency to form fractal structure on small scales. In the next part of the chapter we provide quantitative meaning to the above statements and substantiate them with analytic derivations. We conclude by explaining the physical meaning and measurability of the divergence in the size of the string. We also speculate on the possible qualitative differences between the fundamental strings and the strings of large-N_{color} QCD.

In the light-cone gauge, the transverse coordinates of the string are free fields with mode expansions

$$X^i(\sigma) = X^i_{\text{cm}} + \sum_{n>0} [X^i_n \cos(n\sigma) + \bar{X}^i_n \sin(n\sigma)] \tag{1}$$

The wave function for each transverse coordinate in the ground state of the string has the product form (dropping the superscript i)

$$\Psi(X(\sigma)) = \prod_n \left\{ \left(\frac{\omega_n}{2\pi}\right)^{1/2} \exp[-\omega_n(X_n^2 + \bar{X}_n^2)/4] \right\} \tag{2}$$

with $\omega_n = n$. Squaring this gives a probability distribution for the transverse position of the string. To carry this out in practice it is necessary to truncate the mode expansion at some maximum wave number N. This is one of the ways of introducing a cutoff in the parameter space of the string. Passage to the continuum limit is achieved as $N \to \infty$.

Another cutoff procedure can be defined where string is replaced by $2N+1$ discrete mass points connected by identical springs. The normal modes are such that the positions of the mass points are given by equation (1) evaluated at discrete values of the parameter $\sigma = 2\pi m/(2N+1)$, where m labels the mass points. Then the string wave function is equation (2) with N frequencies

$$\omega_n = \frac{2N+1}{\pi} \sin\left(\frac{\pi n}{2N+1}\right) \tag{3}$$

With both cutoff prescriptions a string configuration is determined by a sequence of values of X^i_n and $\bar{X}^i_n$, with $n = 1, \ldots, N$ and $i = 1, \ldots, D-2$, sampled with probability

$$P(X_n^i) = \left(\frac{\omega_n}{2\pi}\right)^{1/2} \exp[-\omega_n (X_n^i)^2/2] \tag{4}$$

and similarly for $\bar{X}_n^i$.

For the first cutoff procedure each such configuration defines a parametrized curve in $(D-2)$-dimensional space. By necessity we show projection of string onto two transverse dimensions. In practice each run consists of choosing $100 \times (D-2)$ random numbers from their respective probability distributions. For each run we compute curves with $N = 10, 20, 30, 40, 50$. For convenience, let us adopt the following method: as we proceed, for example, from $N = 10$ to $N = 20$, the coefficients of the normal modes with the first 10 wave numbers are kept the same as for the $N = 10$ "snapshot." Similarly, we proceed from $N = 20$ to 30 to 40 to 50, always retaining the previous set of coefficients. Therefore, for each run, increasing N corresponds to observing the same string with improved resolution. The "snapshots" generated by two such runs are shown in Figs. 1 and 2. For one of the runs we also show graphs of total transverse length (Fig. 3) and average transverse line curvature (Fig. 4) as a function of N. In Fig. 5 we show curvature as a function of length along the string for cases with $N = 10, 20$. We find the following qualitative features:

1. Slow growth of the occupied region with N. We will show that the rms radius of the region $\sim (\log N)^{1/2}$.

2. The plots of total string length versus N appear to be linear. We will provide an analytic derivation of this effect.

3. The transverse curvature averaged over the string appears to be approximately independent of the cutoff. We will show analytically that the expectation value of curvature is completely cutoff independent.

4. The growth in length with increasing N is achieved by repetition of similar smooth structures. In fact, a piece of string of given length at $N = 20$ looks similar to a piece of the same length at $N = 10$ (cf. Fig. 5).

5. The slow growth of the occupied volume together with the linear growth of length means that there is a strong tendency for the string to pass through the same small region many times. It is obvious that, as the cutoff is removed, the string fills space densely: there is a point on the string arbitrarily close to any point in space.

In order to elaborate on point (4) and show that no "accidents" occur as we proceed to high values of the cutoff, we have plotted the section of the string confined between $\sigma = 0$ and $4\pi/N$ for $N = 20$ and $N = 500$ (Fig. 6). The remarkable similarity between the two can be qualitatively regarded as a statement of conformal invariance in our approach.

All the above features, except for (3), can also be observed with the discrete regularization of the string. In Fig. 7 we show a typical picture at

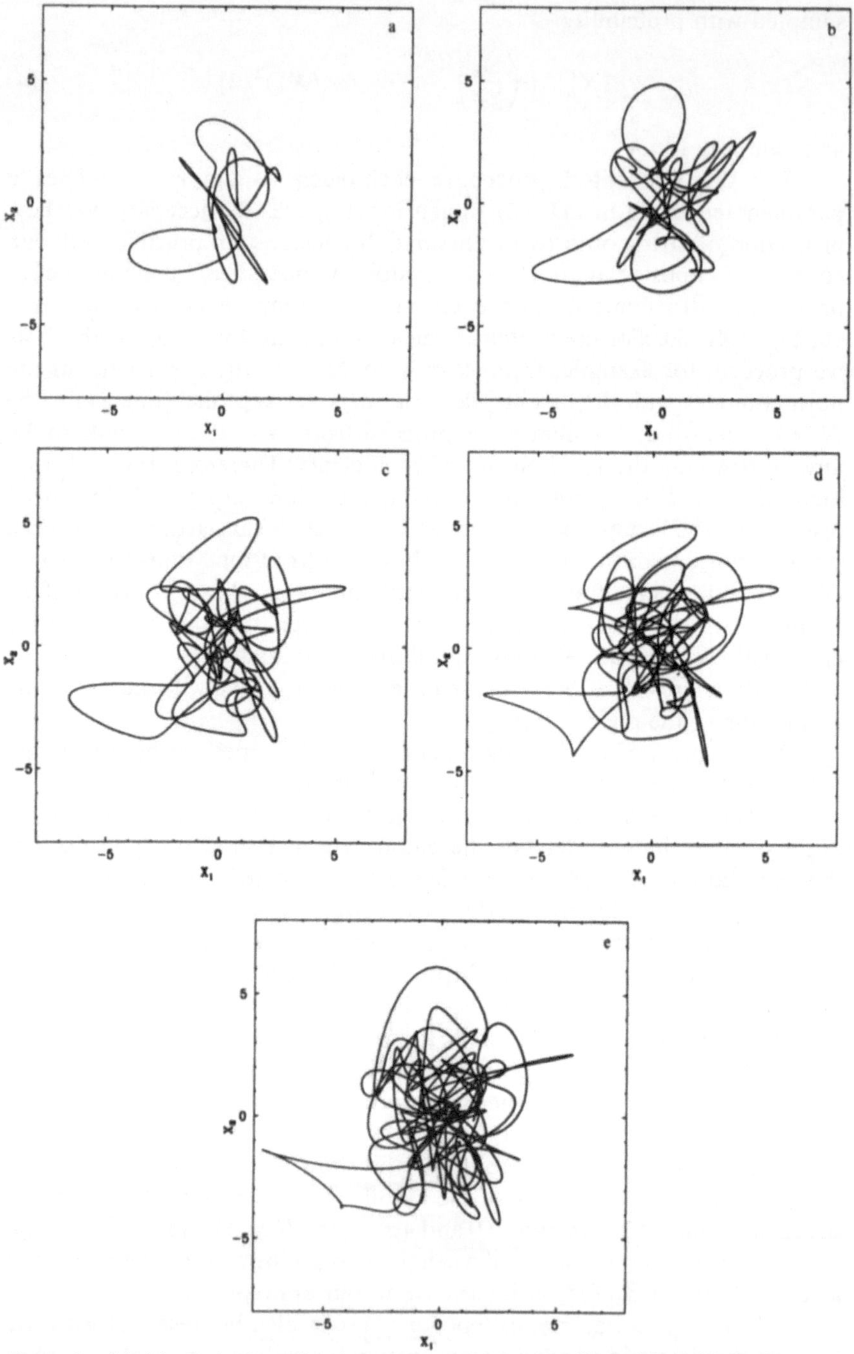

Figure 1. (a)–(e): Projection of string onto two transverse dimensions with mode cutoff $N = 10, 20, 30, 40, 50$, respectively.

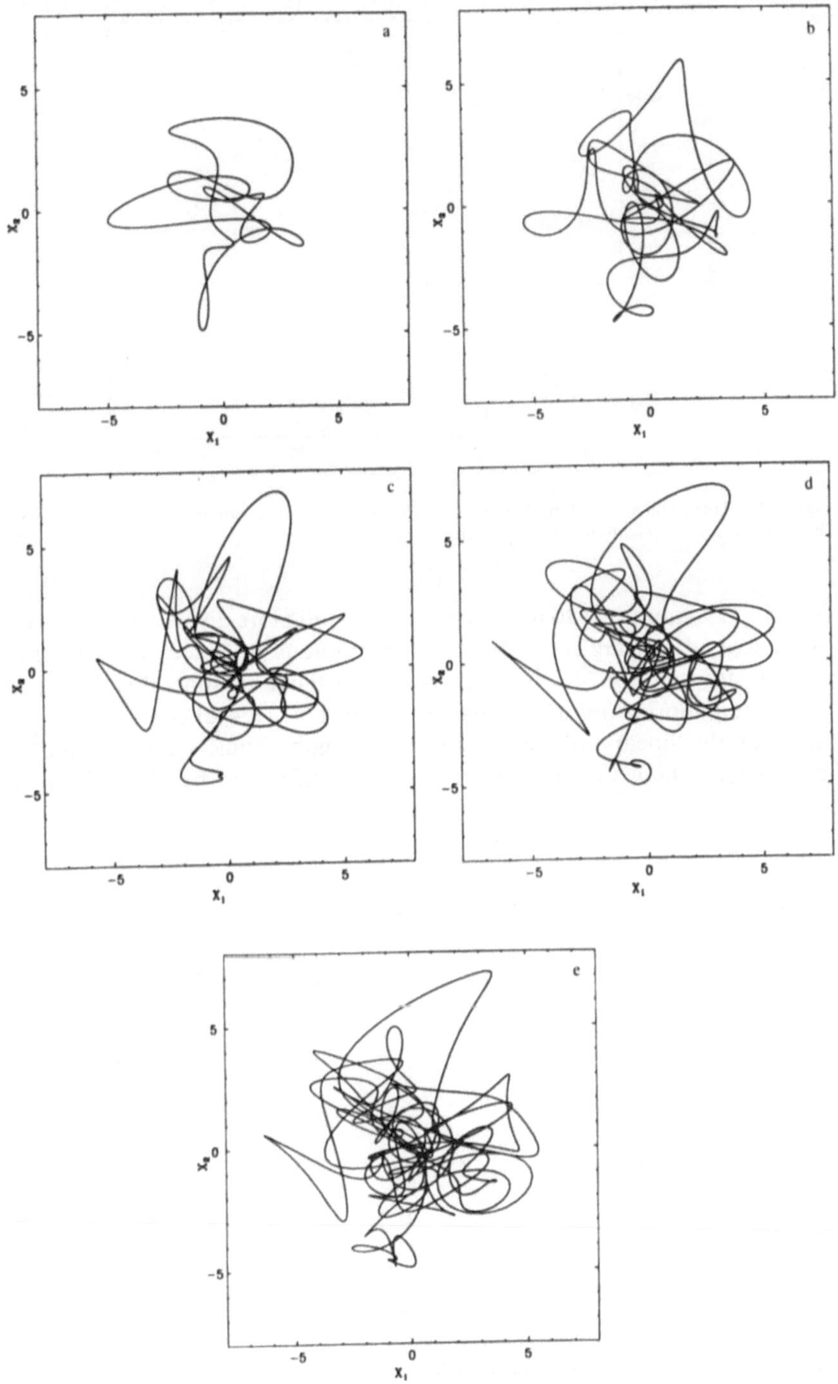

Figure 2. Another run, analogous to Fig. 1.

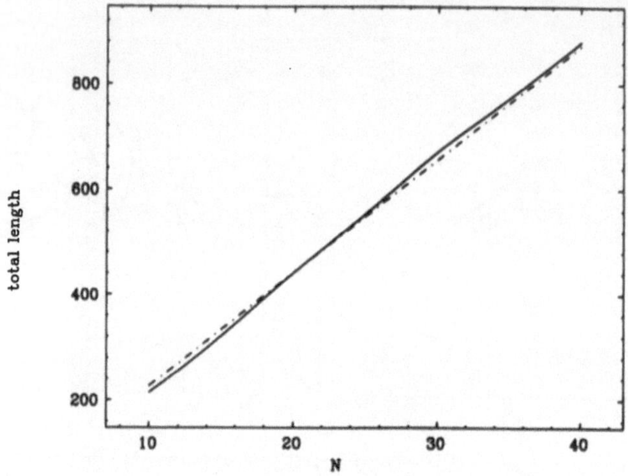

Figure 3. Total transverse length versus mode cutoff for one run in $D = 26$. Broken line shows the analytic result for the expectation value, equation (16).

$N = 50$. It is important to note that, as more and more discrete points crowd the σ axis, the string never becomes continuous in space. We will show analytically that, as $N \to \infty$, the average distance in space between each pair of neighboring points approaches a constant. This, of course, is responsible for the linear growth of the total length. Thus, all the important information about the spatial properties of string can be obtained in the

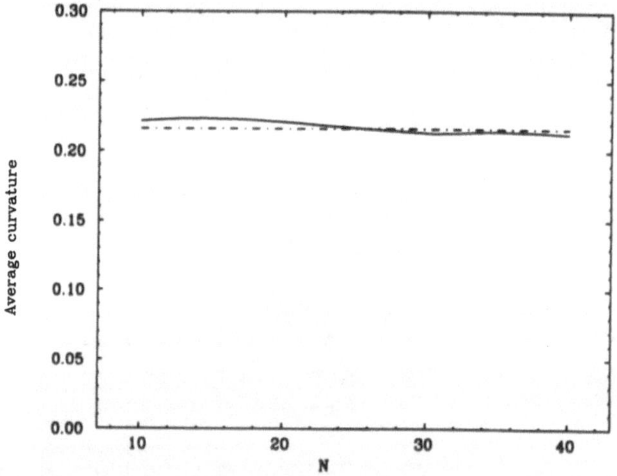

Figure 4. Transverse extrinsic curvature averaged over the string versus mode cutoff for one run in $D = 26$. Broken line shows the analytic result for the expectation value, equation (27).

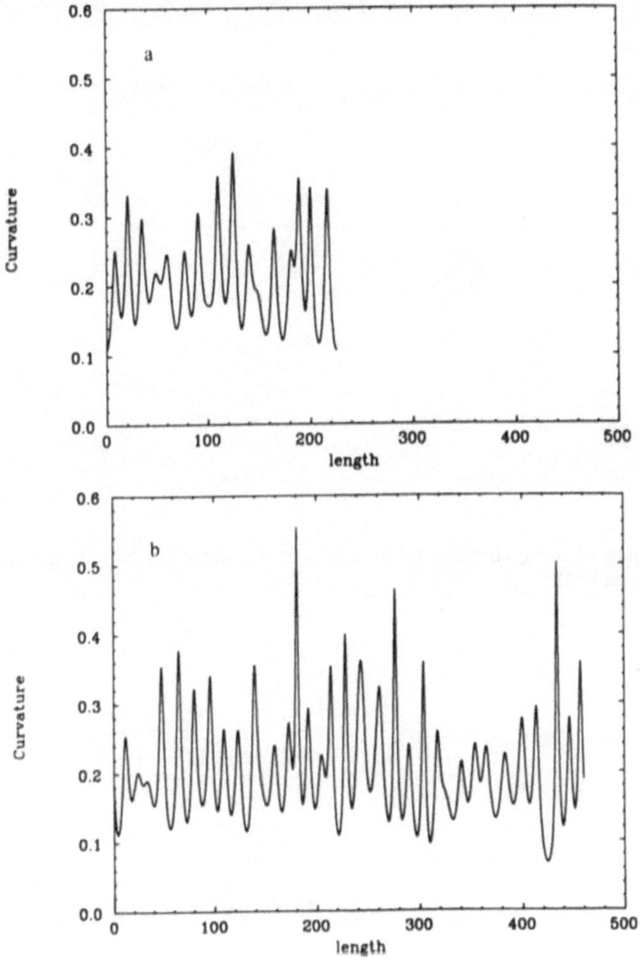

Figure 5. Transverse extrinsic curvature as a function of length along the string, $D = 24$, for (a) $N = 10$ and (b) $N = 20$.

regularization where the string never becomes continuous in space. This fact is essential for treatment of strings in discretized space [4].

Let us now give analytic derivations of some qualitative conclusions reached above. We begin with the growth of the volume occupied by the string with the cutoff N. Define r to be the rms distance of a point on the string to its center of mass:

$$r^2 = \langle (\mathbf{X}(\sigma) - \mathbf{X}_{\text{cm}})^2 \rangle \tag{5}$$

Since there is no preferred point on the closed string, we can arbitrarily set

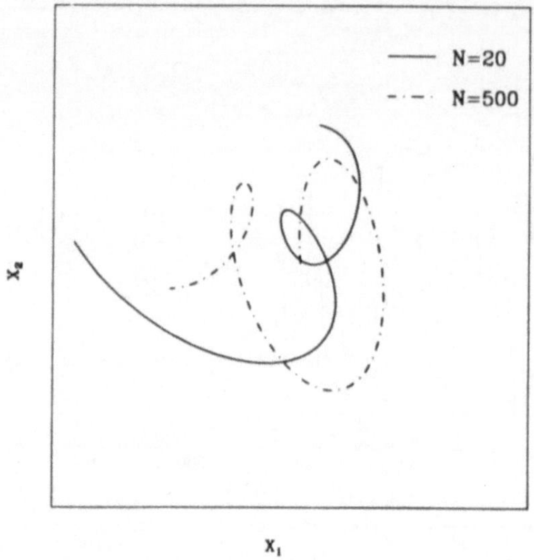

Figure 6. Section of string confined between $\sigma = 0$ and $4\pi/N$ for $N = 20$ (solid line) and $N = 500$ (dashed line).

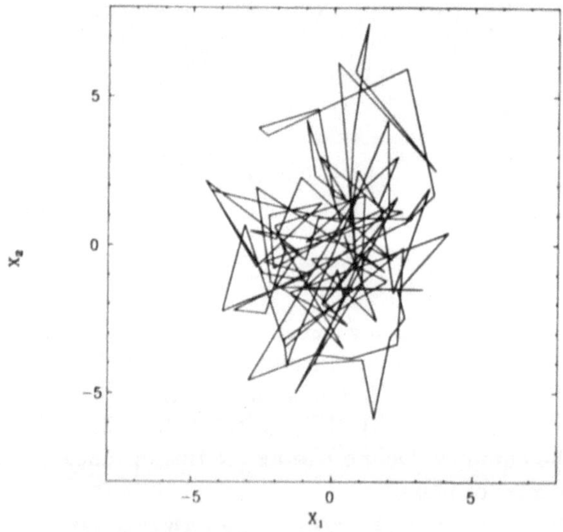

Figure 7. A typical configuration of the ground state of 101 mass points connected by springs, $N_{\text{max}} = 50$.

$\sigma = 0$:

$$r^2 = (D-2)\left\langle\left(\sum_{n=1}^{N} X_n\right)^2\right\rangle$$

$$= (D-2)\sum_{n=1}^{N}\langle X_n^2\rangle = (D-2)\sum_{n=1}^{N}\frac{1}{n} \tag{6}$$

The rms radius of the tachyon grows with the mode cutoff as $(\log N)^{1/2}$. It follows that the rms volume of the transverse region occupied by string $\sim (\log N)^{(D-2)/2}$.

To find the dependence of average length on the cutoff, we start with

$$\langle L\rangle = \int_0^{2\pi}\langle v\rangle\, d\sigma \tag{7}$$

where

$$v(\sigma) = \left[\left(\frac{dX^i}{d\sigma}\right)^2\right]^{1/2} \tag{8}$$

By translation invariance in σ

$$\langle L\rangle = 2\pi\langle v(\sigma = 0)\rangle \tag{9}$$

For each transverse direction

$$\frac{dX^i}{d\sigma}(\sigma = 0) = \sum_{n=1}^{N} n\bar{X}_n^i \tag{10}$$

Using the fact that each $\bar{X}_n^i$ is Gaussian distributed, it is easy to show that $dX^i/d\sigma(\sigma = 0)$ is Gaussian distributed with variance

$$\Sigma^2 = \sum_1^N n = \frac{N(N+1)}{2} \tag{11}$$

Therefore, $\mathbf{v} = d\mathbf{X}/d\sigma$ is distributed according to

$$P(\mathbf{v}) \sim \exp\left(-\frac{\mathbf{v}^2}{2\Sigma^2}\right) d^{D-2}\mathbf{v} \tag{12}$$

As a result, the distribution for the length of $\mathbf{v}$ is

$$P(v) \sim \exp\left(-\frac{v^2}{2\Sigma^2}\right) v^{D-3}dv \tag{13}$$

It follows that

$$\langle v\rangle = \frac{\int_0^\infty \exp(-v^2/2\Sigma^2)v^{D-2}\,dv}{\int_0^\infty \exp(-v^2/2\Sigma^2)v^{D-3}\,dv} \sim \Sigma \sim N \tag{14}$$

The slope of the linear growth determined by above expression depends on dimensionality. For example, if the number of transverse dimensions is an even number ($D-2=2k$) then (14) yields

$$\langle L\rangle = \frac{(2k-1)!\pi^{3/2}}{4^{k-1}[(k-1)!]^2}[N+1/2+O(1/N)] \tag{15}$$

In particular, in $D=26$ we find

$$\langle L\rangle \approx 21.54(N+1/2) \tag{16}$$

As shown in Fig. 3, the data for any given run agree well with this linear dependence. This shows that the standard deviation is small compared with the average length.

It is also interesting to study equation (15) in the limit of a large number of dimensions. Using the Stirling formula for the factorial we find that the slope of growth of transverse length with N is given by

$$\langle L\rangle/N \approx \pi(2D-4)^{1/2} + O(1/(D-2)^{1/2}) \tag{17}$$

as D becomes large. In $D=26$ this predicts the slope of 21.76, which is very close to the exact number (16).

Similar analytic results can be derived in the regularization where string is replaced by a collection of mass points connected by springs. For example, the length is

$$L_d = \sum |\mathbf{X}_{(m+1)} - \mathbf{X}_{(m)}| \tag{18}$$

where the subscript labels the mass points. Using translation invariance, we obtain

$$\langle L_d\rangle = (2N+1)\langle|\mathbf{X}_{(2)} - \mathbf{X}_{(1)}|\rangle \tag{19}$$

After a few steps analogous to the ones for the continuous regularization we find

$$\langle L_d\rangle = \frac{(2k-1)!(8\pi)^{1/2}}{4^{k-1}[(k-1)!]^2}[N+1/2+O(1/N)] \tag{20}$$

Note that, with the definition of length (18), the slope of linear growth in the discrete regularization differs slightly from (15) found in the continuous regularization. However, the linearity of growth and other properties important for our physical conclusions are unaffected.

Let us now investigate the extrinsic line curvature of the string in the regularization where the string is kept continuous. It is conveniently expressed as

$$\kappa = \frac{|\mathbf{a}_\perp|}{v^2} \tag{21}$$

where $\mathbf{a}_\perp$ is the component of $\mathbf{a} = d^2\mathbf{X}/d\sigma^2$ normal to $\mathbf{v} = d\mathbf{X}/d\sigma$. Since

$$\frac{d^2X^i}{d\sigma^2}(\sigma=0) = -\sum_{n=1}^{N} n^2 X_n^i \tag{22}$$

equation (10) implies that **a** and **v** are uncorrelated. Therefore,

$$\langle\kappa\rangle = \langle|\mathbf{a}_\perp|\rangle\left\langle\frac{1}{v^2}\right\rangle \tag{23}$$

From the distribution (13) we find

$$\left\langle\frac{1}{v^2}\right\rangle = \frac{1}{2(k-1)\Sigma^2} \tag{24}$$

It is important to note that this diverges in $D = 4$. Since **a** and **v** are not correlated, $\langle|\mathbf{a}_\perp|\rangle$ is effectively the average length of the vector (22) in $D - 3$ dimensions. Denoting $|\mathbf{a}_\perp| = a$ we find the probability distribution

$$P(a) \sim \exp\left(-\frac{a^2}{2\tilde{\tilde{\Sigma}}^2}\right) a^{D-4}\, da \tag{25}$$

with

$$\tilde{\tilde{\Sigma}}^2 = \sum_1^N n^3 = [\tfrac{1}{2}N(N+1)]^2 = \Sigma^4 \tag{26}$$

It follows that $\langle\kappa\rangle \sim \tilde{\tilde{\Sigma}}/\Sigma^2$ is independent of the cutoff! On the other hand, as shown in Fig. 5, the overwhelming majority of the ensemble of strings has curvature oscillating along the string. The number of oscillations is proportional to N. The oscillatory behavior is characteristic of all other observables: for example, length per unit σ. Oscillations occur because the volume occupied by string grows only as a power of log N while the length of string grows as N: on the average the string has to "turn around" $O(N)$ times. Therefore, position $X^i(\sigma)$ and all functions of position and its derivatives oscillate $O(N)$ times along the string.

After a short calculation we find

$$\langle\kappa\rangle = \frac{[(k-2)!]^2 2^{2k-3}}{(2k-3)!\sqrt{2\pi}} \tag{27}$$

which in $D = 26$ yields $\langle\kappa\rangle = 0.216$. This is in good agreement with Fig. 4, which shows the data for a sample run. In the limit of large dimensionality (27) reduces to

$$\kappa \approx \frac{1}{(D-2)^{1/2}} \tag{28}$$

This confirms the intuitive expectation that increasing dimensionality makes the string smoother.

On the lower end of the range of dimensionalities the average curvature diverges in $D = 4$, [cf. equation (24)]. We believe there is a simple intuitive reason for this, which we proceed to explain. There are two kinds of singular points that can occur on a string: a kink, where the tangent vector $d\mathbf{X}/d\sigma$ is discontinuous, and a cusp, where it vanishes. In other words, at a cusp string turns back onto itself. From equation (21) we see that at a kink the curvature has an integrable (δ-function) singularity, while at a cusp it has a nonintegrable singularity. Our study of projections of strings onto two transverse dimensions indicates that cusps are fairly likely to occur there. We find that most of these cusps are projections of smooth configurations in higher dimensions. Therefore, we conjecture that the relatively high likelihood of cusps in $D = 4$ is responsible for the divergence in the average curvature.

Let us also discuss briefly another important observable characterizing string geometry: correlation of unit tangent vectors

$$\langle \mathbf{t}(\sigma) \cdot \mathbf{t}(\sigma') \rangle \tag{29}$$

where

$$\mathbf{t}(\sigma) = \mathbf{X}'/v(\sigma) \tag{30}$$

To estimate (29) we replace it by

$$\langle \mathbf{X}'(\sigma) \cdot \mathbf{X}'(\sigma') \rangle \langle v \rangle^{-2} \sim -\frac{1}{N^2(\sigma - \sigma')^2} \tag{31}$$

It follows that the correlation of unit tangents falls off quadratically with the length of string between the two points. As the cutoff is removed, the unit tangents are uncorrelated at any finite σ separations because then the length separations are infinite.

One may wonder whether any of the strange effects we have discussed are observable. In particular, the infinite rms radius of all "stringy" particles seems very unphysical. However, it does lead to an observable effect. Consider scattering of a high-energy string from a string at rest. The interaction is mediated by string exchange. In the light-cone frame of the fast string of energy E the lifetime of the interaction is of order $\tau = 1/E$. Oscillations with frequency $> 1/\tau$ average to zero. Thus, we retain a number of modes $\sim E$. This introduces mode cutoff and gives an observable particle radius $\sim(\log E)^{1/2}$. As the resolution is improved, the string "expands." This phenomenon leads to the well-known Regge behavior of scattering cross sections satisfied by the dual amplitudes [2]. Thus the effect is indeed observable and presents no obvious difficulty for scattering of strings by strings.

On the other hand, imagine that the string is being scattered by a local external field. This situation is analogous to the electromagnetic probing of hadrons. In this case the interaction is instantaneous and therefore the string must appear infinite. It is precisely for this reason that the fundamental strings cannot be consistently coupled to arbitrary external fields. That is why they are a theory either of everything or of nothing.

Let us speculate now on how the fundamental strings might differ from the large-N_{color} QCD strings. Our results indicate that the fundamental strings are smooth. This should be contrasted with the expected behavior of QCD strings. In the limit $N_{\text{color}} \to \infty$ Migdal and Makeenko derived an exact lattice string equation [3]. If QCD had an ultraviolet fixed point then the string would be microscopically self-similar. QCD being asymptotically free is likely to make strings even rougher.

Another difference between the QCD and fundamental strings involves the spatial distribution of the longitudinal momentum p^+. For a hadron, this could be measured by interaction with external gravitational field. The result is a form factor $F(q^2)$. For a fundamental string an analogous form factor can be obtained by observing that the distribution of p^+ is measured by the vertex operator $\partial_\alpha X^+ \partial^\alpha X^+ \exp(iq \cdot X)$, where α is the world-sheet index. In the light-cone gauge $X^+ = \tau$ and the form factor reduces to

$$F(q^2) = \left\langle \int d\sigma \exp[iq \cdot X(\sigma)] \right\rangle \sim \exp(-q^2 \log N) \qquad (32)$$

In the limit $N \to \infty$ the form factor is nonvanishing only at $q = 0$. It follows that p^+ is smeared uniformly all over space. This peculiar property applies not only to the ground state of the string but also to any finitely excited state. For any such state the change in the wave function relative to the ground state concerns only a finite number of normal modes and becomes negligible in the limit $N \to \infty$. The strange behavior of the gravitational form factors is possibly connected with the existence of the graviton: at least for the massless spin-2 state it could be foreseen on the basis of general principles. A theorem by Weinberg and Witten [5] states that in a Lorentz invariant theory with a Lorentz invariant energy-momentum tensor the gravitational form factor of a massless spin-2 particle must satisfy $F(q^2 \neq 0) = 0$. It seems that string theory uses its infinite zero-point fluctuations to allow the existence of gravitons.

ACKNOWLEDGMENTS. We thank Marvin Weinstein for his help at the early stages of this work. LS acknowledges support from the NSF under grant No. PHY 812280. The work of MK and IK was supported by the Department of Energy under contract No. DE-AC03-76SF00515.

REFERENCES

1. M. Green, J. Schwartz, and E. Witten, *Superstring Theory*, Cambridge University Press, Cambridge, 1986.
2. L. Susskind, *Phys. Rev. D 1*, 1182 (1970).
3. A. A. Migdal, *Phys. Rep. 102*, 199 (1983).
4. I. Klebanov and L. Susskind, *Nucl. Phys. B 309*, 175 (1988).
5. S. Weinberg and E. Witten, *Phys. Lett. B 96*, 59 (1980).

Chapter 11

Covariantized Light-Cone String Field Theory

Taichiro Kugo

1. MOTIVATION

String field theory (SFT) is certainly a powerful and basic framework in which to reveal the nonperturbative dynamics as well as the underlying symmetry principles of string. Many active investigations have actually been done and are being done in that direction, and the corresponding developments have been achieved. Nevertheless we still have severe difficulties in constructing SFTs in a satisfactory manner.

There are two major approaches to covariant SFTs; one is the approach based on the joining-splitting-type interaction vertex, which was first suggested by Siegel [1] and fully developed by Hata, Itoh, Kugo, Kunitomo, and Ogawa (HIKKO) [2-5] and partly by Neveu and West [6]. Another is the one based on the midpoint interaction vertex, which was presented by Witten [7-9]. We call these HIKKO's and Witten's theories, respectively.

HIKKO's theory gives quite a consistent SFT as a classical field theory (i.e., at tree level) for both cases of open and closed string, although it contains an additional unphysical parameter α called "string length." At the loop levels, however, the presence of this parameter α causes trouble

TAICHIRO KUGO • Department of Physics, Kyoto University, Kyoto 606, Japan.

in yielding an infinity factor in the amplitude [4]:

$$\mathrm{Im}\, T^{N\text{-loop}} \propto \left[\int_{-\infty}^{\infty} d\alpha\, 1\right]^N$$

In addition, the amplitudes obtained after renormalizing this factor $[\int d\alpha\, 1]$ into the loop expansion parameter $\hbar$ or g, still contain infinite multiple overcounting concerning the modular invariance. The latter difficulty is suspected to come from the former since the factorization of the ∞ factor $[\int d\alpha]^N$ is very subtle and dangerous.

Witten's theory is, on the other hand, free from this α-problem. However, Witten's theory works only for open string and seems very difficult to extend to closed string [10, 11] and, hence, to heterotic string in particular.

We should thus solve the α-problem in HIKKO's approach. Either in HIKKO or Witten, the nonzero modes are treated perfectly well. Namely, the extra unphysical modes $\alpha_n^{\pm} \equiv (\alpha_n^0 \pm \alpha_n^{25})/\sqrt{2}$, which have to appear to achieve the manifest covariance, are associated with the Faddeev–Popov ghost modes c_n, $\bar{c}_n$ to form a quartet (= a pair of BRS doublets) as [12, 13]

$$\boldsymbol{\alpha}_n = (\alpha_n^1, \alpha_n^2, \ldots, \alpha_n^{24})$$

physical

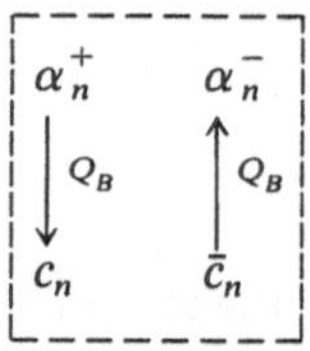

quartet

(1)

Just as in the Yang–Mills case, these quartet modes are sufficiently well controlled not to appear in physical subspace by the "quartet" decoupling mechanism [14]. The zero modes, on the other hand, appear as follows in HIKKO's case:

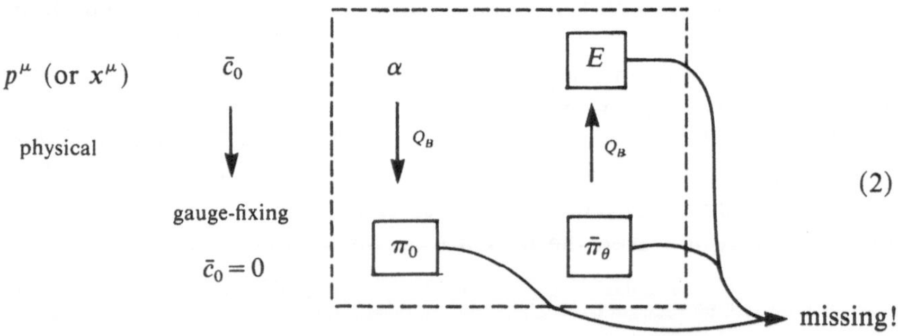

(2)

Namely, the center-of-mass momentum p^μ is physical and okay, and the unique ghost zero mode $\bar{c}_0$ is unphysical but can be eliminated as a gauge-fixing condition $\bar{c}_0 = 0$. The string length parameter α is, however, unphysical but nevertheless appears alone and does not decouple accordingly. This is the source of the α problem. One way out of it is therefore to introduce additionally three unphysical variables, one bosonic E plus two fermionic π_0 and $\bar{\pi}_\theta$, to complete the "quartet" together with α as indicated in (2). Then we can expect that the contribution of α will be canceled by the "quartet" (or Parisi–Sourlas [15]) mechanism. This is exactly what happens in the "covariantized light-cone" SFT, which we shall discuss in this chapter.

2. COVARIANTIZED LIGHT-CONE SFTs: INTRODUCTION

The first covariantized light-cone SFT was given in 1984 by Siegel in the first [16] of his series of three papers [1], entitled "Covariantly Second-Quantized String." That paper gave a gauge-fixed (and hence BRS-invariant) version of the covariantized light-cone SFT, which we refer to as Siegel's ortho-symplectic-group, $OSp(d, 2/2)$, invariant theory. Later, in the end of 1986, Neveu and West [17, 18] succeeded in finding the gauge-invariant version [see also Uehara [19]] and discussed the gauge-fixing procedure leading to the original Siegel's theory.

As we shall explain shortly, once the light-cone gauge SFT is known, Siegel's original gauge-fixed version, $OSp(d, 2/2)$ invariant theory, follows immediately through a clear prescription given by himself. On the other hand, Neveu and West constructed the gauge-invariant action in a very heuristic manner and the procedure how to obtain it was left unclear. One of the main purposes of this chapter is to present a general procedure to obtain gauge-invariant action directly from the light-cone gauge SFT.* We shall state this procedure as clearly and generally as possible so that its extension to the superstring may not be difficult.

The basic ingredients of our procedure are the following two, as we will see in Section 4 in detail: First is the important proposal made by Siegel and Zwiebach [24] that the BRS transformation Q_B should be identified with an OSp extended Lorentz transformation $\mathbf{M}^{-c}$ in the $-$ and c(ghost) plane. Second is the introduction of an additional Grassmann coordinate θ as a BRS partner of the proper time τ, which was first done by Uehara [19] explicitly and noted also by Siegel and Zwiebach [25]. This θ, or its conjugate $\bar{\pi}_\theta$, will play the role of one of the missing zero-mode ghosts in (2).

* The approach that has recently been pursued by Siegel and Zwiebach [20] is slightly different from ours; they demand the local gauge invariance under the full $OSp(1, 1/2)$ group, while we require the invariance only under one generator of that group as a gauge symmetry. Yet other approaches to covariantized light-cone SFTs can be found in Refs. 21–23.

2.1. Siegel's $OSp(d, 2/2)$ Invariant Theory [16]

Siegel's procedure for covariantizing the light-cone gauge SFT is very simple. One has just to replace any (transverse) $O(d-2)$ vector X^i with metric δ_{ij} by $OSp(d-1, 1/2)$ vector $X^M = (X^\mu, c, \bar{c})$, consisting of bosonic Lorentz coordinate $X^\mu = (X^i, X^+, X^-)$ and Grassmann ghost coordinates c and $\bar{c}$, with OSp metric η_{MN} [given in (23) below]. Namely,

	Light-cone SFT		Siegel's OSp SFT	
Coordinates	$x^+ = \tau, x^-$	$\to$	τ, x^-	
	$X^i(\sigma)$	$\to$	$X^M(\sigma) = (X^\mu(\sigma), c(\sigma), \bar{c}(\sigma))$	
Equivalent Momentum	$p^- = E, 2p^+ = \alpha$	$\to$	E, α	(3)
	$P^i(\sigma)$	$\to$	$P^M(\sigma) = (P^\mu(\sigma), -\pi(\sigma), \bar{\pi}(\sigma))$	
Metric	$O(d-2)\ \delta_{ij}$	$\to$	$OSp(d-1, 1/2)\ \eta_{MN}$	

Once this replacement is performed in the light-cone SFT, the resultant theory is Siegel's OSp theory. Since the starting light-cone SFT has Lorentz symmetry $O(d-1, 1)$ in which $O(d-2)$ is a manifest linear symmetry realized by the transverse vector X^i, the resultant Siegel theory is clearly invariant under the extended Lorentz transformation $OSp(d, 2/2)$ and the $OSp(d-1, 1/2)$ subgroup is a manifest linear symmetry realized by X^M:

	Linear symmetry	S-matrix symmetry	
Light-cone SFT	$O(d-2)$	$O(d-1, 1)$	(4)
Siegel's OSp SFT	$OSp(d-1, 1/2)$	$OSp(d, 2/2)$	

Because of this linear $OSp(d-1, 1/2)$ symmetry, the effects of the added two bosonic coordinates $X^\pm$ and two fermionic coordinates c and $\bar{c}$ in going from X^i to $X^M = (X^i, X^+, X^-, c, \bar{c})$ cancel with each other in any loop integrations. This is the Parisi–Sourlas mechanism [15], as a result of which the Green's functions in the light-cone gauge SFT and in Siegel's OSp SFT become the same functions; that is, suppose that a certain Green's function is given by

$$G(\alpha, E, \mathbf{p}^2), \qquad \mathbf{p} = (p^i) \tag{5}$$

in the light-cone gauge SFT, where $\mathbf{p}^2$ is understood to be a generic notation representing the $O(d-2)$ invariants $\mathbf{p}_r \cdot \mathbf{p}_s = \sum_{i=1}^{d-2} p_r^i p_s^i$ and $\mathbf{p}_r \cdot \boldsymbol{\varepsilon}_s$ of arbitrary external momenta $\mathbf{p}_r$ and polarization vectors $\boldsymbol{\varepsilon}_r$, then the corresponding Green's function in Siegel's OSp SFT is given by

$$G(\alpha, E, p^2 + 2i\pi\bar{\pi}), \qquad p^2 \equiv p^\mu p_\mu = \mathbf{p}^2 - 2p^+p^- \tag{6}$$

with the same function G as in (5). Further, if we consider the S-matrix elements S, which we know are invariant under the larger group $O(d-1,1)$ or $OSp(d,2/2)$ in the two theories, respectively, we immediately see that the Green's function relation, (5) $\leftrightarrow$ (6), implies the following S-matrix relation:

$$\begin{array}{ccc} \text{Light-cone SFT} & & \text{Siegel's } OSp \text{ SFT} \\ S(-\alpha E+\mathbf{p}^2) & \leftrightarrow & S(-\alpha E+p^2+2i\pi\bar{\pi}) \end{array} \tag{7}$$

Namely, the function S is common to the two theories. Therefore if we put

$$E=0, \qquad \pi=\bar{\pi}=0 \tag{8}$$

in Siegel's OSp SFT, we obtain the amplitudes $S(p^2)$, which coincide with the correct light-cone gauge's amplitudes $S(-\alpha E+\mathbf{p}^2)$ provided that we identify $p^\mu=(2p^+,p^-,p^i)$ in Siegel's theory with (α,E,p^i) in light-cone gauge SFT. Note that in Siegel's theory the Lorentz transformation is thus identified with the $O(d-1,1)$ rotating p^μ and hence is a manifest symmetry.

The constraint (8) is, however, merely a condition put *by hand* by Siegel. Dissatisfied with this point, probably, Siegel himself left this theory and pursued another approach in his later papers in the above-mentioned series [1].

2.2. Neveu and West's Work and a Severe Criticism

One of the important contributions of Neveu and West [17, 18] is, besides the construction of gauge-invariant action, the suggestion that the Siegel's by-hand condition (8) may come out automatically from a more natural physical state condition; that is, they imposed

$$Q_B\phi^{\text{phys}}=0 \tag{9}$$

for physical states taking $Q_B=M^{-c}$, an OSp Lorentz generator, as proposed by Siegel and Zwiebach [24], and argued that the physical on-shell states ϕ^{phys} satisfying both (9) and the on-shell condition $(i\alpha\,\partial_\tau-L)\phi^{\text{phys}}=0$, realize

$$E=\pi=\bar{\pi}=0, \qquad \alpha\text{-independent}, \qquad L\phi^{\text{phys}}=0 \tag{10}$$

aside from a trivial state of the form $Q_B\chi$. Uehara [19] also analyzed the BRS cohomology for (9) more neatly and obtained a similar result.

There arises, however, a serious question about their claim: "Does physical state really realize (10)?" If so, then, from $\pi=\bar{\pi}=0$, the physical state ϕ^{phys} is proportional to π and $\bar{\pi}$, $\phi^{\text{phys}}=\pi\bar{\pi}\tilde{\phi}$, and hence must have *zero norm*:

$$\langle\phi^{\text{phys}}|\phi^{\text{phys}}\rangle=\int d\pi\,d\bar{\pi}|\phi^{\text{phys}}|^2\propto\int d\pi\,d\bar{\pi}\;\pi\bar{\pi}\pi\bar{\pi}=0^2! \tag{11}$$

[Or equivalently, in the coordinate (c and $\bar{c}$ conjugate to $\bar{\pi}$ and π) representation, $\pi = \bar{\pi} = 0$ implies that ϕ^{phys} is independent of c and $\bar{c}$ and $\langle \phi^{\text{phys}} | \phi^{\text{phys}} \rangle \propto \int dc\, d\bar{c}\, 1 = 0^2$ follows.] This is indeed a severe criticism, since the zero-norm property means that, if we substitute a physical state, for instance, Yang–Mills state

$$|\phi^{\text{phys}}\rangle = \pi\bar{\pi} A_\mu(x) \alpha^{\mu\dagger}_{n=1} |0\rangle$$

into the Neveu–West action $S[\phi]$, then we obtain a vanishing $S(A_\mu) = 0$!

On the other hand, however, they also claim that the physical state is α-independent and realizes $E = 0$. This implies $\phi^{\text{phys}} \propto \delta(E)$ as for the (α, E) dependence and hence the contribution of (α, E) sector to the norm *diverges*:

$$\int d\alpha\, dE |\delta(E)|^2 = \delta(0) \cdot \int d\alpha\, 1 = \infty^2 \tag{12}$$

Thus we actually have an indefinite expression $(0 \times \infty)^2$ for the physical state norm. Can we avoid such an indefinite expression? Are their physical states really correct? We shall give a complete answer to these questions, which is another main purpose of this chapter.

3. LIGHT-CONE GAUGE STRING FIELD THEORY

Let us recapitulate here the light-cone gauge SFT [26–28] very briefly for the purpose of later use.

The action of the light-cone gauge SFT is given by [27–29]

$$S_{\text{LC}} = \int d\tau [\phi^\dagger (i\partial_\tau - L_0^\perp/\alpha)\phi + \tfrac{2}{3} g \phi^3 + \tfrac{2}{4} g^2 \phi^4]$$

$$\phi^\dagger \left(i\partial_\tau - \frac{L_0^\perp}{\alpha} \right) \phi \equiv \int d\mathbf{1}\, \text{tr} \langle \phi(\mathbf{1}, \tau) | (i\partial_\tau - L_0^\perp/\alpha) | \phi(\mathbf{1}, \tau) \rangle \tag{13}$$

$$\phi^n \equiv \int d\mathbf{1} \cdots d\mathbf{n}\, \text{tr} \langle \phi(\mathbf{1}, \tau) | \cdots \langle \phi(\mathbf{n}, \tau) | | V_{\text{LC}}^{(n)}(\mathbf{1}, \ldots, \mathbf{n}) \rangle \qquad (n = 3, 4)$$

where $\mathbf{r}$ and $d\mathbf{r}$ denote the zero-mode variables of string r and their integrations,

$$\mathbf{r} \equiv (\mathbf{p}_r = (p_r^1, \ldots, p_r^{24}), \alpha_r), \qquad d\mathbf{r} \equiv (2\pi)^{-25}\, d^{24}\mathbf{p}_r \alpha_r\, d\alpha_r \tag{14}$$

and $L_0^\perp$ is the $n = 0$ component of the Virasoro operator

$$L_n^\perp = \sum_{m=-\infty}^{\infty} \tfrac{1}{2} {:}\boldsymbol{\alpha}_{n+m} \cdot \boldsymbol{\alpha}_{-m}{:} - \delta_{n,0}, \qquad \boldsymbol{\alpha}_0 = \mathbf{p} \tag{15}$$

with $\mathbf{a}\cdot\mathbf{b}$ generally denoting the summation over the transverse directions $\sum_{j=1}^{24} a^j b^j$. The three-string vertex $|V_{\rm LC}^{(3)}\rangle$ is given as

$$|V_{\rm LC}^{(3)}\rangle=\mu^{(3)}(\alpha_1,\alpha_2,\alpha_3)\exp\left\{\frac{1}{2}\sum_{r,s=1}^{3}\sum_{n,m=0}^{\infty}\bar{N}_{nm}^{rs}\boldsymbol{\alpha}_{-n}^{(r)}\cdot\boldsymbol{\alpha}_{-m}^{(s)}\right\}|0\rangle$$

$$\times(1/\alpha_1\alpha_2\alpha_3)(2\pi)^{25}\delta^{24}\left(\sum_r\mathbf{p}_r\right)\delta\left(\sum_r\alpha_r\right)\tag{16}$$

$$\mu^{(3)}=\exp\left(-\tau_0\sum_{r=1}^{3}\frac{1}{\alpha_r}\right),\qquad \tau_0=\sum_{r=1}^{3}\alpha_r\ln|\alpha_r|$$

in terms of the Fourier components $\bar{N}_{nm}^{rs}$ of the Neumann function defined on the corresponding three-string diagram. The four-string vertex $|V_{\rm LC}^{(4)}\rangle$ is also given by a similar expression [28, 29].

The Lorentz transformation property of this system has been studied by many authors [24, 22, 19] since Goddard, Goldstone, Rebbi, and Thorn (GGRT) [30] and Mandelstam [26]. The full expression for the nontrivial Lorentz transformation $\delta_L\phi\equiv[i\epsilon_{j-}\mathbf{M}^{j-},\phi]$ has been given only recently [31, 32]:

$$\delta_L|\phi\rangle=i\epsilon_{j-}[M^{j-}|\phi\rangle-gX^j(\sigma_{\rm int})|\phi*\phi\rangle-g^2X^j(\sigma_{\rm int})|\phi\circ\phi\circ\phi\rangle]\tag{17}$$

$$M^{j-}=ip^j\frac{\partial}{\partial\alpha}+\frac{x^j}{\alpha}L_0^{\perp}-\frac{i}{\alpha}\sum_{m=1}^{\infty}\left(\frac{\alpha_{-m}^jL_m^{\perp}-L_{-m}^{\perp}\alpha_m^j}{m}\right)\tag{18a}$$

$$X^j(\sigma_{\rm int})|\phi*\phi(\mathbf{3})\rangle=\int d\mathbf{1}\,d\mathbf{2}\langle\phi(\mathbf{1},\tau)|\langle\phi(\mathbf{2},\tau)|X^j(\sigma_{\rm int})|V_{\rm LC}^{(3)}(\mathbf{1},\mathbf{2},\mathbf{3})\rangle\tag{18b}$$

$$X^j(\sigma_{\rm int})|\phi\circ\phi\circ\phi(\mathbf{4})\rangle$$
$$=\int d\mathbf{1}\,d\mathbf{2}\,d\mathbf{3}\langle\phi(\mathbf{1},\tau)|\langle\phi(\mathbf{2},\tau)|\langle\phi(\mathbf{3},\tau)|X^j(\sigma_{\rm int})|V_{\rm LC}^{(4)}(\mathbf{1},\ldots,\mathbf{4})\rangle\tag{18c}$$

Based on the α-symmetry [4] and BRS-invariance [3] proved by HIKKO in their covariant SFT, the present author [31] has shown the invariance of the light-cone gauge action (13) under the Lorentz transformation (17),

$$\delta_L S_{\rm LC}=0\tag{19}$$

as well as the closure of the Lorentz transformation algebra:

$$[\delta_L(\epsilon),\delta_L(\epsilon')]=0\quad\text{or}\quad[\mathbf{M}^{i-},\mathbf{M}^{j-}]=0\tag{20}$$

4. FROM LIGHT-CONE TO COVARIANT SFT: GENERAL PROCEDURE

Once we know the light-cone SFT, we can immediately construct a covariant (gauge-invariant) SFT. Let us explain this procedure in a general manner so that its extension may be easy to the other SFTs than the present bosonic 26-dimensional one.

In this procedure, we "extend" the light-cone gauge string field ϕ to the covariant one Φ:

$$\phi[X^i(\sigma), \alpha, \tau] \rightarrow \Phi[X^M(\sigma), \alpha, \tau, \theta] \tag{21}$$

Here two extensions are performed: (1) *OSp* extension $X^i(\sigma) \rightarrow X^M(\sigma)$ and (2) superextension of proper time $\tau \rightarrow (\tau, \theta)$, which we now explain separately.

4.1. *OSp* extension

This part of extension is the same as Siegel's original one [16] explained in Section 2.1. The $O(d-2)$ vector $X^i(\sigma)$ is replaced by the $OSp(d-1, 1/2)$ vector $X^M(\sigma)$ by adding two (one positive-metric and one negative-metric) bosonic coordinates and two fermionic coordinates

$$X^i(\sigma) \rightarrow X^M(\sigma) \equiv (X^\mu(\sigma), c(\sigma), \bar{c}(\sigma)) \tag{22}$$

so that the original $O(d-2)$ symmetry with metric δ_{ij} is extended to the $OSp(d-1, 1/2)$ symmetry with metric η_{MN}:

$$\delta_{ij} \rightarrow \eta_{MN} = \begin{array}{c} \\ \mu \\ c \\ \bar{c} \end{array} \!\! \begin{array}{c} \begin{array}{ccc} \nu & c & \bar{c} \end{array} \\ \left(\begin{array}{c|cc} \eta_{\mu\nu} & \multicolumn{2}{c}{0} \\ \hline & 0 & -i \\ 0 & i & 0 \end{array}\right) \end{array} = \eta^{MN} \tag{23}$$

In terms of the zero-mode coordinate x (or momentum p) and nonzero mode oscillators α_n, this extension (22) is written as

$$\begin{aligned} x^i &\rightarrow x^M \equiv (x^\mu, c_0, \bar{c}_0) \\ p^i = \alpha_0^i &\rightarrow p^M \equiv (p^\mu, -\pi_0, \bar{\pi}_0) = \alpha_0^M \\ \alpha_n^i &\rightarrow \alpha_n^M \equiv (\alpha_n^\mu, -\gamma_n, \bar{\gamma}_n) \end{aligned} \tag{24}$$

and the commutation relations become

$$\begin{aligned} [x^i, p^j] = i\delta^{ij} &\rightarrow [x^M, p^N\} = i\eta^{MN} \\ [\alpha_n^i, \alpha_m^j] = n\delta^{ij}\delta_{n+m,0} &\rightarrow [\alpha_n^M, \alpha_m^N\} = n\eta^{MN}\delta_{n+m,0} \end{aligned} \tag{25}$$

where the graded commutator $[A, B\}$ denotes anticommutator when A and B are both fermionic quantities and commutator otherwise. Equation (25) means, in particular, that

$$\{c_0, \bar{\pi}_0\} = \{\bar{c}_0, \pi_0\} = 1, \qquad \{\gamma_n, \bar{\gamma}_m\} = in\delta_{n+m,0} \tag{26}$$

The Virasoro operators $L_n^\perp$ (15) are accordingly replaced by

$$\begin{aligned} L_n &= \sum_{m=-\infty}^{\infty} \tfrac{1}{2}:\alpha^M_{n+m}\eta_{MN}\alpha^N_{-m}:-\delta_{n,0} \\ &= \sum_{m=-\infty}^{\infty} :(\tfrac{1}{2}\alpha_{n+m}\cdot\alpha_{-m} + i\gamma_{n+m}\bar{\gamma}_{-m}):-\delta_{n,0} \end{aligned} \tag{27}$$

Comparing these commutation relations (25) and the Virasoro operators (27) with those in HIKKO's covariant SFT, we see that the oscillators α_n^M here are just identical with $(\alpha_n^\mu, \gamma_n \equiv in\alpha c_n, \bar{\gamma}_n \equiv \alpha^{-1}\bar{c}_n)$ in HIKKO's theory. This correspondence, however, breaks down in the zero-mode sector $\alpha_0^M = (p^\mu, -\pi_0, \bar{\pi}_0)$; although p^μ and $\bar{\pi}_0$ here still have their counterparts p^μ and $\bar{\gamma}_0 = \alpha^{-1}\bar{c}_0^{\text{HIKKO}}$ in HIKKO's theory, the ghost zero mode π_0 has no counterpart since γ_0 vanishes by definition $\gamma_n = in\alpha c_n$ in HIKKO's case. Therefore π_0 is the extra ghost zero-mode necessary for the *OSp* symmetry, which was missing in HIKKO's theory.

The BRS operator Q_B is identified with the *OSp*-extended Lorentz transformation operator M^{c-}, rotation on the c(ghost) and $-$ plane, as first pointed out by Siegel and Zwiebach [24]. Here M^{c-} is obtained by making the *OSp* extension to the original $O(d-1,1)$ Lorentz generator M^{j-} given in (18a), and by taking the ghost direction $N = c$ in the $OSp(d, 2/2)$ extended Lorentz generator M^{N-}. Thus, clearly from (18a) with $x^c = c_0, p^c = -\pi_0$ and $\alpha_n^c = -\gamma_n$, we have [22]

$$-M^{c-} \equiv Q_B = -\frac{c_0}{\alpha}L_0 + i\pi_0\frac{\partial}{\partial\alpha} - \frac{i}{\alpha}\sum_{n=1}^{\infty}\left(\frac{\gamma_{-n}L_n - L_{-n}\gamma_n}{n}\right) \tag{28a}$$

where the Virasoro operators L_n are of course the *OSp* extended ones (27). Separating the terms containing the extra ghost zero mode $\gamma_0 = \pi_0$ out of L_n and L_0, we easily see that the rest part just coincides with the Kato-Ogawa BRS operator Q_B^{KO} [12]:

$$Q_B = Q_B^{\text{KO}} + \frac{i\pi_0}{\alpha}\left(N_{\text{FP}} + \alpha\frac{\partial}{\partial\alpha} + 1\right) \tag{28b}$$

where the Kato-Ogawa ghost variables $c_n^{\text{KO}}(n \neq 0)$, $\bar{c}_n^{\text{KO}}$ have been identified with the present ones as

$$\begin{aligned} &\gamma_n = in\alpha c_n^{\text{KO}} \qquad \text{for } n \neq 0, \qquad c_0 = \frac{\partial}{\partial\bar{\pi}_0} = \alpha\frac{\partial}{\partial\bar{c}_0^{\text{KO}}} = \alpha c_0^{\text{KO}} \\ &\bar{\gamma}_n = \alpha^{-1}\bar{c}_n^{\text{KO}} \qquad \text{for all } n \end{aligned} \tag{29}$$

and N_{FP} is the ghost number operator given by

$$N_{\text{FP}} = \pi_0\frac{\partial}{\partial\pi_0} - \bar{\pi}_0\frac{\partial}{\partial\bar{\pi}_0} + i\sum_{n=1}^{\infty}\frac{1}{n}(\gamma_n^\dagger\bar{\gamma}_n + \bar{\gamma}_n^\dagger\gamma_n) \tag{30}$$

The full nonlinear BRS transformation δ_B is thus identified with the full *OSp*-extended Lorentz transformation $\mathbf{M}^{-c}$ accordingly; hence, from (17) we obtain

$$\delta_B \Phi = Q_B \Phi + g\Phi * \Phi + g^2 \Phi \circ \Phi \circ \Phi \tag{31}$$

where we have absorbed the prefactor $X^c(\sigma_{\text{int}}) = c(\sigma_{\text{int}})$ into the definition of the vertex so that we have, from (18) with (16),

$$|\Phi * \Psi(3)\rangle = \epsilon_\Phi \epsilon_\Psi \int d1\, d2\, \langle \Phi(1)| \langle \Psi(2)| \, |V^{(3)}(1,2,3)\rangle$$

$$|V^{(3)}(1,2,3)\rangle = \frac{\mu^{(3)}}{\alpha_1\alpha_2\alpha_3} c(\sigma_{\text{int}}) \exp\left[\sum_{r,s=1}^{3} \sum_{n,m=0}^{\infty} \bar{N}^{rs}_{nm} (\tfrac{1}{2}\alpha^{(r)}_{-n} \cdot \alpha^{(s)}_{-m} + i\gamma^{(r)}_{-n}\bar{\gamma}^{(s)}_{-m}) \right]$$
$$\times |0\rangle \delta(\alpha, p) \tag{32}$$

$$\delta(\alpha, p) \equiv (2\pi)^{27} \delta(\textstyle\sum_r \alpha_r) \delta^{26}\left(\sum_r p_r\right) \delta\left(\sum_r \bar{\pi}^{(r)}_0\right) \delta\left(\sum_r \pi^{(r)}_0\right)$$

$$dr \equiv (2\pi)^{-27} \alpha_r\, d\alpha_r\, d^{26}p_r\, d\pi^{(r)}_0\, d\bar{\pi}^{(r)}_0$$

and similar expression for $\Phi \circ \Psi \circ \Lambda$ using the four-string vertex. Directly from the algebra (20), it follows that the *OSp*-extended full Lorentz transformation $\mathbf{M}^{N-}$ satisfies

$$[\mathbf{M}^{N-}, \mathbf{M}^{M-}\} = 0 \tag{33}$$

hence the full BRS transformation (31) $\delta_B = \delta_{\mathbf{M}^{-c}}$ being nilpotent:

$$(\delta_B)^2 \Phi = 0 \tag{34}$$

This nilpotency $(\delta_B)^2 = 0$ in fact holds in a stronger form; namely, we have

$$O(g^0){:}\; Q_B^2 = 0 \tag{35a}$$

$$O(g^1){:}\; Q_B(\Phi * \Psi) = Q_B\Phi * \Psi + (-1)^\Phi \Phi * Q_B \Psi \tag{35b}$$

$$O(g^2){:}\; (\Phi * \Psi) * \Lambda - \Phi * (\Psi * \Lambda)$$
$$= (-1)^{\Phi+\Psi+\Lambda+1} [Q_B(\Phi \circ \Psi \circ \Lambda) - Q_B\Phi \circ \Psi \circ \Lambda$$
$$- (-1)^\Phi \Phi \circ Q_B \Psi \circ \Lambda - (-1)^{\Phi+\Psi} \Phi \circ \Psi \circ Q_B \Lambda] \tag{35c}$$

and so on, which are just the same identities as those in HIKKO [3, 5]. These identities in our case result immediately from the corresponding identities in the light-cone gauge SFT [(53), (55), and (56) in Ref. 31].

4.2. Introduction of τ, θ

Note that the extra ghost zero-mode π_0 appears in the BRS operator Q_B (28) essentially in the form

$$i\pi_0 \frac{\partial}{\partial \alpha}$$

This implies that π_0 is playing the role of BRS partner of the string length parameter α. This is good since α should be an unphysical parameter in our approach.

We have yet another unphysical parameter, proper time τ, which we have not touched upon in the above Section 4.1. To make τ really unphysical, we have to introduce a new Grassmann coordinate θ to make up a BRS doublet (τ, θ).

To add the variables (τ, θ) as a BRS doublet, we now replace the above BRS operator Q_B in (28) by [19]

$$Q_B \to Q_B - i\theta \frac{\partial}{\partial \tau} \qquad \left[\text{or } Q_B - E\frac{\partial}{\partial \bar{\pi}_\theta}\right] \tag{36}$$

and simultaneously we change the δ function $\delta(\alpha, p)$ and the zero-mode integration dr in the vertex (32) as

$$\delta(\alpha, p) \to \delta(\alpha, p)\delta(\tau_1 - \tau_3)\delta(\tau_2 - \tau_3)\delta(\theta_1 - \theta_3)\delta(\theta_2 - \theta_3)$$
$$\left[\text{or } \delta(\alpha, p)2\pi\delta\left(\sum_r E_r\right)\delta\left(\sum_r \bar{\pi}_\theta^{(r)}\right)\right] \tag{37}$$

$$dr \to (2\pi)^{-28}\alpha_r\, d\alpha_r\, d^{26}p_r\, d\pi_0^{(r)}\, d\bar{\pi}_0^{(r)}\, dE_r\, d\bar{\pi}_\theta^{(r)} \tag{38}$$

Note that (37) means that we have taken interactions "instantaneous" with respect to the time variables τ and θ. Here, in (36)–(37), E and $\bar{\pi}_\theta$ are momentum variables conjugate to τ and θ, respectively, and the quantities given in the bracket stand for the expression in the "momentum" representation.

Even with these changes, all the previous identities (35), and hence the nilpotency (34), $(\delta_B)^2 = 0$, in particular, still remain to hold: First $Q_B^2 = 0$ is trivial since the original Q_B and the added term $i\theta\partial/\partial\tau$ are commutative with each other and nilpotent separately. Second the distributive law (35b), or equivalently $\sum_{r=1}^3 Q_B^{(r)}|V^{(3)}\rangle = 0$, remains valid since

$$\left(\sum_r \theta_r \frac{\partial}{\partial \tau_r}\right)\delta(\tau_1 - \tau_3)\delta(\tau_2 - \tau_3)\delta(\theta_1 - \theta_3)\delta(\theta_2 - \theta_3) = 0 \tag{39}$$

All the other associative laws (or Jacobi identity in the closed string case) clearly remain valid since the additional variables τ and θ appear only in the δ functions, indicating simply that the interactions are instantaneous.

4.3. Gauge-Invariant Action

We have thus constructed the covariantized BRS operator Q_B, $*$-product and $(\circ\circ)$-product which satisfy the same identities (35) as in HIKKO. Therefore, the gauge-invariant action S is constructed such that $\delta S/\delta\Phi$ is proportional to the full BRS transformation $\delta_B\Phi$, just as in HIKKO's case [2, 3]:

$$S = \Phi \cdot Q_B\Phi + \tfrac{2}{3}g\Phi \cdot (\Phi * \Phi) + \tfrac{2}{4}g^2\Phi \cdot (\Phi \circ \Phi \circ \Phi) \tag{40}$$

The gauge transformation is given by

$$\delta\Phi = Q_B\Lambda + g(\Phi * \Lambda - \Lambda * \Phi) + g^2(\Phi \circ \Phi \circ \Lambda - \Phi \circ \Lambda \circ \Phi + \Lambda \circ \Phi \circ \Phi) \tag{41}$$

The invariance of (40) under (41) is a result of only the nilpotency $(\delta_B)^2 = 0$, cyclic symmetry of the vertices, and partial integrability [3].

The action (40) turns out to coincide with that given by Neveu and West [18] if we write $\Phi = \chi + \theta\bar{\chi}$ and express the action in terms of χ and $\bar{\chi}$ by integrating away the θ variable,* although they gave the gauge transformation and the gauge invariance proof only up to $O(g^1)$.

5. GAUGE-FIXED BRS-INVARIANT ACTION AND PHYSICAL STATES

The gauge-invariant action (40) takes quite the same form as HIKKO's, and hence we can find the gauge-fixed one by the same procedure [33, 34].

5.1. BRS-Invariant Action for $\bar{\pi}_0|\Phi\rangle = 0$ Gauge

We now consider the case of imposing a gauge-fixing condition

$$\bar{\pi}_0|\Phi\rangle = 0 \tag{42}$$

Corresponding to this, we write Q_B [in (36) with (28)] and Φ as follows, making explicit the dependence on $\bar{\pi}_0$ and $c_0 = \partial/\partial\bar{\pi}_0$:

$$Q_B = -\frac{c_0}{\alpha}L_0' + (M - ic_0\pi_0)\frac{\bar{\pi}_0}{\alpha} + \tilde{Q}_B \tag{43}$$

$$\tilde{Q}_B = \tilde{Q}_B^{\mathrm{KO}} + \frac{i\pi_0}{\alpha}\left(N_{\mathrm{FP}}' + \alpha\frac{\partial}{\partial\alpha}\right) - E\frac{\partial}{\partial\bar{\pi}_\theta} \tag{44a}$$

$$M = \sum_{n=1}^{\infty}\frac{2}{n}\gamma_{-n}\gamma_n \tag{44b}$$

$$\Phi = -\bar{\pi}_0\phi + \psi \tag{45}$$

* The author was informed first by B. Zwiebach of the fact that the Neveu–West gauge invariant action may be cast into the standard $\Phi \cdot Q_B\Phi$ form if one uses the additional variable θ. This is commented on also in Ref. 25.

where $\tilde{Q}_B^{\mathrm{KO}}$ is the Kato-Ogawa $\tilde{Q}_B$ containing no ghost zero-modes, and the prime on L_0' and N_{FP}' means the omission of the ghost zero mode parts. Following HIKKO, the guage-fixed BRS-invariant action corresponding to the gauge (42) is simply given by the same action as the gauge-invariant one (40) provided that we set the ψ component in (45) equal to zero and abandon the ghost number constraint on ϕ; namely, putting $\Phi = -\bar{\pi}_0\phi$ and integrating away the $\bar{\pi}_0$ variable in (40), we obtain

$$S_{\text{gauge-fixed}} = [S]_{\psi=0} = \phi \cdot \frac{-L_0'}{\alpha}\phi + \tfrac{2}{3}g\phi^3 + \tfrac{2}{4}g^2\phi^4 \tag{46}$$

The BRS-transformation $\hat{\delta}_B\phi$ under which this action (46) is invariant is also given by setting $\psi = 0$ in the original BRS transformation (41):

$$\hat{\delta}_B\phi = \tilde{Q}_B\phi + gw\phi * \phi + g^2u\phi \circ \phi \circ \phi \tag{47}$$

Again (46) and (47) are formally the same as HIKKO, but the vertices $|v^{(n)}\rangle$ $(n = 3, 4)$, factors w and u, and the zero-mode integral dr differ from HIKKO's at the parts containing π_0, E, and $\bar{\pi}_\theta$ modes:

$$|v^{(n)}\rangle = \frac{1}{(\prod_r \alpha_r)} \exp\left(i \sum_{r,s=1}^{n} \pi_0^{(r)} \prod_{m=1}^{\infty} \bar{N}_{0m}^{rs}\bar{\gamma}_{-m}^{(s)} \right) |v_{\mathrm{HIKKO}}^{(n)}\rangle \delta\left(\sum_r E_r\right) \delta\left(\sum_r \bar{\pi}_\theta^{(r)}\right)$$

$$w^{(r)} = w_{\mathrm{HIKKO}}^{(r)} + i(\tau_0/\alpha_r^2)\pi_0^{(r)}, \qquad u^{(r)} = u_{\mathrm{HIKKO}}^{(r)} + \frac{1}{\alpha_r}\sum_{s=1}^{4} \bar{N}_{00}^{(4)sr}\pi_0^{(s)} \tag{48}$$

$$dr = \alpha_r (dr)_{\mathrm{HIKKO}} \cdot (2\pi)^{-1}\, dE_r\, d\bar{\pi}_\theta^{(r)}$$

In perturbation theory, the *physical on-shell* states ϕ_{phys} are specified by the linearized versions of BRS transformation (47) and equation of motion $\delta S_{\text{gauge-fixed}}/\delta\phi = 0$:

$$\tilde{Q}_B\phi_{\mathrm{phys}} = 0 \tag{49}$$

$$L_0'\phi_{\mathrm{phys}} = 0 \tag{50}$$

5.2. Analysis of $\tilde{Q}_B\phi_{\mathrm{phys}} = 0$

First note that $\partial/\partial\alpha$ in $\tilde{Q}_B$ (44a) is the derivative $(\partial/\partial\alpha)_\gamma$ with γ_n, $\bar{\gamma}_n$ kept fixed. We regard henceforth the Kato-Ogawa oscillators c_n^{KO} and $\bar{c}_n^{\mathrm{KO}}$ $(n \neq 0)$ in (29) instead of γ_n and $\bar{\gamma}_n$ as α independent, and then the α derivative $(\partial/\partial\alpha)_c$ becomes [35]

$$\begin{aligned}\left(\frac{\partial}{\partial\alpha}\right)_c &= \left(\frac{\partial\gamma}{\partial\alpha}\right)_c \frac{\partial}{\partial\gamma} + \left(\frac{\partial\bar{\gamma}}{\partial\alpha}\right)_c \frac{\partial}{\partial\bar{\gamma}} + \left(\frac{\partial\alpha}{\partial\alpha}\right)_c \left(\frac{\partial}{\partial\alpha}\right)_\gamma \\ &= \frac{1}{\alpha}\left(\gamma\frac{\partial}{\partial\gamma} - \bar{\gamma}\frac{\partial}{\partial\bar{\gamma}}\right) + \left(\frac{\partial}{\partial\alpha}\right)_\gamma = \frac{1}{\alpha}N_{\mathrm{FP}}' + \left(\frac{\partial}{\partial\alpha}\right)_\gamma\end{aligned} \tag{51}$$

So the BRS operator $\tilde{Q}_B$ now takes the simple form

$$\tilde{Q}_B = Q_B^{(1)} + Q_B^{(2)}$$

$$Q_B^{(1)} = \tilde{Q}_B^{\mathrm{KO}}, \qquad Q_B^{(2)} = i\pi_0 \frac{\partial}{\partial \alpha} - E\frac{\partial}{\partial \bar{\pi}_\theta} \tag{52}$$

$Q_B^{(1)}$ and $Q_B^{(2)}$ contain mutually independent variables and are nilpotent separately:

$$(Q_B^{(1)})^2 = 2ML_0' = 0 \qquad \text{(on-shell)}, \qquad (Q_B^{(2)})^2 = 0 \tag{53}$$

Generally in such a case, one can show that [36]

$$(Q_B^{(1)} + Q_B^{(2)})\phi_{\text{phys}} = 0 \Rightarrow \phi_{\text{phys}} = \phi_{\text{phys}}^{(1)(2)} + (Q_B^{(1)} + Q_B^{(2)})\chi$$

with (54)

$$Q_B^{(1)}\phi_{\text{phys}}^{(1)(2)} = Q_B^{(2)}\phi_{\text{phys}}^{(1)(2)} = 0$$

That is, the physical states are annihilated by $Q_B^{(1)}$ and $Q_B^{(2)}$ simultaneously, aside from a "null" state (trivial cohomology) part of the form $(Q_B^{(1)} + Q_B^{(2)})\chi$.

Thus it is sufficient to analyze the physical states satisfying

$$\left(i\pi_0 \frac{\partial}{\partial \alpha} - E\frac{\partial}{\partial \bar{\pi}_\theta}\right)\phi_{\text{phys}} = 0 \tag{55a}$$

$$\tilde{Q}_B^{\mathrm{KO}}\phi_{\text{phys}} = 0 \tag{55b}$$

As we show in Appendix A, equation (55a) implies that ϕ_{phys} can depend on the $2+2$ variables $(\alpha, E; \pi_0, \bar{\pi}_\theta)$ only through the $OSp(1, 1/2)$ invariant $\alpha E - i\pi_0\bar{\pi}_\theta$, again aside from the "null" state $Q_B^{(2)}\chi$. Since plane waves e^{iax} span a complete set in the function space of single variable $x = -i(\alpha E - i\pi_0\bar{\pi}_\theta)$, the solutions of (55) now take the form

$$\phi_{\text{phys}} = e^{a(\alpha E - i\pi_0\bar{\pi}_\theta)}\phi^{\mathrm{KO}} \qquad (\tilde{Q}_B^{\mathrm{KO}}\phi^{\mathrm{KO}} = 0) \tag{56}$$

where ϕ^{KO} is the Kato-Ogawa physical states containing no unphysical zero modes. Note that $e^{a(\alpha E - i\pi_0\bar{\pi}_\theta)}$ is a Gaussian wave function (by Wick rotation).

Important remarks are in order on the physical state (56). First the states with different values of a differ from one another only by null states $Q_B^{(2)}\chi$; indeed,

$$\begin{aligned} e^{a(\alpha E - i\pi_0\bar{\pi}_\theta)} - e^{b(\alpha E - i\pi_0\bar{\pi}_\theta)} &= \int_b^a dx(\alpha E - i\pi_0\bar{\pi}_\theta)\, e^{x(\alpha E - i\pi_0\bar{\pi}_\theta)} \\ &= Q_B^{(2)}\left(\alpha\bar{\pi}_\theta \int_b^a dx\, e^{x(\alpha E - i\pi_0\bar{\pi}_\theta)}\right) \end{aligned} \tag{57}$$

Therefore there exists only one genuine physical state in the function space of $(\alpha, E, \pi_0, \bar{\pi}_\theta)$. Second, in accordance with this observation also, the norm of the physical states $e^{a(\alpha E - i\pi_0\bar{\pi}_\theta)}$ is 1 independently of a:

$$\int \frac{d\alpha\, dE}{2\pi i}\, d\bar{\pi}_\theta i\, d\pi_0\, e^{a(\alpha E - i\pi_0\bar{\pi}_\theta)} = \frac{1}{a}\cdot a = 1! \tag{58}$$

Note here that the bosonic Gaussian integration has yielded the factor $1/a$, which is canceled by the factor a coming from fermionic Gaussian integration. Owing to the a independence, one may of course choose the $a = 0$ (or $a = \infty$) state as a representative of physical state. But then one would encounter the indefinite expression $\infty \times 0$, which was just the difficulty of Neveu and West's physical state as explained in Section 2. It is therefore now clear that this difficulty simply resulted from the inadequacy of their particular choice $a = 0$ (or $a = \infty$) for the physical state. There appears no difficulty if we use the physical state (56) with finite a.

This gauge is interesting because it resembles the original HIKKO theory very much.

In this gauge-fixed theory, all the tree level on-shell physical amplitudes are easily seen to coincide correctly with those of light-cone gauge SFT as follows: First one can directly check that the $*$ product of two string functions of the particular $OSp(1,1/2)$ Gaussian form $e^{a(\alpha E - i\pi_0\bar{\pi}_\theta)}$ is again of such a form, provided that a is proportional to any *conserved* quantity (like a component of momentum p^μ). So, if we put all the external lines to be physical states of the form (56), with a set equal to a conserved quantity p, we see in the case of tree level diagrams that all the internal lines also take the form $e^{a(\alpha E - i\pi_0\bar{\pi}_\theta)}$. Thus the extra four zero-mode variables $(\alpha, E, \pi_0, \bar{\pi}_\theta)$ can be integrated away explicitly and the resultant amplitudes turn out to be the same as those of HIKKO's theory with α set equal to p.* The latter is known to reproduce the dual amplitude at the tree level.

At the loop order levels, however, this gauge encounters the difficulty of $\infty \times 0$: Since, in this gauge, the propagator $1/L_0'$ contains none of the zero-mode variables α, E, π_0, and $\bar{\pi}_\theta$, and, in particular, E and $\bar{\pi}_\theta$ appear in the vertices (48) only in the δ functions $\delta(\sum_r E_r)\delta(\sum_r \bar{\pi}_\theta^{(r)})$, the integrations over the loop momenta E_l and $\bar{\pi}_\theta^{(l)}$ are factored out to yield

$$\prod_{l=1}^{L}\left[\int dE_l\, 1\right]\cdot\left[\int d\bar{\pi}_\theta^{(l)}\, 1\right] = (\infty \times 0)^L$$

Thus this gauge is singular. We therefore examine another gauge next.

* It is also known that the $\alpha = p^+$ HIKKO model (i.e., HIKKO's theory with α set equal to p^+) itself reproduces the same amplitudes as light-cone gauge SFT at any loop order level.

6. ANOTHER GAUGE-FIXED ACTION: SIEGEL'S ORIGINAL ONE

The gauge that leads to Siegel's original $OSp(d, 2/2)$ invariant action [16] as a gauge-fixed action is

$$(\bar{\pi}_0 - \bar{\pi}_\theta)|\Phi\rangle = 0 \tag{59}$$

This is essentially the gauge discussed by Neveu and West [18], although their arguments were much less transparent because of the lack of the θ variable.

For convenience, to discuss the gauge (59), let us make a change of variables $(\bar{\pi}_0, \bar{\pi}_\theta)$ into

$$\begin{aligned} \bar{\pi}_0^{\text{new}} &= \bar{\pi}_0 - \bar{\pi}_\theta \\ \bar{\pi}_\theta^{\text{new}} &= \bar{\pi}_\theta \end{aligned} \tag{60a}$$

or equivalently, in "coordinate" variables,

$$\begin{aligned} c_0^{\text{new}} &\equiv \partial/\partial\bar{\pi}_0^{\text{new}} = \partial/\partial\bar{\pi}_0 = c_0 \\ \theta^{\text{new}} &\equiv \partial/\partial\bar{\pi}_\theta^{\text{new}} = \partial/\partial\bar{\pi}_\theta + \partial/\partial\bar{\pi}_0 = \theta + c_0 \end{aligned} \tag{60b}$$

Then the BRS operator (43) with (44) is rewritten in terms of these new variables as

$$Q_B = \frac{c_0^{\text{new}}}{\alpha}(\alpha E - L_0^{OSp}) + (M - ic_0^{\text{new}}\pi_0)\frac{\bar{\pi}_0^{\text{new}}}{\alpha} + \tilde{Q}_B^{\text{new}} \tag{61}$$

$$L_0^{OSp} = L_0' + i\pi_0\bar{\pi}_\theta^{\text{new}} \tag{62a}$$

$$\tilde{Q}_B^{\text{new}} = \tilde{Q}_B^{\text{KO}} + M\frac{\bar{\pi}_\theta^{\text{new}}}{\alpha} - E\frac{\partial}{\partial\bar{\pi}_\theta^{\text{new}}} + i\pi_0\frac{\partial}{\partial\alpha} \tag{62b}$$

where $\partial/\partial\alpha$ is again the α derivative with c_n^{KO} and $\bar{c}_n^{\text{KO}}$ $(n \neq 0)$ fixed. Hereafter we use only new variables and omit the superscripts "new" on the new variables accordingly.

As before, writing $\Phi = -\bar{\pi}_0\phi + \psi$, setting $\psi = 0$, relaxing the ghost number constraint on ϕ, and integrating away the $\bar{\pi}_0$ variable in the gauge-invariant action (40), we obtain the gauge-fixed action in this gauge (59) as

$$S_{\text{gauge-fixed}} = \phi\cdot\left(E - \frac{1}{\alpha}L_0^{OSp}\right)\phi + \tfrac{2}{3}g\phi^3 + \tfrac{2}{4}g^2\phi^4 \tag{63}$$

which is invariant under the BRS transformation

$$\hat{\delta}_B\phi = \tilde{Q}_B^{\text{new}}\phi + gw^{\text{new}}\phi * \phi + g^2u^{\text{new}}\phi\circ\phi\circ\phi \tag{64}$$

obtained from (31) by setting ψ equal to zero. (We omit the detailed expressions for w^{new}, u^{new}, etc., here.)

By construction, it is clear that this gauge-fixed action (63) is just identical with Siegel's original *OSp* action, which results directly from the light-cone gauge SFT action (13) by the following replacement of the $O(d-2)$ vectors by $OSp(d-1, 1/2)$ ones:

$$p^i \to (p^\mu, -\pi_0, \bar{\pi}_\theta)$$
$$\alpha_n^i \to (\alpha_n^\mu, -\gamma_n, \bar{\gamma}_n)$$

So the action (63) has actually a very large symmetry, $OSp(d, 2/2)$ extended Lorentz invariance, in which only the $OSp(d-1, 1/2)$ subgroup is linearly realized. Our BRS transformation (64) obtained above is, however, not identical with the Lorentz transformation $\mathbf{M}^{-c}$ in the $(-, c)$ plane. It coincides with the latter only on the mass shell (i.e., if the equation of motion is used). (Actually $\hat{\delta}_B$ is nilpotent only on the mass shell, while $\mathbf{M}^{-c}$ is nilpotent off-shell.)

Again we specify the physical on-shell external states by the linearized versions of BRS transformation $\hat{\delta}_B$ (64) and equation of motion:

$$\tilde{Q}_B^{\text{new}} \phi_{\text{phys}} = 0 \tag{65a}$$
$$(\alpha E - L_0^{OSp})\phi_{\text{phys}} = 0 \tag{65b}$$

This case, the form (62b) of $\tilde{Q}_B^{\text{new}}$, is a bit more complicated than before and so the analysis of equation (65a) becomes harder than the previous gauge. It is, however, clear that the state

$$\phi_{\text{phys}} = e^{a(\alpha E - i\pi_0\bar{\pi}_\theta)} \cdot \delta(p^2 - \alpha E + i\pi_0\bar{\pi}_\theta + M^2)\phi^{\text{DDF}} \tag{66}$$

satisfies (65) and hence is a representative of physical state, where ϕ^{DDF} is a state constructed by DDF modes alone and M^2 is the mass square operator or eigenvalue. This state (66) has a well-defined norm as far as $a \neq 0, \neq\infty$, again.

Finally we now show that the S matrix for the physical on-shell states in this gauge agrees with that in the light-cone gauge SFT. As explained in Section 2, the S matrix for the external states possessing *definite momenta* $P_r \equiv (p_r^\mu, \alpha_r, E_r, \pi_0^{(r)}, \bar{\pi}_\theta^{(r)})$ is given by the same function as in the light-cone gauge SFT:

$$\underset{\text{This SFT}}{S(P^2 = p_\mu p^\mu - \alpha E + i\bar{\pi}_0\pi_\theta)} \xleftrightarrow{\text{Same fn.S}} \underset{\text{Light-cone SFT}}{S(p^2 = -\alpha E + \mathbf{p}^2)} \tag{67}$$

where P^2 (p^2) are generic notations denoting $OSp(d, 2/2)[O(d-1, 1)]$ invariants $P_r \cdot P_s$ $(p_r \cdot p_s)$ *and* $P_r \cdot \mathcal{E}_s$ $(p_r \cdot \epsilon_s)$ for momenta P_r (p_r) and polarization $\mathcal{E}_r$ (ϵ_r). Our physical states take the form (66) and do not realize $E = \pi_0 = \bar{\pi}_\theta = 0$, unlike Siegel's case. Nevertheless, equation (67) is sufficient for concluding the desired result. Indeed, we fortunately have the following generalized Parisi–Sourlas formula which holds for arbitrary *invariant* function f of n $OSp(1,1/2)$ vectors $(k_r^+, k_r^-, \kappa_r, \bar{\kappa}_r)$ $(r = 1, 2, \ldots, n)$:

$$\int \prod_{r=1}^{n} \left(\frac{d^2 k_r^{\pm}}{2\pi i} d\bar{\kappa}_r i\, d\kappa_r \right) f(k_r^+ k_s^- + k_r^- k_s^+ - i\kappa_r \bar{\kappa}_s + i\bar{\kappa}_r \kappa_s) = f(0) \quad (68)$$

This is proved in Appendix B. The S matrix for our physical on-shell states (66) is given by

$$\int \prod_{r=1}^{n} \left[\frac{d\alpha_r\, dE_r}{2\pi i} d\bar{\pi}_\theta^{(r)} i\, d\pi_0^{(r)} \phi_{(r)}^{\mathrm{DDF}}(p_r, \epsilon_r) \right.$$
$$\left. \times\, \delta(p_r^2 - \alpha_r E_r + i\pi_0^{(r)} \bar{\pi}_\theta^{(r)} + m_r^2)\, e^{a(\alpha_r E_r - i\pi_0^{(r)} \bar{\pi}_\theta^{(r)})} \right] S(P^2) \quad (69)$$

Here the integrand is an invariant function under $OSp(1,1/2)$ rotating the vectors $(\alpha, E, \pi_0, \bar{\pi}_\theta)$ since $S(P^2)$ is so under the larger group $OSp(d, 2/2)$ and the other factors depend only on the invariants $\alpha E - i\pi_0 \bar{\pi}_\theta$. So we can apply the formula (68) to (69) with $(k_r^{\pm}, \kappa_r, \bar{\kappa}_r) = (\alpha_r, E_r, \pi_0^{(r)}, \bar{\pi}_\theta^{(r)})$, and the result of integrations in (69) become equivalent to simply setting $\alpha_r = E_r = \pi_0^{(r)} = \bar{\pi}_\theta^{(r)} = 0$. (In this argument, important is also the condition of physical polarization that the polarization vector ε_r have no nonzero components in the $(\alpha, E, \pi_0, \bar{\pi}_\theta)$ directions.) Thus we again obtain the same S matrix $S(p^\mu p_\mu)$ as in the light-cone gauge SFT.

APPENDIX A: SOLUTIONS TO EQUATION (55a)

Let us show in this appendix that any solution to the equation

$$Q\phi(\alpha, E, \pi, \bar{\pi}) = 0 \qquad \text{with } Q = i\pi \frac{\partial}{\partial \alpha} - E \frac{\partial}{\partial \bar{\pi}} \quad (A.1)$$

is given by an $OSp(1,1/2)$ invariant function f aside from a trivial null state of the form $Q\chi$:

$$\phi = f(\alpha E - i\pi\bar{\pi}) + Q\chi(\alpha, E, \pi, \bar{\pi}) \quad (A.2)$$

We see first that

$$\phi = F(\alpha E - i\pi\bar{\pi}, E, \pi) + (\text{null state}) \quad (A.3)$$

is the most general solution of (A.1). However, considering the Taylor expansion of F with respect to E and π around $E = \pi = 0$, we easily understand that F can be rewritten into the form

$$F(\alpha E - i\pi\bar{\pi}, 0, 0) + \pi G(\alpha E - i\pi\bar{\pi}, E) + EH(\alpha E - i\pi\bar{\pi}, E) \quad (A.4)$$

But the first term is the desired $OSp(1,1/2)$ invariant function and the second and third terms can be written as a null state as follows:

$$Q[-i\alpha G(\alpha E - i\pi\bar{\pi}, E) - \bar{\pi}H(\alpha E - i\pi\bar{\pi}, E)] \tag{A.5}$$

Therefore (A.2) is proved.

APPENDIX B: PROOF OF THE GENERALIZED PARISI–SOURLAS FORMULA (68)

The original Parisi–Sourlas formula [15] is the one for the case of one $OSp(1,1/2)$ vector $(k^+, k^-, \kappa, \bar{\kappa})$, which is derived as follows:

$$\begin{aligned}
&\int \frac{d^2k^\pm}{2\pi}\, d\bar{\kappa}\, d\kappa\, f(k^+k^- - i\kappa\bar{\kappa}) \\
&\qquad = \int \frac{d^2k^\pm}{2\pi i}\, d\bar{\kappa} i\, d\kappa [f(k^+k^-) - i\kappa\bar{\kappa} f'(k^+k^-)] \\
&\qquad = \int \frac{d^2k^\pm}{2\pi i} f'(k^+k^-) = \int_0^\infty dk^2 \frac{d}{d(-k^2)} f(-k^2) = f(0)
\end{aligned} \tag{B.1}$$

where we have assumed $f(-\infty) = 0$ and used $k^\pm = k_0 \pm k_1 = ik_2 \pm k_1$, $dk^+\, dk^- = 2i\, dk_1\, dk_2 = 2\pi i\, dk^2$. [The equation (58) is just a special example of this formula.]

Now we prove the generalized formula (68) for the case of n $OSp(1,1/2)$ vectors $K_r \equiv (k_r^+, k_r^-, \kappa_r, \bar{\kappa}_r)$ $(r = 1, 2, \ldots, n)$. The proof goes by induction. The $n = 1$ case holds by (B.1). We prove that (68) holds for arbitrary n, assuming that it holds for the $n-1$ case. For that purpose let us first perform the integration over the nth momentum $K_n \equiv K \equiv (k^+, k^-, \kappa, \bar{\kappa})$, which generally looks like

$$\begin{gathered}
\int dK f(K^2, P\cdot K, Q^2) \\
K^2 = k^+k^- - i\kappa\bar{\kappa}, \qquad dK = d^2k^\pm d\bar{\kappa}\, d\kappa/2\pi \\
P\cdot K = p^+k^- + p^-k^+ - i\pi\bar{\kappa} + i\bar{\pi}\kappa
\end{gathered} \tag{B.2}$$

where $P \equiv (p^+, p^-, \pi, \bar{\pi})$ is an $OSp(1,1/2)$ momentum given by a certain linear combination of $K_1, K_2, \ldots, K_{n-1}$, and Q^2 denote generically the invariants made of $K_1, \ldots, K_{n-1}$ alone.

The integral (B.2) is evaluated by separating the $P \cdot K$ dependent and independent pieces in the integrand:

$$(\text{B.2}) = \int dK f(K^2, 0, Q^2) + \int dK[f(K^2, P\cdot K, Q^2) - f(K^2, 0, Q^2)] \tag{B.3}$$

The results of integrals yield, of course, $OSp(1, 1/2)$ invariant functions of the rest of the $n-1$ variables $K_1, K_2, \cdots, K_{n-1}$. The second term of (B.3) is, however, seen to give a vanishing contribution when the integrations $\int d^{n-1}K$ over $K_1, K_2, \ldots, K_{n-1}$ are performed. This is seen as follows. Since the integrand of the second term contains at least a factor $P\cdot K$, the result of the $\int dK$ integration contains at least a factor P^2 by the $OSp(1, 1/2)$ invariance, and hence takes the form $P^2 g(Q^2)$ with a certain function g of generic invariants Q^2 of $K_1, K_2, \ldots, K_{n-1}$. But the integral $\int d^{n-1}KP^2 g(Q^2)$ gives $[P^2 g(Q^2)]_{K_1=\cdots=K_{n-1}=0}$ by the induction assumption for the $n-1$ case and vanishes owing to the explicit factor P^2. Thus, keeping only the first term in (B.3) and using (B.1) again, we find

$$(\text{B.2}) = \int dK f(K^2, o0, Q^2) = f(0, 0, Q^2) \tag{B.4}$$

implying that the effect of the integration over the nth momentum K is equivalent to settting $K = 0$. Repeating this procedure we obtain the desired formula (68).

ACKNOWLEDGMENTS. This work was supported in part by a Grant-in-Aid for Scientific Research from the Ministry of Education, Science, and Culture (No. 63540214). The author would like to thank H. Hata, H. Kunitomo, and K. Suehiro for fruitful discussions and critical comments. He is also grateful to P. C. West, B. Zwiebach, and Nouri-Maghadam for valuable discussions in the early stages of this work.

REFERENCES

1. W. Siegel, *Phys. Lett. 149B*, 157, 162 (1984); *151B*, 391, 396 (1985).
2. H. Hata, K. Itoh, T. Kugo, H. Kunitomo, and K. Ogawa, *Phys. Lett. 172B*, 186, 195 (1986); *175B*, 138 (1986); *Nucl. Phys. B 283*, 433 (1987); *Prog. Theor. Phys. 77*, 443 (1987).
3. H. Hata, K. Itoh, T. Kugo, H. Kunitomo, and K. Ogawa, *Phys. Rev. D 34*, 2360 (1986); *35*, 1318 (1987).
4. H. Hata, K. Itoh, T. Kugo, H. Kunimoto, and K. Ogawa, *Phys. Rev. D 35*, 1356 (1987).
5. For a review, see T. Kugo, *Lectures delivered at 25th Course of the International School of Subnuclear Physics on The SuperWorld II*, Erice 1987, Plenum Press, New York, to appear.
6. A. Neveu and P. C. West, *Phys. Lett. 168B*, 192 (1986); *Nucl. Phys. B 278*, 601 (1986).
7. E. Witten, *Nucl. Phys. B 268*, 253 (1986).

8. S. B. Giddings, *Nucl. Phys. B 278*, 79 (1986); S. B. Giddings and E. Martainec, *Nucl. Phys. B 278*, 91 (1986); S. B. Giddings, E. Martainec, and E. Witten, *Phys. Lett. 176B*, 362 (1986); D. J. Gross and A. Jevicki, *Nucl. Phys. B 283*, 1 (1987); *B287*, 1225 (1987); *293*, 29 (1987); S. Samuel, *Phys. Lett. 181B*, 255 (1986); *Nucl. Phys. B 296*, 187 (1988); K. Itoh, K. Ogawa, and K. Suehiro, *Nucl. Phys. B 289*, 127 (1987); K. Suehiro, *Nucl. Phys. B 296*, 333 (1987); *Prog. Theor. Phys. 78*, 1151 (1987); E. Cremmer, A. Schwimmer, and C. B. Thorn, *Phys. Lett. 179B*, 57 (1986).
9. E. Witten, *Nucl. Phys. B 276*, 291 (1986).
10. S. B. Giddings and E. Martainec, *Nucl. Phys. B 278*, 91 (1986); H. Hata, K. Itoh, T. Kugo, H. Kunitomo, and K. Ogawa, *Phys. Rev. D 35*, 1318 (1987).
11. See, however, e.g., A. Strominger, Lectures delivered at ICTP School on Superstring, Trieste, Italy, April 1987.
12. M. Kato and K. Ogawa, *Nucl. Phys. B 212*, 443 (1983).
13. See also K. Fujikawa, *Phys. Rev. D 25*, 443 (1982); S. Hwang, *Phys. Rev. D 28*, 2614 (1983).
14. T. Kugo and I. Ojima, *Phys. Lett. 73B*, 459 (1978); *Prog. Theor. Phys. Suppl. 66*, 1 (1979).
15. G. Parisi and N. Sourlas, *Phys. Rev. Lett. 43*, 744 (1979).
16. W. Siegel, *Phys. Lett. 142B*, 276 (1984).
17. A. Neveu and P. C. West, *Phys. Lett. 182B*, 343 (1987).
18. A. Neveu and P. C. West, *Nucl. Phys. B293*, 266 (1987).
19. S. Uehara, *Phys. Lett. 190B*, 76 (1987); King's College preprint 87-0412 (May 1987).
20. W. Siegel and B. Zwiebach, *Nucl. Phys. B 282*, 125 (1987); *Phys. Lett. 184B*, 325 (1987); *Nucl. Phys. B 288*, 332 (1987); see also, W. Siegel, *Nucl. Phys. B 284*, 632 (1987); L. Baulieu, W. Siegel, and B. Zwiebach, *Nucl. Phys. B 287*, 93 (1987).
21. H. Aoyama, *Nucl. Phys. B 299*, 379 (1988).
22. A. K. H. Bengtsson and N. Linden, *Phys. Lett. 187B*, 289 (1987).
23. H. Aratyn, R. Ingermanson, and A. J. Niemi, *Phys. Lett. 194B*, 506 (1987); *195B*, 149, 155 (1987); A. J. Niemi, Berkeley preprints UCB-PTH-87-28 (June 1987); UCB-PTH-87-29 (July 1987); UCB-PTH-87-32 (Aug. 1987).
24. W. Siegel and B. Zwiebach, *Nucl. Phys. B 282*, 125 (1987).
25. W. Siegel and B. Zwiebach, *Nucl. Phys. B 288*, 332 (1987).
26. S. Mandelstam, *Nucl. Phys. B 64*, 205 (1973); *69*, 77 (1974); *83*, 413 (1974).
27. M. Kaku and K. Kikkawa, *Phys. Rev. D 10*, 1110, 1823 (1974).
28. E. Cremmer and J. L. Gervais, *Nucl. Phys. B 90*, 1653 (1975).
29. M. B. Green and J. H. Schwarz, *Nucl. Phys. B 218*, 43 (1983); *243*, 475 (1984); M. B. Green, J. H. Schwarz, and L. Brink, *Nucl. Phys. B 219*, 437 (1983); T. Yoneya, *Phys. Rev. Lett. 55*, 1828 (1985).
30. P. Goddard, J. Goldstone, C. Rebbi, and C. B. Thorne, *Nucl. Phys. B 56*, 109 (1973).
31. T. Kugo, *Prog. Theor. Phys. 78*, 690 (1987).
32. S.-J. Sin, LBL preprint LBL-23715 (June 1987).
33. H. Hata, K. Itoh, T. Kugo, H. Kunitomo, and K. Ogawa, *Nucl. Phys. B 283*, 433 (1987).
34. C. B. Thorne, *Nucl. Phys. B 287*, 61 (1987); M. Bochicchio, *Phys. Lett. 193B*, 31 (1987).
35. W. Siegel and B. Zwiebach, *Nucl. Phys. B 299*, 206 (1988).
36. T. Kugo and S. Uehara, *Prog. Theor. Phys. 64*, 1395 (1980).

Chapter 12

Topology, Superspace, and Anomalies

Burt A. Ovrut

In the first part of this chapter, I would like to discuss a topological approach to heterotic superdiffeomorphic and Lorentz anomalies developed in collaboration with J. Louis and R. Garreis [1]. This work is carried out using the superfield formulation of (1, 0) supergravity [2] and, hence, world-sheet supersymmetry is manifest. Although topology, and in particular cohomology, have been successfully applied [3] to the study of anomalies involving component fields, previous attempts to apply cohomology to superspace [4] were beset with difficulties. These difficulties arose, primarily, from the lack of a Stokes theorem for superspace. However, it was shown in Ref. 1 that for (1, 0) superspace there is a modified notion of Stokes theorem which enables topological techniques to be employed. Using these techniques, a satisfactory superspace theory of gauge and gravitational anomalies can be developed. It is interesting to note that this modified notion of Stokes theorem is intimately related to the choice of torsion constraints [2] in (1, 0) superspace.

We begin by briefly reviewing the formalism of (1, 0) supermanifolds. A point in superspace is written as

$$z^M = (x^+, x^-, \theta^{+1}) \tag{1}$$

Burt A. Ovrut • Department of Physics, University of Pennsylvania, Philadelphia, Pennsylvania 19104-6396.

The geometry is determined by the supervielbeins $E_M{}^A$, $E_A{}^M$ and the Lorentz connection $\phi_{MA}{}^B$. The associated gauge groups are as follows:

1. Superdiffeomorphisms—parameters ξ^M;
2. Lorentz transformations—parameter L and generator

$$\lambda_A{}^B = \left(\begin{array}{cc|c} 1 & & \\ & -1 & \\ \hline & & 1/2 \end{array}\right) \tag{2}$$

Under superdiffeomorphisms

$$\begin{aligned} \delta_D E_M{}^A &= -\xi^L \partial_L E_M{}^A - (\partial_M \xi^L) E_L{}^A \\ \delta_D E_A{}^M &= -\xi^L \partial_L E_A{}^M + E_A{}^L (\partial_L \xi^M) \\ \delta_D \phi_M &= -\xi^L \partial_L \phi_M - (\partial_M \xi^L) \phi_L \end{aligned} \tag{3}$$

Under Lorentz transformations

$$\begin{aligned} \delta_L E_M{}^A &= L E_M{}^B \lambda_B{}^A \\ \delta_L E_A{}^M &= -L \lambda_A{}^B E_B{}^M \\ \delta_L \phi_M &= -\partial_M L \end{aligned} \tag{4}$$

It is clear from (3) and (4) that, thus far, the supergravity multiplet is highly reducible. The torsion superfields, $T_{BC}{}^A$, and the Lorentz curvature superfields $R_{CDA}{}^B$ $(= R_{CD}\lambda_A{}^B)$ can be constructed from the vielbein and connection. The supergravity multiplet can be rendered irreducible by imposing the following constraints:

$$\begin{aligned} T_{+1+1}{}^a &= -2i\delta^{+a} \\ T_{bc}{}^a &= T_{+1+1}{}^{+1} = 0 \\ T_{-+1}{}^a &= 0 \end{aligned} \tag{5}$$

The Bianchi identities then imply

$$\begin{aligned} T_{a+1}{}^{+1} &= T_{++1}{}^a = 0 \\ R_{++1} &= R_{+1+1} = 0 \\ R_{-+1} &= -2iT_{+-}{}^{+1} \\ R_{+-} &= 2\mathcal{D}_{+1} T_{+-}{}^{+1} \end{aligned} \tag{6}$$

Some of the consequences of (5) and (6) are

1. ϕ_M are determined from $E_M{}^A$, $E_A{}^M$
2. $E_+{}^M$ are determined from $E_-{}^M$, $E_{+1}{}^M$.

Define $H_A{}^B$ by

$$\delta E_M{}^A = E_M{}^B H_B{}^A \tag{7}$$

Then constraints (5) are left invariant for independent $H_{+1}{}^A$, $H_-{}^A$ as long as

$$H_-{}^{+1} = -\frac{i}{2}(\mathscr{D}_{+1}H_-{}^+ - \mathscr{D}_-H_{+1}{}^+)$$
$$H_+{}^+ = -i(\mathscr{D}_{+1}H_{+1}{}^+ - 2iH_{+1}{}^{+1}) \tag{8}$$
$$H_+{}^- = -i\mathscr{D}_{+1}H_{+1}{}^-$$

Super-Weyl transformations are defined as those transformations that leave the torsion constraints invariant and that satisfy

$$\delta_W E_M{}^a = lE_M{}^a \tag{9}$$

Under super-Weyl transformations

$$\delta_W E_M{}^a = lE_M{}^a$$
$$\delta_W E_M{}^{+1} = \frac{l}{2}E_M{}^{+1} - iE_M{}^+(\partial_{+1}l)$$
$$\delta_W E_a{}^M = -lE_a{}^M + i(\partial_{+1}l)\delta_{a+}E_{+1}{}^M \tag{10}$$
$$\delta_W E_{+1}{}^M = -\frac{l}{2}E_{+1}{}^M$$
$$\delta_W \phi_M = N_a E_N{}^a(\partial_a l) + E_M{}^{+1}(\partial_{+1}l)$$

The density superfield is

$$E_{+1}{}^{-1} = [\det(E_a{}^m)^{-1}][E_{+1}{}^{+1} - E_{+1}{}^m(E_a{}^m)^{-1}E_a{}^{+1}] \tag{11}$$

Under the above transformations

$$\delta_D E_{+1}^{-1} = -(-1)^m\partial_M(\xi^M E_{+1}^{-1})$$
$$\delta_L E_{+1}^{-1} = -\frac{L}{2}E_{+1}^{-1} \tag{12}$$
$$\delta_W E_{+1}^{-1} = \frac{3l}{2}E_{+1}^{-1}$$

Denote $dx^+\,dx^-\,d\theta^{+1}$ by dz. Then the superspace integration measure is $dz\,E_{+1}^{-1}$. The $(1, 0)$ supergravity Einstein action is given by

$$S_{SG} = \int dz\,E_{+1}^{-1}2T_{+-}{}^{+1}$$
$$= \int dz\,E_{+1}^{-1}i(\mathscr{D}_-\phi_{+1} - \mathscr{D}_{+1}\phi_-) \tag{13}$$
$$= 0$$

Having reviewed the formalism of (1, 0) superspace we can now construct the sigma model for the heterotic superstring [5]. The matter superfields are

1. $X^\mu - \mu = 0, \ldots, D-1$
2. $\psi^I_{-1} - I = 1, \ldots, N$

Note that -1 is equivalent to $-+1$. Under the above transformations

$$\delta_D X^\mu = -\xi^L \partial_L X, \qquad \delta_D \psi^I_{-1} = -\xi^L \partial_L \psi^I_{-1}$$

$$\delta_L X^\mu = 0, \qquad \delta_L \psi^I_{-1} = \frac{L}{2} \psi^I_{-1} \tag{14}$$

$$\delta_W X^\mu = 0, \qquad \delta_W \psi^I_{-1} = -\frac{l}{2} \psi^I_{-1}$$

The most general action that is superdiffeomorphic, Lorentz, and super-Weyl invariant is

$$S = S_X + S_\psi \tag{15}$$

where

$$\begin{aligned} S_X &= \int dz\, E^{-1}_{+1}[-i(\mathscr{D}_{+1} X^\mu)(\mathscr{D}_- X_\mu)] \\ S_\psi &= \int dz\, E^{-1}_{+1}[-\psi^I_{-1}(\mathscr{D}_{+1}\psi^I_{-1})] \end{aligned} \tag{16}$$

We have taken $G_{\mu\nu} = \eta_{\mu\nu}$, $B_{\mu\nu} = 0$, $g_{IJ} = \delta_{IJ}$, and $A_{+1K}{}^I = 0$ in (16).

We now discuss the superdiffeomorphic anomaly associated with the heterotic string. Prior to gauge fixing the effective action, $W[E_A{}^M]$, is given by

$$e^{iW} = N \int [dX^\mu][d\psi^I_{-1}]\, e^{i(S_X + S_\psi)} \tag{17}$$

The superdiffeomorphic variation of W can be written as

$$\delta_D W = \int dz\, E^{-1}_{+1} \Lambda_P{}^M G_{D-1M}{}^P \tag{18}$$

where

$$\Lambda^M_P = \partial_P \xi^M \tag{19}$$

Note that $\delta_D W$ vanishes if and only if $G_{D-1M}{}^P$ vanishes. Therefore, $G_{D-1M}{}^P$ is the superdiffeomorphic anomaly. What is the functional structure of $G_{D-1M}{}^P$? Remarkably, this can be determined topologically. This is done in three steps.

1. *BRST Transformations.* Let ξ be a constant, anticommuting parameter and define superdiffeomorphic ghost superfields by $\xi^M = \xi C^M$. Note that $C^{\pm}$ and C^{+1} are anticommuting and commuting superfields, respectively. The BRST transformations of $E_A{}^M$, X^μ, and ψ^I_{-1} are obtained from superdiffeomorphisms by replacing ξ^M with ξC^M. Define BRST generator, Σ_D, by

$$\delta_D \phi_i = \xi \Sigma_D \phi_i \tag{20}$$

Note that Σ_D is an anticommuting operator. Hence,

$$\Sigma_D^2 = 0 \tag{21}$$

It follows from the δ_D variations above that

$$\begin{aligned} \Sigma_D E_A{}^M &= -C^L \partial_L E_A{}^M + (-1)^a E_A{}^L (\partial_L C^M) \\ \Sigma_D X^\mu &= -C^L \partial_L X^\mu \\ \Sigma_D \psi^I_{-1} &= -C^L \partial_L \psi^I_{-1} \end{aligned} \tag{22}$$

Define

$$C_P{}^M = (-1)^p \partial_P C^M \tag{23}$$

Then

$$\Sigma_D C_P{}^M = -C^L \partial_L C_P{}^M - (-1)^{p+l} C_P{}^L C_L{}^M \tag{24}$$

Also

$$\Sigma_D E^{-1}_{+1} = -\partial_L (C^L E^{-1}_{+1}) \tag{25}$$

2. *Wess-Zumino Consistency Condition.* Replacing ξ^M by ξC^M in $\delta_D W$ implies

$$\Sigma_D W = -\int dz\, E^{-1}_{+1} C_P{}^M G_{D-1M}{}^P \tag{26}$$

Operating with Σ_D yields

$$\int dz\, \Sigma_D (E^{-1}_{+1} C_P{}^M G_{D-1M}{}^P) = 0 \tag{27}$$

Can Σ_D be brought through E^{-1}_{+1}? Decompose $\Sigma_D = \Sigma_l + \Sigma_g$, where

$$\begin{aligned} \Sigma_l E_A{}^M &= -C^L \partial_L E_A{}^M + (-1)^a E_A{}^L (\partial_L C^M), \qquad \Sigma_g E_A{}^M = 0 \\ \Sigma_l X^\mu &= -C^L \partial_L X^\mu, \qquad \Sigma_g X^\mu = 0 \\ \Sigma_l \psi^I_{-1} &= -C^L \partial_L \psi^I_{-1}, \qquad \Sigma_g \psi^I_{-1} = 0 \\ \Sigma_l C_P{}^M &= -C^L \partial_L C_P{}^M, \qquad \Sigma_g C_P{}^M = -(-1)^{p+l} C_P{}^L C_L{}^M \end{aligned} \tag{28}$$

Note that

$$\Sigma_l E_{+1}^{-1} = -\partial_L(C^L E_{+1}^{-1}), \qquad \Sigma_g E_{+1}^{-1} = 0 \tag{29}$$

It can be shown that $\Sigma_g^2 = 0$. Assume that $G_{D-1M}{}^P$ behaves like a Lie algebra valued scalar superfield. That is,

$$\Sigma_l G_{D-1M}{}^P = -C^L \partial_L G_{D-1M}{}^P \tag{30}$$

Then

$$\int dz\, \Sigma_l(E_{+1}^{-1} C_P{}^M G_{D-1M}{}^P) = -\int dz\, \partial_P(C^P E_{+1}^{-1} C_M{}^L G_{D-1L}{}^M)$$
$$= 0 \tag{31}$$

Therefore

$$\int dz\, E_{+1}^{-1} \Sigma_g(C_P{}^M G_{D-1M}{}^P) = 0 \tag{32}$$

This is the Wess–Zumino consistency condition for the superdiffeomorphic anomaly. Note that (32) is linear in the ghost fields and $G_{D-1M}{}^P$ has Lorentz charge $-+1$.

3. *Topological Solution of the Consistency Condition.* Let $\Gamma_{NM}{}^R$ be the superfield Christoffel connection. Define $\Gamma_M{}^R = dy^N \Gamma_{NM}{}^R$. Differential dy^N is chosen so that $\Gamma_M{}^R$ has the same statistics as $C_M{}^R$. Under BRST transformations take

$$\begin{aligned} \Sigma_l \Gamma_M{}^R &= dy^N(-1)^n(C^L \partial_L \Gamma_{NM}{}^R + C_N{}^L \Gamma_{LM}{}^R) \\ \Sigma_g \Gamma_M{}^R &= dC_M{}^R + (-1)^{1+m+l}(C_M{}^L \Gamma_L{}^R + \Gamma_M{}^L C_L{}^R) \end{aligned} \tag{33}$$

The curvature two-form is defined by

$$R_P{}^Q = d\Gamma_P{}^Q + (-1)^{1+p+r} \Gamma_P{}^R \Gamma_R{}^Q \tag{34}$$

Consider the three-form

$$\omega_3^0(\Gamma) = (-1)^q \Gamma_Q{}^P R_P{}^Q + \tfrac{1}{3}(-1)^{r+p+q} \Gamma_Q{}^P \Gamma_P{}^R \Gamma_R{}^Q \tag{35}$$

It can be shown that

$$(-1)^q R_Q{}^P R_P{}^Q = d\omega_3^0(\Gamma) \tag{36}$$

Now define $\tilde{\Gamma}_M{}^R = \Gamma_M{}^R - C_M{}^R$. Furthermore, let $\Delta_D = d + \Sigma_g$. Note that $\Delta_D^2 = 0$. The associated curvature two-form is

$$\tilde{R}_P{}^Q = \Delta_D \tilde{\Gamma}_P{}^Q + (-1)^{1+p+r} \tilde{\Gamma}_P{}^R \tilde{\Gamma}_R{}^Q \tag{37}$$

Since the algebra of $\Delta_D, \tilde{\Gamma}_P{}^Q$ is the same as $d, \Gamma_P{}^Q$ implies

$$(-1)^q \tilde{R}_Q{}^P \tilde{R}_P{}^Q = \Delta_D \omega_3(\tilde{\Gamma}) \tag{38}$$

where

$$\omega_3(\tilde{\Gamma}) = (-1)^q \tilde{\Gamma}_Q{}^P \tilde{R}_P{}^Q + \tfrac{1}{3}(-1)^{r+p+q} \tilde{\Gamma}_Q{}^P \tilde{\Gamma}_P{}^R \tilde{\Gamma}_R{}^Q \tag{39}$$

However, one can show that $\tilde{R}_P{}^Q = R_P{}^Q$. Therefore

$$d\omega_3^0(\tilde{\Gamma}) = \Delta_D \omega_3(\tilde{\Gamma}) \tag{40}$$

Now $\omega_3(\tilde{\Gamma})$ can be expanded as $\omega_3(\tilde{\Gamma}) = \omega_3^0 + \omega_2^1 + \omega_1^2 + \omega_0^3$, where, for example

$$\omega_2^1 = -(-1)^q C_Q{}^P d\Gamma_P{}^Q \tag{41}$$

Comparing the left- and right-hand sides of (40) and equating terms of identical form and ghost number yields the descent equations. The equation of interest is

$$\Sigma_g \omega_2^1 = -d\omega_1^2 \tag{42}$$

One cannot integrate superforms over superspace. Therefore, consider the component form of (42)

$$\Sigma_g \omega_{2AB}^1 = \mathscr{D}_A \omega_{1B}^2 - (-1)^{ab} \mathscr{D}_B \omega_{1A}^2 + T_{AB}{}^D \omega_{1D}^2 \tag{43}$$

Integrating over superspace yields

$$\int dz\, E_{+1}^{-1} \Sigma_g \omega_{2AB}^1 = \int dz\, E_{+1}^{-1} T_{AB}{}^D \omega_{1D}^2 \tag{44}$$

Note that the torsions $T_{AB}{}^D$ are an obstruction to Stokes' theorem. However, recall that $G_{D-1M}{}^P$ has Lorentz charges $-+1$ and that $T_{-+1}{}^D = 0$ for any D. Therefore

$$\int dz\, E_{+1}^{-1} \Sigma_g \omega_{2-+1}^1 = 0 \tag{45}$$

and, hence, ω_{2-+1}^1 is a nontrivial solution of the Wess–Zumino consistency condition. The $-+1$ component of ω_2^1 is

$$\omega_{2-+1}^1 = (-1)^{1+p} C_Q{}^P E_{-1}{}^{NM} \partial_M \Gamma_{NP}{}^Q \tag{46}$$

where $E_{-1}{}^{NM} = (-1)^n E_-{}^N E_{+1}{}^M - E_{+1}{}^N E_-{}^M$. Comparing (45) with (32) implies

$$G_{D-1M}{}^P \propto (-1)^m E_{-1}{}^{NL} \partial_L \Gamma_{NM}{}^P \tag{47}$$

The constant of proportionality can be found by a one-loop supergraph calculation. The result is

$$G_{D-1M}{}^P = -\frac{1}{168\pi}(D-N)(-1)^m E_{-1}{}^{NL} \partial_L \Gamma_{NM}{}^P \tag{48}$$

The Lorentz anomaly also has a topological solution. We find that

$$G_{L-1B}{}^A \propto (-1)^b E_{-1}{}^{NM} \partial_M \phi_{NB}{}^A \tag{49}$$

The superdiffeomorphic and Lorentz anomalies are closely related. To see this define superfields $\mathcal{H}_M{}^N$ by

$$(e^{\mathcal{H}})_M{}^N = E_A{}^M \delta_A{}^N \tag{50}$$

and let

$$\Gamma_t = (e^{-t\mathcal{H}})\Gamma(e^{t\mathcal{H}}) - (e^{-t\mathcal{H}})d(e^{t\mathcal{H}}) \tag{51}$$

where $0 \leq t \leq 1$. Then the Wess–Zumino term for the superdiffeomorphic anomaly is

$$S[E, \Gamma] = \int_0^1 dt \int dz\, E_{+1}^{-1} \mathcal{H}_Q{}^P G_{D-1}[\Gamma_t]_P{}^Q \tag{52}$$

where

$$G_{D-1}[\Gamma_t]_P{}^Q = -\frac{1}{168\pi}(D-N)(-1)^m E_{-1}{}^{NL} \partial_L \Gamma_{tNM}{}^P \tag{53}$$

Decompose $\delta_D = \mathcal{L}_\xi + T_\Lambda$ where $\Lambda_M{}^N = \partial_M \xi^N$. It follows that

$$\begin{aligned} \mathcal{L}_\xi \Gamma_{tNM}{}^R &= -\xi^L \partial_L \Gamma_{tNM}{}^R - (\partial_N \xi^L)\Gamma_{tLM}{}^R \\ T_\Lambda \Gamma_t &= [\Gamma_t, \Lambda_t] - d\Lambda_t \end{aligned} \tag{54}$$

where

$$\Lambda_t = (e^{-t\mathcal{H}})\Lambda(e^{t\mathcal{H}}) + (e^{-t\mathcal{H}})T_\Lambda(e^{t\mathcal{H}}) \tag{55}$$

Also note that $\Gamma_0 = \Gamma$, $\Lambda_0 = \Lambda$, and $\Lambda_1 = 0$. Consider $\delta_D S[E, \Gamma]$. After a long calculation we find

$$\delta_D S[E, \Gamma] = -\int dz\, E_{+1}^{-1} \Lambda_Q{}^P G_{D-1}[\Gamma]_P{}^Q \tag{56}$$

It follows that

$$\delta_D(W + S[E, \Gamma]) = 0 \tag{57}$$

Hence, by adding Wess–Zumino term (52) to the action the superdiffeomorphic anomaly can be made to vanish. However, there is now a Lorentz anomaly, as we now show. The relation between the Christoffel and Lorentz connections is

$$\Gamma_{NL}{}^M = -(-1)^{n(a+l)}(E_L{}^A \partial_N E_A{}^M - E_L{}^A \phi_{NA}{}^B E_B{}^M) \tag{58}$$

Defining

$$\phi_\tau = (e^{-\tau\mathcal{H}})\phi(e^{\tau\mathcal{H}}) - (e^{-\tau\mathcal{H}})d(e^{\tau\mathcal{H}}) \tag{59}$$

where $\tau = t - 1$. Then

$$\Gamma_{tP}{}^Q = \delta_P{}^A \delta_B{}^Q \phi_{\tau A}{}^B \tag{60}$$

The superdiffeomorphic Wess–Zumino term can now be written as

$$S[E,\Gamma] = -\int_{-1}^{0} d\tau \int dz\, E_{+1}^{-1} \mathcal{H}_B{}^A G_{L-1}[\phi_\tau]_A{}^B \tag{61}$$

One can now calculate $\delta_L S[E,\Gamma]$. We find that

$$\delta_L(W + S[E,\Gamma]) = \int dz\, E_{+1}^{-1} \mathcal{H}_A{}^B G_{L-1}[\phi]_B{}^A \tag{62}$$

where the Lorentz anomaly is given by

$$G_{L-1}[\phi]_B{}^A = \frac{1}{168\pi}(D-N)(-1)^a E_{-1}{}^{NM} \partial_M \phi_{NB}{}^A \tag{63}$$

Since there is no longer a superdiffeomorphic anomaly we can now fix this gauge, thus introducing ghost superfields into the action. The contribution of these ghosts to the Lorentz anomaly can be calculated with supergraphs. We find that the entire Lorentz anomaly is

$$G_{L-1B}{}^A = \frac{1}{168\pi}(D+N+22)(-1)^a E_{-1}{}^{NM} \partial_M \phi_{NB}{}^A \tag{64}$$

Finally, the super-Weyl anomaly can also be calculated topologically. Setting $D - N + 22 = 0$, we find that the condition that the super-Weyl anomaly vanish is $D - 10 = 0$. Hence, we find the well-known result that the heterotic string has no superdiffeomorphic, Lorentz, or super-Weyl anomalies as long as $D = 10$ and $N = 32$.

In the second part of this chapter I would like to discuss a formalism that allows a calculation of the Schwinger term in the conformal Virasoro algebra, and the associated seagulls, directly from the Weyl anomaly. This can be done for both the bosonic string [6] and the superstring [7], but, for simplicity, I will emphasize the former. This work was done in collaboration with J. Louis, R. Garreis, and J. Wess.

The anomalous part of the one-loop effective action of the bosonic string can be shown to be

$$W^{(1)}[g] = -\frac{1}{96\pi}(D-26)\int d^2x\,(-g)^{1/2} R^{(2)} \frac{1}{\nabla^2} R^{(2)} \tag{65}$$

Recalling that the stress-energy tensor T_{mn} is defined by

$$T_{mn} = 2\frac{\delta W}{\delta g^{mn}} \tag{66}$$

it is not hard to show that the superdiffeomorphic and Weyl anomalies are given by

$$\begin{aligned} G_D &= \nabla^m T_{mn} \\ G_W &= T^m{}_m \end{aligned} \tag{67}$$

respectively. Since (65) is invariant under diffeomorphisms, G_D must vanish. What about a Weyl anomaly? It follows from (65), (66), and (67) that

$$G_W = -\frac{1}{24\pi}(D-26)R^{(2)} \tag{68}$$

and, hence, there is in general a nonzero Weyl anomaly. This anomaly vanishes if and only if

$$D = 26 \tag{69}$$

which is the critical dimension for the bosonic string. What is the relationship between the Weyl anomaly and the anomalous Schwinger term in the conformal algebra? Recall that

$$e^{iW[g]} = \int [db][dc][dX^\mu]\, e^{i(S+S_{GT})} \tag{70}$$

where

$$W[g] = W^{(0)}[g] + W^{(1)}[g] \tag{71}$$

$W^{(0)}[g]$ is nonanomalous and can be ignored, whereas $W^{(1)}[g]$ is given by (65). Let ξ^m be a vector field satisfying

$$\nabla^m \xi^n + \nabla^n \xi^m = \Lambda g^{mn} \tag{72}$$

Such vector fields are called conformal Killing vectors (CKVs). The conformal currents are given by

$$J_m = T_{mn}\xi^n \tag{73}$$

Start with equation (70). Functionally differentiate with respect to $g^{pq}(y)$, multiply by CKV $\xi^q(y)$, operate with ∇_y^p, functionally differentiate with respect to $g^{rs}(z)$, and multiply by CKV $\xi^s(z)$. Then, to lowest order in h_{mn}, we find that

$$\frac{-i\Lambda(y)}{2}\xi^s(z)\frac{\delta G_w(y)}{\delta g^{rs}(z)} = \nabla_y^p \int [db][dc][dX^\mu]\Big[J_p(y)J_r(z) - i\xi^s(z)\frac{\delta J_p(y)}{\delta g^{rs}(z)}\Big]\, e^{i(S+S_{GT})}\, e^{-iW[g]} \tag{74}$$

Define

$$\langle T^*(\hat{J}_p(y)\hat{J}_r(z))\rangle = \int [db][dc][dX^\mu] J_p(y)J_r(z)\, e^{i(S+S_{GT})}\, e^{-iW[g]} \tag{75}$$

where T^* is the covariant T-star product and the caret notation identifies quantum operators. Also, let

$$\Xi_{pq,rs}(y)\xi^q(y)\delta(y-z) = \int [db][dc][dX^\mu]\frac{\delta J_p(y)}{\delta g^{rs}(z)}\, e^{i(S+S_{GT})}\, e^{-iW[g]} \tag{76}$$

Recall that

$$\langle T^*(\hat{J}_p(y)\hat{J}_r(z))\rangle = \langle T(\hat{J}_p(y)\hat{J}_r(z))\rangle + \xi^q(y)\xi^s(z)\tau_{pq,rs}(y,z) \tag{77}$$

where T is the time-ordered product and the $\tau_{pq,rs}$ functions are "seagull" terms. In the Bjorken–Johnson–Low limit one can show

$$\nabla^p_t T(\hat{J}_p(y)\hat{J}_r(z)) = g^{p0}(y)\delta(y^0-z^0)[\hat{J}_p(y),\hat{J}_r(z)] \tag{78}$$

Define

$$\tau'_{pq,rs}(y,z) = \tau_{pq,rs}(y,z) - i\Xi_{pq,rs}(y)\delta(y-z) \tag{79}$$

Then, using (75)–(79), equation (74) becomes

$$\frac{-i}{2}[\partial_{yl}\xi^l(y)]\xi^s(z)\frac{\delta G_W(y)}{\delta g^{rs}(z)} = -\delta(y^0-z^0)\langle[\hat{J}_0(y),\hat{J}_r(z)]\rangle$$
$$+\eta^{pt}\partial_{yt}[\xi^q(y)\xi^s(z)\tau'_{pq,rs}(y,z)] \tag{80}$$

For the closed string a complete set of CKVs is

$$\xi^\pm{}_m(x^\pm) = \frac{1}{2\sqrt{2}}\, e^{im2\sqrt{2}x^\pm} \tag{81}$$

where m is any integer. Define the associated conformal currents and conformal generators by

$$\begin{aligned}\hat{J}^{(\pm)}{}_{(m)p} &= \hat{T}_{p\pm}\xi^\pm{}_m\\ \hat{L}^{(\pm)}{}_m &= \int dx^1\, \hat{J}^{(\pm)}{}_{m(0)}\end{aligned} \tag{82}$$

respectively. Henceforth, we consider the (+) terms only. The most general form for the first term on the right-hand side of (80) [with index r in (80) set equal to zero] is

$$\delta(y^0-z^0)\langle[\hat{J}^{(+)}{}_{(m)0}(y),\hat{J}^{(+)}{}_{(n)0}(z)]\rangle$$
$$= \xi^+{}_m(y)\xi^+{}_n(z)\sum_{r=0}^{\infty} S^{(+,+)r}(y)\left(\frac{\partial}{\partial y^1}\right)^r\delta(y-z) \tag{83}$$

Equation (80) can now be written as

$$\frac{-i}{2}[\partial_{y+}\xi^+{}_m(y)]\xi^+{}_n(z)\frac{\delta G_W(y)}{\delta g^{0+}(z)}$$
$$= -\xi^+{}_m(y)\xi^+{}_n(z)\sum_{r=0}^{\infty} S^{(+,+)r}(y)\left(\frac{\partial}{\partial y^1}\right)^r\delta(y-z)$$
$$+\eta^{pt}\partial_{yt}[\xi^+{}_m(y)\xi^+{}_n(z)\tau'{}_{p+,0+}(y,z)] \tag{84}$$

G_W is the Weyl anomaly given in (68). Hence $\delta G_W(y)/\delta g^{0+}(z)$ can be evaluated and is found to be

$$\frac{\delta G_W(y)}{\delta g^{0+}(z)} = \frac{\sqrt{2}}{48\pi}(D-26)(\partial_0\partial_1 - \partial_1^2)|_y \delta(y-z) \tag{85}$$

Putting (85) into (84) and comparing left- and right-hand sides yields the following results: Seagulls

$$\begin{aligned} \tau'_{0+,0+}(y,z) &= \frac{-i}{48\pi}(D-26)\partial_{y1}^2\delta(y-z) \\ \tau'_{1+,0+}(y,z) &= \frac{-i}{48\pi}(D-26)(\partial_0\partial_1 + 2\partial_1^2)|_y\delta(y-z) \end{aligned} \tag{86}$$

Schwinger terms:

$$\begin{aligned} S^{(+,+)3}(y) &= \frac{-i}{24\pi}(D-26) \\ S^{(+,+)r}(y) &= 0, \qquad r \neq 3 \end{aligned} \tag{87}$$

Equation (83) now becomes

$$\begin{aligned} &\delta(y^0 - z^0)\langle[\hat{J}^{(+)}{}_{(m)0}(y), \hat{J}^{(+)}{}_{(n)0}(z)]\rangle \\ &= \frac{-i}{24\pi}(D-26)\xi^+{}_m(y)\xi^+{}_n(z)\partial_{y1}^3\delta(y-z) \end{aligned} \tag{88}$$

Integrating (88) over $\int_{-\infty}^{+\infty} dz^0 \int_0^\pi dz^1\, dy^1$ and using (82) implies

$$\langle[\hat{L}^{(+)}{}_m, \hat{L}^{(+)}{}_n]\rangle = \frac{(D-26)}{24\pi} m^3 \delta_{m+n,0} \tag{89}$$

It follows that

$$[\hat{L}^{(+)}{}_m, \hat{L}^{(+)}{}_n] = (m-n)\hat{L}^+{}_{m+n} + \frac{(D-26)}{24\pi} m^3 \delta_{m+n,0}\hat{1} \tag{90}$$

The second term on the right-hand side is the anomalous Schwinger term. (Note that the m^3 can be turned into the more standard expression $m^3 - m$ by a trivial redefinition of $\hat{L}^{(+)}{}_0$.) A similar result holds for the $\hat{L}^{(-)}{}_m$ generators. We see, therefore, that there is a direct relationship between the Weyl anomaly and the anomalous Schwinger term in the conformal algebra.

ACKNOWLEDGMENTS. Work supported in part by the Department of Energy under contract No. DOE-AC02-76-ERO-3071 and NATO under grant No. 86/0684.

REFERENCES

1. R. Garreis, J. Louis, and B. Ovrut, *Phys. Lett. 198B*, 189 (1987); *Nucl. Phys. B 306*, 567 (1988).
2. M. Evans and B. Ovrut, *Phys. Lett. 174B*, 177 (1986); *175B* 145 (1986); *184B*, 153 (1987); *186B*, 134 (1987); *Phys. Rev. D 35*, 3045 (1987); R. Brooks, F. Muhammad, and S. Gates, *Nucl. Phys. B 268*, 599 (1986); G. Moore and P. Nelson, *ibid. 274*, 509 (1986).
3. See B. Zumino, Chiral Anomalies and Differential Geometry, Relativity Groups and Topology II, Les Houches (1983); and L. Baulieu, *Phys. Rep. C129*, 1 (1985), and references therein.
4. G. Girardi, R. Grimm, and R. Stora, *Phys. Lett. 156B* 203 (1985); L. Bonora, P. Pasti, and M. Tonin, *Phys. Lett. 156B*, 341 (1985); *Nucl. Phys. B 252*, 458 (1985).
5. D. Gross, J. Harvey, E. Martinec, and R. Rohm, *Phys. Rev. Lett. 54*, 502 (1985); *Nucl. Phys. B 256*, 253 (1985); *267*, 75 (1986).
6. J. Louis, R. Garreis, B. Ovrut, and J. Wess, *Phys. Lett. 199B*, 57 (1987).
7. J. C. Lee, J. Louis, and B. Ovrut, University of Pennsylvania preprint UPR-0345T (1987).

Chapter 13

Field and String Quantization in Curved Space-Times

Norma Sánchez

1. CONTEXTUAL BACKGROUND

Perhaps the main challenge in theoretical physics today is unification of all interactions including gravity. At present, string theories appear as the best candidates to achieve such a unification. However, many technical and conceptual problems remain *and* a quantum theory of gravity is still not available. Continuous effort over the last quarter of a century has demonstrated the many difficulties encountered in repeated attempts to construct such a theory and has also indicated some of the particular properties that an eventual complete theory will have to possess. The amount of work in that direction can be by now presented in two different sets, which have mainly evolved (and remain) separated: (1) *conceptual unification* (introduction of the uncertainty principle in general relativity, the interpretation problem, quantum field theory (QFT) in curved space-time and by accelerated observers, Hawking radiation and its consequences, "wave function of the universe", . . .); (2) *grand unification* (the unification of all interactions including gravity from the particle physics point of view. Gravity is considered as a massless spin-2 particle (the graviton, supergravities, Kaluza-Klein theories, and the more successful superstrings).

NORMA SÁNCHEZ • Observatoire de Paris, Section de Meudon, 92195 Meudon Principal Cedex, France.

Whatever the final theory of the world will be, if it is to be a theory of everything, we would like to know what new understanding it will give us about the singularities of classical general relativity. If string theory is to provide a theory of quantum gravity, it should give us a proper theory (not yet existent) for describing the ultimate state of quantum black holes and the initial (say, very early) state of the universe, that is, a theory describing the physics (and the geometry) at Planck energies and lengths. Till now gravity has not completely been incorporated in string theory: strings are most frequently formulated in a *flat* space-time. Gravity appears through massless spin-2 particles (graviton). One disposes only of partial results for strings in curved backgrounds; these mainly concern the problem of consistency (validity of quantum conformal invariance) through the vanishing of the beta functions. The nonlinear quantum string dynamics in curved space-time has only been studied in the slowly varying approximation for the geometry (background field method) where the field propagator is essentially taken as the flat space Feynman propagator. Clearly, such approximations are useless for the study of strings in *strong* curvature regimes where quantum gravity effects are important. Our aim is to properly understand strings in the context of quantum gravity. As a first step in this program we propose studying QST (*quantum string theory*) *in curved space-times.* (Of course the main goal should be to extract the particle spectrum and the space-time itself from string theory, but, as is known, one is very far from doing that explicitly.) There are different kinds of effects to be considered here: *ground state* and *thermal effects* (associated with the fact that in general relativity there are no preferred reference frames, and with the possibility of having different choices of time), and *curvature effects,* which will modify the mass formula, critical dimension, and scattering amplitudes of strings.

2. FIELDS IN CURVED SPACE

The formulation of quantum field theory in nontrivial (curved or flat) space-times has given new fundamental features with respect to the usual understanding of QFT in trivial (Minkowski flat) space-time, viz., (1) the possibility for a given field theory to have different alternative well-defined Fock spaces (different "sectors" of the theory); (2) the presence of "intrinsic" statistical features (temperature, entropy) arising from the nontrivial structure (geometry, topology) of the space-time and not from a superimposed statistical description of the quantum matter fields. Relevant examples are QFT on the Rindler manifold and its analytic mapping extensions [1, 2], black holes and cosmological (de Sitter) space-times. Quantum field theory that has been developed for curvilinear (accelerated) coordinates in flat space in a way that can be directly generalized to curved space-time is a useful step for a physical and mathematical discussion of the full theory. The genuine coordinate independence, which is so familiar in the classical theory of general relativity, is not a particular property of gravity but a fundamental principle prevalent in all descriptions of physical laws. On the other hand, the apparent difference that results from the treatment of a quantum field theory in a variety of coordinate systems (in either curved or flat space-time) is not a coordinate effect at all, but is a consequence of the fact that physically different quantum states are correctly described by the quantum theory as being physically distinct. "Canonical" states for different coordinate systems are physically different (each timelike vector field leads to a separate indication of what constitutes a definition of positive frequency).

Some time ago the present author [2] proposed a new approach to QFT in curved space-times based on analytic (holomorphic) mappings. The mappings relating some manifold to its global analytic extension are an essential ingredient in the discussion of QFT on curved manifolds and its thermal properties. This approach allows (1) studying the vacuum and thermal properties of the quantum theory entirely in terms of the properties of holomorphic functions; (2) classifying the vacuum spectra (particle production rates and their associated temperature and entropy) according to the singularities of the mappings (the singularities here describing the asymptotic regions of the space-time). Consider a two-dimensional space-time as embedded in a higher four-dimensional one with metric tensor g. If g has removable singularities (event horizons), the coordinates (without horizons) in which the manifold has its maximal analytic extension can be found from mappings of the form

$$x_k \pm t_k = f(r^* \pm t)$$

where (x_k, t_k) are Kruskal (maximal) type coordinates and (r^*, t) are of Schwarzschild type. For the most important metrics in general relativity, the presence of isometry groups allows us to perform the maximal analytic extension of the four-dimensional manifold M through the extension of a relevant two-dimensional manifold containing the time axis and a suitable spatial coordinate. This two-dimensional manifold can be taken as the fiber of the manifold M considered as an appropriate fiber bundle. Examples of this situation are the Taub–NUT (T-NUT) and Kerr–Newman (KN) metrics, which can be also generalized to be solutions of the Einstein equations with nonzero cosmological term, i.e., the Taub–NUT-de Sitter and Kerr–Newman-de Sitter families. After analytic continuation $t_k = i\tau_k$ $(t = i\tau)$, the mapping

$$u_k = f(u')$$

($u_k = X_k \pm i\tau_k$, $u' = r^* \pm i\tau$) is holomorphic (or antiholomorphic). These mappings should satisfy the boundary conditions

$$u_{k\pm} = f(\pm\infty)$$

$u_{k\pm}$ are the images of the event horizons in u'. Thus, in Schwarzschild space-time $r^* \equiv r + r_s \ln(r/r_s - 1) = -\infty$, (i.e., $r = r_s \equiv$ Schwarzschild radius) corresponds to $u_{k\pm} = 0$ (i.e., $x_k = \pm t_k$). For the Kerr–Newman or Taub–NUT families, f is the *exponential mapping*, i.e.,

$$f(u') = e^{2\pi\beta^{-1}u'}$$

where

$$\beta = \begin{cases} 8\pi[M + M^2(M^2 + l^2 + Q^2]^{-1/2}] & \text{(KN)} \\ \pi\left\{\dfrac{M(M+1) + 2[L^2 - M(M^2 + L^2)^{1/2}]}{L(M^2 + L^2)^{1/2}}\right. & \text{(T-NUT)} \end{cases}$$

The temperature (T) of these solutions is a topological invariant: It is given by the differential winding number of the mapping f, i.e.,

$$T = \frac{1}{2\pi}\frac{\partial}{\partial\tau}\,\mathrm{Im}\log f(r^* + i\tau)$$

If f is the exponential mapping, τ is periodic ($0 \leq \tau \leq \beta$) and $T = \beta^{-1}$ is the Hawking temperature. In the Euclidean regime, these solutions exist as gravitational instantons (i.e., complete solutions of the Einstein equations with ++++ signature). The Euclidean action $I = \beta^2/16\pi$ is finite, interpreted as the intrinsic entropy of the solutions and related to its Euler number. [The presence of event horizons is a necessary condition to have nonzero entropy and a true (global) thermal equilibrium over the whole space-time, but they are not necessary to have finite temperature and asymptotic thermal equilibrium.]

3. STRINGS IN RINDLER SPACE

Much of the present interest in string theories comes from the hope that they may provide a finite theory of quantum gravity. It seems natural in this context to investigate the quantization of strings in a curved space-time. As a first step in this program we study the quantization of a string in a uniformly accelerated space (D-dimensional Rindler space). Although this is a flat space-time, it possesses a space-time structure including an event horizon, similar to a black-hole manifold. In order to quantize the string properly in this manifold, a horizon regularization is needed. This regularization can be introduced through the definition of the Rindler coordinates as follows:

$$x_1 \pm x_0 + \varepsilon = e^{\alpha(X^1 \pm X^0)}$$
$$x^i = X^i, \qquad i = 2, \ldots, D \tag{1}$$

where (x^1, x^0) and (X^1, X^0) are, respectively, Minkowskian (Kruskak-like) and Rindler (Schwarzschild-like) coordinates. α is a constant defining the proper acceleration of the Rindler observers (α is equal to the surface gravity in the black-hole case) and ε is an infinitesimal parameter of the order of the Planck length. The horizon is now at a finite distance $|X^1 \pm X^0| \sim (1/\alpha)\ln(1/\varepsilon)$. This regularization reflects the fact that a classical description of the geometry is no longer valid at distances of the order of the Planck length.

We quantize the string in a light-cone gauge where the light-cone variables are $(x^1 \pm x^0)$ and $(X^1 \pm X^0)$. This choice is particularly convenient here since the acceleration points in the x^1 direction. We recall that for QFT in accelerated frames (in flat and curved space-times), different timelike vector fields lead to inequivalent positive frequency modes [1–3]. In the light-cone gauge, positive frequency modes with respect to the timelike variable (τ) on the string world sheet are physical states when τ is identified with the appropriate time variable in the physical space-time. We associate inertial and accelerated particle states of the string to positive frequency modes with respect to x^0 and X^0, respectively. We find that the accelerated frequencies of the accelerated modes differ in a large factor

$$\lambda_0 = \frac{2\pi\alpha}{\ln(2\pi/\varepsilon + 1)} \tag{2}$$

from the inertial ones. Physically this factor reflects the indefinite increasing of the string length when it approaches the event horizon (Fig. 1). We develop the string dynamics in Rindler space and write the corresponding constraints. The longitudinal coordinates X^0 and X^1 obey nonlinear equations of motion but, as in the inertial case, they can be eliminated in terms of the transverse coordinates which are the independent dynamical variables. Two possible situations appear depending on whether the center of mass has a uniform speed or a uniform acceleration. The mass formulas are derived in each case. The Poincaré invariance of the flat Minkowski space-time has a nonlinear realization in terms of the Rindler coordinates. We prove that the passage from the inertial to accelerated modes of the string is a canonical transformation both at the classical and quantum levels. We also check that an explicit realization of the Poincaré algebra can be constructed in terms of the accelerated modes, provided (at the quantum level) one sets $D = 26$ and uses symmetric ordering of the operators. The

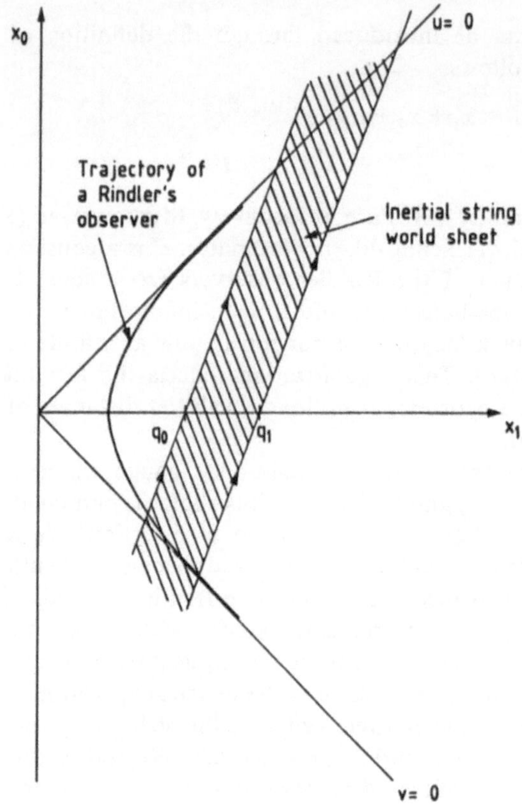

Figure 1. World-sheet of the inertial string in Minkowski coordinates. $u = 0$ and $v = 0$ are the horizons of Rindler space.

explicit expression of the Bogoliubov coefficients relating the inertial and Rindler modes of the string is found. The expectation value of the Rindler number mode operator in the ground state (tachyon) of the string is computed. This follows a thermal distribution [equation (62)] with temperature

$$T_s = \frac{\alpha}{2\pi} \tag{3}$$

This is the same Hawking–Unruh value that appears in the field-theoretical context. However, if one measures the frequency in dimensionless units $(1, 2, \ldots)$ instead of multiples of λ_0 [equation (2)], the temperature of the ground state is a (very large) pure number

$$T_0 = \frac{T_s}{\lambda_0} = \frac{1}{4\pi^2} \ln\left(\frac{2\pi}{\varepsilon} + 1\right) \tag{4}$$

We find the expectation value of the accelerated mass M' operator in the (tachyon) ground state; it turns out to be a large positive number

$$M' = 2\pi T_0 \tag{5}$$

The major features of the string quantization in Rindler space also hold for a string in a Schwarzschild space-time. An appropriate light-cone gauge can be introduced for left or right movers in the null Kruskal and Schwarzschild coordinates. Now, equation (2) for λ_0 gives the relation between the Kruskal and Schwarzschild frequencies and T_s and T_0 [equations (3)-(4)] give the temperatures characterizing the ground state of the string in the black hole manifold with

$$\varepsilon \sim \left(\frac{M_{\rm Pl}}{M}\right)^{(D-2)/(D-3)} \quad \text{and} \quad \alpha \equiv K \sim \left(\frac{M_{\rm Pl}}{M}\right)^{(D-2)/(D-3)}$$

Here $M_{\rm Pl}$ and M stand for the Planck mass and the black-hole mass, respectively. It can be noticed that the Hagedorn temperature in this context $(1/\sqrt{\alpha'} \sim M_{\rm Pl})$ is always $\gtrsim T_s \sim M_{\rm Pl}(M_{\rm Pl}/M)^{1/(D-3)}$ since a basic requirement for the present semiclassical treatment is $M \gtrsim M_{\rm Pl}$. The investigations presented here can also be extended to the case of fermionic strings and to more general (nonuniform) accelerations described by analytic (holomorphic) mappings as in the approach of Ref. 2.

4. HORIZON REGULARIZATION IN RINDLER SPACE—INTRODUCTION OF THE ε PARAMETER

Let us discuss now some features of Rindler space that are important for our study of strings in this space. The Rindler transformation maps the right-hand wedge $x^1 \geq |x^0|$ of Minkowski space onto the whole Rindler space $-\infty \leq X^1, X^0 \leq +\infty$ (the whole Minkowski space can be covered using four different Rindler patches). As is known, a quantum field in Rindler space is in a thermal state with temperature $T = \alpha/(2\pi)$. In addition, ultraviolet divergences arise in the free energy and entropy of quantum fields from the existence of a horizon at $x^1 = |x^0|$ (i.e., $X^1 = -\infty$) in the space-time. The same problem appears in the case of a four-dimensional black hole. Let us illustrate this phenomenon by considering a free massive scalar field Ψ. In D-dimensional Rindler space, the positive frequency modes are

$$\Psi = \frac{1}{(2\pi)^{(D-2)/2}} e^{i(-\lambda X^0 + k^i X^i)} \Phi(X^1), \qquad \lambda > 0, \qquad k^i \in \mathbb{R}$$

where $\phi(X^1)$ satisfies

$$\left[\frac{d^2}{dX^{12}} + \lambda^2 - (m^2 + k^{i2})\alpha^2 e^{2\alpha X^1}\right]\Phi(X^1) = 0$$

The total number ($\mathcal{N}$) of wave modes with frequency less than λ can be computed in the semiclassical approximation for $\lambda \gtrsim \alpha$. This is enough to study the ultraviolet behavior of the quantities interesting us. In the WKB approximation $\mathcal{N}_\lambda$ is given by

$$\pi\mathcal{N}_\lambda = \int_{-H}^{a(\lambda)} dX^1 \int \prod_{i=1}^{D-2} \frac{dk^i}{2\pi} [\lambda^2 - (m^2 + k^{i^2})\alpha^2 e^{2\alpha X^1}]^{1/2} \tag{7}$$

The integration is taken over the values of k^i and X^1 for which the argument of the square root is positive. Here $a(\lambda)$ is the classical turning point

$$e^{\alpha a(\lambda)} = \frac{\lambda}{m\alpha}$$

and H is a large cutoff ($H \gtrsim 1/\alpha$) on the negative Rindler coordinate X^1. This shifts the horizon by replacing the light-cone $x^1 = |x_0|$ as a boundary of Rindler space-time by the hyperbola

$$(x^1)^2 - (x^0)^2 = e^{-2\alpha H} \tag{8}$$

This regularization takes into account the fact that a classical description of the geometry is no longer valid at distances of order of the Planck length. Thus

$$e^{-2\alpha H} \sim l_p$$

Evaluating $\mathcal{N}_\lambda$ for large H it yields

$$\pi\mathcal{N}_\lambda = \frac{1}{4} \frac{(\lambda/\alpha)^{D-1}}{(4\pi)^{(D-3)/2}(D-2)} e^{\alpha H(D-2)}[1 + O(e^{-2\alpha H})] \tag{9}$$

Then the free energy and the entropy at temperature T:

$$F = -\int_0^\infty \mathcal{N}_\lambda \frac{d\lambda}{(e^{\lambda/T} - 1)}, \qquad S = -\frac{\partial F}{\partial T}$$

are equal to

$$F = -\frac{\Gamma(D)\xi(D)T}{(4\pi)^{(D-1)/2}(D-2)} \alpha^{1-D} e^{\alpha H(D-2)}[1 + O(e^{-2\alpha H})] \tag{10}$$

$$S = -\frac{D}{T} F > 0$$

In Rindler space, $T = \alpha/2\pi$ and so

$$F = -\frac{\Gamma(D)\xi(D)}{(4\pi)^{D-1/2}\pi^D} \frac{\alpha}{(D-2)} e^{\alpha H(D-2)} \tag{11}$$

$$S = \frac{\Gamma(D)\xi(D)}{(4\pi)^{D-1/2}\pi^D} \frac{2\pi D}{(D-2)} e^{\alpha H(D-2)} \tag{12}$$

We explicitly see that F and S need the ultraviolet cutoff H to be finite. This is equivalent to considering the following mapping defining the accelerated coordinates:

$$\begin{aligned} x^1 - x^0 + \varepsilon &= e^{\alpha(X^1 - X^0)} \\ x^1 + x^0 + \varepsilon &= e^{\alpha(X^1 + X^0)} \end{aligned} \tag{13}$$

where $\varepsilon = e^{-\alpha H}, -H \leq X^1 \leq +\infty$.

5. ANALYTIC MAPPINGS AND THE HAWKING–UNRUH EFFECT IN STRING THEORY

As it is known, the string action in a D-dimensional (generically curved) space-time is given by

$$S = \frac{1}{2\pi\alpha'} \int d\sigma\, d\tau\, \sqrt{g} g^{\alpha\beta} G_{AB}(X) \partial_\alpha X^A \partial_\beta X^B \tag{14}$$

$G_{AB}(X)$ and $g_{\alpha\beta}(\sigma, \tau)$ stand for the space-time and the world-sheet metrics, respectively, namely,

$$\begin{aligned} dS^2 &= G_{AB}(X)\, dX^A\, dX^B, \qquad 0 \leq A, B \leq D-1 \\ d\mathcal{S}^2 &= g_{\alpha\beta}(x)\, dx^\alpha\, dx^\beta, \qquad 0 \leq \alpha, \beta \leq 1 \end{aligned} \tag{15}$$

$(2\pi\alpha')^{-1}$ is the string tension (we will take here $\alpha' = 1$).

The equations of motion and the constraints are

$$\Box^2 X^A + \Gamma^A_{BC}(X) \partial_\alpha X^B \partial_\beta X^C g^{\alpha\beta} = 0 \tag{16}$$

$$T_{\alpha\beta} = G_{AB} \partial_\alpha X^A \partial_\beta X^B - \tfrac{1}{2} g_{\alpha\beta} \partial_\sigma X^A \partial^\sigma X^B = 0 \tag{17}$$

$\Gamma^A_{BC}(X)$ stand for the Christoffel connections of the metric G_{AB}. As is well known, the action (14) is invariant under reparametrizations of the world-sheet coordinates

$$\delta X^\mu = \varepsilon^\alpha(\xi) \partial_\alpha X^\mu(\xi), \qquad \xi^\alpha = (\sigma, \tau)$$

under which the two-dimensional metric $g_{\alpha\beta}$ goes to

$$\delta g_{\alpha\beta}(\xi) = \varepsilon^\gamma \partial_\gamma g_{\alpha\beta} + (\partial_\alpha \varepsilon^\alpha) g_{\alpha\beta} + (\partial_\beta \varepsilon^\gamma) g_{\alpha\beta}$$

This reparametrization invariance allows us to choose $g_{\alpha\beta}$ in the conformal form (so-called "conformal gauge"), namely,

$$\begin{aligned} g_{\alpha\beta} &= \Lambda(\xi) \eta_{\alpha\beta} \\ d\mathcal{S}^2 &= \Lambda(\sigma, \tau)(d\sigma^2 - d\tau^2) \end{aligned} \tag{18}$$

The choice of the conformal gauge still allows the reparametrizations

$$\begin{aligned} x_+ + \varepsilon &= f(x'_+ + l) \\ x_- + \varepsilon &= g(x'_- + l) \end{aligned} \tag{19}$$

(where $x_\pm = \sigma \pm \tau$, $x'_\pm = \sigma' \pm \tau'$ and we have introduced the constants ε, l whose meaning will be clear in the sequel). In this way, we can choose the light-cone gauge:

$$V \equiv X_0 + X_1 = p_+ x_+ \tag{20}$$

where the proportionality constant $p_+ > 0$ is the momentum of the center of mass of the string.

An important step in field as well as string quantization is the definition of positive frequency states and its associated ground state. It is by now well known, in QFT (in either flat or curved space-times) that "canonical states" for different coordinate systems are physically different (each time-like vector field leads to a separate indication of what constitutes a positive frequency) [1, 2]. There are different ways (and there is thus an ambiguity) by choosing such a basis. This makes it possible for a given field theory to have different alternative well-defined Fock spaces (different "sectors" of the theory). To illustrate this feature in string theory, let us consider the simplest case $G_{AB} = \eta_{AB}$, i.e.,

$$dS^2 = -dX^{0^2} + dX^{1^2} + \cdots + dX^{D-1^2} \tag{21}$$

which means both that the space-time is flat and that the string is described in an inertial frame. Usually the string is described in this frame where

$$\partial_{x+}\partial_{x-}X^A = 0, \qquad A = 0, 1, \ldots, D-1 \tag{22}$$

and then positive frequency modes are defined with respect to the inertial time X^0. In the light-cone gauge, X^0 is proportional to τ and therefore the modes

$$\overleftarrow{\varphi}_n = \frac{1}{2(\pi|n|)^{1/2}}\, e^{-inx_+}, \qquad \vec{\varphi}_n \frac{1}{2(\pi|n|)^{1/2}}\, e^{inx_-} \tag{23}$$

define the (inertial) particle states of the string.

Obviously, if we consider the (σ', τ') parametrization of the string world sheet [equation (19)], i.e.,

$$\begin{aligned} \sigma' + l &= \tfrac{1}{2}[G(\sigma + \tau + \varepsilon) + F(\sigma - \tau + \varepsilon)], \qquad & F \equiv f^{-1} \\ \tau' &= \tfrac{1}{2}[G(\sigma + \tau + \varepsilon) - F(\sigma - \tau + \varepsilon)], \qquad & G \equiv g^{-1} \end{aligned} \tag{24}$$

for which we have

$$\partial_{x'_+}\partial_{x'_-}X^A = 0 \tag{25}$$

positive frequency modes with respect to τ', namely,

$$\overleftarrow{\phi}_n = \frac{1}{2(\pi|n|)^{1/2}} e^{-i\lambda_n x'_+}, \qquad \overrightarrow{\phi}_n = \frac{1}{2(\pi|n|)^{1/2}} e^{i\lambda_n x'_-} \tag{26}$$

do not define positive frequency modes with respect to X^0, i.e., are not inertial particle states of the string. However, they are positive frequency modes with respect to another (accelerated) time $X^{0\prime}$ defined by

$$\begin{aligned} X_1 - X_0 &= p - f(X'_1 - X'_0) \\ X_1 + X_0 &= p + g(X'_1 + X'_0) \end{aligned} \tag{27}$$

where f is the same as in equation (19). This corresponds to the description of the string in an accelerated reference frame $\{X'_0, X'_1, \ldots, X'_{D-1}\}$, with acceleration

$$\tilde{a} = \frac{1}{(f'g')^{1/2}} \partial_{X'_1}[\ln(f'g')]$$

and metric

$$\begin{aligned} dS^2 = {} & p_+p_- f'(X'_1 - X'_0) g'(X'_1 + X'_0)(-dX'^2_0 + dX'^2_1) \\ & + dX'^2_2 + \cdots + dX'^2_{D-1} \end{aligned} \tag{28}$$

where, for simplicity, we have taken

$$X'_i = X_i \qquad (i = 2, \ldots, D-1)$$

For the description of the string in this frame we can always choose

$$V \equiv X'_1 + X'_0 = x'_+ + l \tag{29}$$

as it follows from equations (19), (20), and (27). Thus, equation (25) defines the accelerated particle states of the string. [The choice (29) is particularly useful here because of the acceleration points in the X^1 direction.] In this frame, the equations of motion of the string are

$$\begin{aligned} &\partial_{x'_+}\partial_{x'_-} U' + \frac{f''}{f'} \partial_{x'_+} U' \partial_{x'_-} U' + \left(\frac{f''}{f'} - \frac{g''}{g'}\right) \partial_{x'_+} U' \partial_{x'_-} V' = 0 \\ &\partial_{x'_+}\partial_{x'_-} V' + \frac{g''}{g'} \partial_{x'_+} V' \partial_{x'_-} V' - \left(\frac{f''}{f'} - \frac{g''}{g'}\right) \partial_{x'_+} U' \partial_{x'_-} V' = 0 \end{aligned} \tag{30}$$

$$\partial_{x'_+}\partial_{x'_-} X'^i = 0 \tag{31}$$

The equations of motion of the physical (transverse degrees) of freedom are the same in both frames. The equations of motion of the longitudinal coordinates (U', V') are different, but it can be shown [3] that, as in the inertial case [4], they can be eliminated in terms of the transverse coordinates by using the constraints

$$
\begin{aligned}
T'_{\pm\pm} &= f'(U')g'(V')\partial_{\pm'}U'\partial_{\pm'}V' + \partial_{\pm'}X^i\partial_{\pm'}X^i \approx 0 \\
T'_{+-} &= 0
\end{aligned}
\tag{32}
$$

The Poincaré invariance of flat space-time has a nonlinear realization in terms of the accelerated coordinates, but an explicit realization of the Poincaré algebra can be constructed in terms of the accelerated modes provided (at the quantum level) one sets $D = 26$ and uses symmetric ordering of the operators [3].

The manifold (σ', τ') defined by equation (24) is the convenient world-sheet parametrization for an accelerated string or a string in an accelerated space-time. This is true in either flat or curved space-time [although in the last case, equation (25) can only be valid asymptotically for $x'_{\pm} \to \pm\infty$].

Let us describe the string in the reference frames (21) and (28) referred to in the sequel as (I) and (A) respectively. We consider here a closed string and take for simplicity $f \equiv g$. The boundary conditions are

$$
\begin{aligned}
X^\mu(0, \tau) &= X^\mu(\pi, \tau) \\
X^\mu(0, \tau') &= X^\mu(L_\varepsilon, \tau')
\end{aligned}
\tag{33}
$$

Notice that

$$
0 < \sigma < \pi, \qquad -\infty < X^0 < +\infty
$$
$$
-\infty < \tau < +\infty \tag{34}
$$

In order to have

$$
0 < \sigma' < L_\varepsilon, \qquad -\infty < X'^0 < +\infty
$$
$$
-\infty < \tau' < +\infty \tag{35}
$$

appropriate boundary conditions on the mappings must be imposed. Condition (33) is automatically satisfied from equations (19) with

$$
\begin{aligned}
L_\varepsilon &= F(\varepsilon + 2\pi) - F(\varepsilon), \qquad F \equiv f^{-1} \\
l &= F(\varepsilon)
\end{aligned}
\tag{36}
$$

Condition (35) on τ' and X'^0 is not particular to strings. It is required to get a consistent quantization (of fields or strings) in accelerated manifolds and to have a complete in (out) basis [2]. On the other hand, the manifold A can cover only a (bounded) region (namely, $|X^1| > |X^0|$) of the original (I) one. Similarly, the manifold (σ', τ') can cover only a domain $(|\sigma| > \tau)$

of the inertial or global world sheet. All these considerations are satisfied by taking mappings such that

$$u_{\pm} = f(\pm\infty) \tag{37}$$

where u_+ (u_-) are constants that can take independently finite or infinite values and $u_+ > u_-$. This means that the inverse mapping $F \equiv f^{-1}$ has singularities at $u = u_{\pm}$, i.e.,

$$F(u_{\pm}) = \pm\infty \tag{38}$$

Singularities of these mappings describe the asymptotic regions of the space-time. Critical points of f, i.e.,

$$f'(\pm\infty) = 0 \tag{39}$$

describe event horizons at $u = u_-$ and $u = u_+$. In this case u_- (u_+) are finite and the manifold A cover the region $u_- < |X^1 \pm X^0| < u_-$ of the global (I) space-time. If $u_{\pm} = \pm\infty$, there are no horizons. In this approach, the well-known Rindler's space corresponds to the mapping

$$f = e^{\alpha U'}, \qquad \alpha = \text{const} \tag{40}$$

In this case, $u_- = 0$ and $u_+ = +\infty$; $u = u_- = 0$ is an event horizon and the manifold covers the right-hand wedge $|U| > 0$. (In order to cover the whole plane, four Rindler's patches are needed.) For the string world sheet we have

$$x_{\pm} + \varepsilon = e^{\alpha(x'_{\pm} + l)} \tag{41}$$

and we see how the presence of the constant ε is necessary in order to have a finite string period in the manifold (σ', τ'). For the Rindler's mapping (41) we have

$$l = \frac{1}{\alpha} \log \varepsilon$$

$$L_{\varepsilon} = \frac{1}{\alpha} \log\left(\frac{2\pi}{\varepsilon} + 1\right) \tag{42}$$

For $\varepsilon \to 0$, there is a stretching effect of the string due to the presence of an event horizon in the world sheet.

Let us now consider the string quantization in the frames (I) and (A). The string coordinates can be split into left and right movers as

$$X^A(\xi) = \tilde{X}^A(\xi_+) + \vec{X}^A(\xi_-) \tag{43}$$

where $\xi_\pm$ stands for $\sigma \pm \tau$ or $\sigma' \pm \tau'$. As usual, left and right modes of the closed string decouple both in the classical and quantum theories. We consider first the left movers. The wave modes are periodic functions of x_+ and x_- with period 2π. The left modes which are only functions of x_+ will be periodic functions of $x'_+ = \sigma' + \tau'$ with period L_e. In the inertial frame (I) we have

$$\begin{aligned} \tilde{X}^i_{(x_+)} &= \frac{q^i}{2} + \frac{p^i}{2} x_+ + \frac{1}{2} \sum_{n\neq 0} \frac{\tilde{a}^i_n}{\sqrt{|n|}} e^{-inx_+} \\ \tilde{U}_{(x_+)} &= q_- + p_- x_+ + \frac{i}{2} \sum_{n\neq 0} \frac{\tilde{u}_n}{n} e^{-inx_+} \\ p_- &= u_0/2 \end{aligned} \tag{44}$$

The independent dynamical variables $\tilde{a}^i_n$ $(n \in z)$, q_i, q_+, and p_+ obey the canonical commutation relations

$$\begin{aligned} &[\tilde{\alpha}^i_n, \tilde{\alpha}^j_m] = n\delta_{n+m}\delta^{ij}, \qquad \tilde{\alpha}^i_n = -i\sqrt{|n|}\,\mathrm{sgn}\,\tilde{a}^i \qquad n \neq 0 \\ &[q^i, \tilde{\alpha}^j_0] = i\delta^{ij}, \qquad \tilde{\alpha}^i_0 = p^i \\ &[q_-, p_-] = i \end{aligned} \tag{45}$$

Then

$$[\tilde{X}^i(x_+), \tilde{p}^i(y_+)] = \frac{\delta^{ij}}{2} \delta(x_+ - y_+)$$

The energy-momentum tensor, the conformal generators and the mass operator are

$$\begin{aligned} T_{++}(x_+) &= \sum_{n\in z} \frac{L_n}{2\pi} e^{-inx_+} \\ L_n &= \tfrac{1}{2}\sum_m \tilde{\alpha}^i_m \tilde{\alpha}^i_{n-m} + p_+ \tilde{u}_n \\ L_0 &= \tfrac{1}{2}\sum_m \tilde{\alpha}^i_m \tilde{\alpha}^i_{-m} + 2p_+p_- \end{aligned} \tag{46}$$

and

$$\begin{aligned} M^2 = p_+p_- - p_i^2 &= \sum_n \frac{n}{2}(a^{i+}_n a^i_n + a^i_n a^{i+}_n) \\ &= \sum_n n a^{i+}_n a^i_n - \frac{(D-2)}{24} \end{aligned} \tag{47}$$

In the accelerated frame (A) we have

$$\overleftarrow{X}^i(x'_+) = \frac{Q^i}{2} + \frac{P^i}{2} x'_+ + \sum_{n \neq 0} \overleftarrow{C}^i_n \overleftarrow{\phi}_n(x'_+)$$
$$\overleftarrow{U}(x'_+) = u_0 + \frac{P_-}{2} x'_+ + \frac{i}{2} \sum_{n \neq 0} \frac{\gamma_n}{n} e^{-i\lambda_n x'_+} \tag{48}$$

with

$$\overleftarrow{X}(x'_+ + L_\varepsilon) - \overleftarrow{X}(x'_+) = \frac{P^i}{2\pi} L_\varepsilon$$
$$\overleftarrow{U}(x'_+ + L_\varepsilon) - \overleftarrow{U}(x'_+) = \frac{P_-}{2} L_\varepsilon \tag{49}$$

and

$$P^i = \frac{2\pi}{L_\varepsilon} pi, \qquad P_- = \frac{2\pi}{L_\varepsilon} p_-$$

The left modes provide the T'_{++} component of the energy-momentum tensor. The accelerated conformal generators (L'_n) follow from the Fourier transform of T'_{++}:

$$T'_{++}(x'_+) = \sum_{-\infty}^{\infty} \frac{L'_n}{L_\varepsilon} e^{-i\lambda_n x'_+}$$
$$L'_n = L'^{\perp}_n + L'^{\parallel}_n \approx 0 \tag{50}$$
$$L'^{\perp}_n = -\frac{1}{L_\varepsilon} \sum_m \overleftarrow{\beta}^i_m \overleftarrow{\beta}^i_{n-m}$$

$$\overleftarrow{C}^{i+}_n = i \operatorname{sgn} \overleftarrow{\beta}^i_n / \sqrt{|n|}, \qquad \overleftarrow{\beta}^i_n = \overleftarrow{\beta}^i_{-n}, \qquad \beta_0 = p^i, \qquad \gamma_0 = p_-$$

We find the following Fock representation

$$[\overleftarrow{C}^i_m, \overleftarrow{C}^j_m] = \delta_{nm}\, \delta^{ij}, [\overleftarrow{\beta}^i_n, \overleftarrow{\beta}^j_m] = n\delta_{n+m,0}\, \delta^{ij}$$
$$[P^i, Q^j_0] = -\left(\frac{2\pi}{L_\varepsilon}\right) i\, \delta^{ij} \tag{51}$$
$$[p_+, u_0] = -i$$

which implies

$$[\overleftarrow{X}^i(\xi), \overleftarrow{P}^j(\xi')] = \frac{i}{2} \delta(\xi - \xi')\, \delta^{ij} \tag{52}$$

For a string whose center of mass follows an accelerated world line in the longitudinal direction X^1, of the type

$$x^1 \pm x^0 \pm a_\pm = A_\pm\, e^{\pm\alpha\tau'}, \qquad a_\pm, A_\pm = \text{const}$$

it is convenient to use the expansion [3]

$$U = U_0 + \frac{1}{f(x'_+)}\left[\tilde{p}_- + \frac{i}{2}\sum_{n\neq 0} r_n\, e^{-i\lambda_n x'_+}\right] \tag{53}$$

for which the mass formula is given by

$$\begin{aligned} M'^2 = (\alpha L_\varepsilon)^2 p_+\tilde{p}_- - p_i^2 &= \sum_1 \frac{n}{2}(C_n^{i+}C_n^i + C_n^i C_n^{i+}) \\ &= \sum_1 nC_n^{i+}C_n^i - \frac{(D-2)}{24} \end{aligned} \tag{54}$$

The operators M^2 and M'^2 [equations (47) and (54)] are different but both have the same eigenvalues.

The same treatment applies to the right movers $\vec{X}^A$ in a gauge defined by

$$\vec{U}' = p_- x_-$$

and with

$$\vec{U}' = x'_- + l$$

i.e.,

$$x_- + \varepsilon = f(x'_- + l)$$

The (I) and (A) Fock representations are related by a Bogoliubov transformation

$$C_m^i = p^i O_m + \sum_{n=1}^{\infty}(A_{mn}a_n^i + B_{mn}a_n^{i+}), \qquad m \neq 0 \tag{55}$$

with Bogoliubov coefficients

$$\begin{aligned} A_{mn} &= \langle \phi_{\lambda m}, \varphi_n\rangle = \frac{1}{2\pi}\left(\frac{n}{m}\right)^{1/2}\int_0^{2\pi} e^{in\sigma - i\lambda m F(\sigma+\varepsilon)}\, d\sigma \\ B_{mn} &= \langle \phi_{\lambda m}, \varphi_n^*\rangle = \frac{1}{2\pi}\left(\frac{n}{m}\right)^{1/2}\int_0^{2\pi} e^{in\sigma + i\lambda m F(\sigma+\varepsilon)}\, d\sigma \end{aligned} \tag{56}$$

and zero-mode contribution

$$O_m = \left\langle \frac{1}{2}\left(x_+ - \frac{2\pi}{L_\varepsilon}x'_+\right), \phi_{\lambda m}\right\rangle$$

The ground state of the string is defined by

$$\begin{aligned} a_n^i|0\rangle &= 0, \qquad \forall n > 1 \\ p^i|0\rangle &= 0 \end{aligned} \tag{57}$$

We also have

$$\begin{aligned} C_m|0'\rangle &= 0, \qquad \forall m > 1 \\ P^i|0'\rangle &= 0 \end{aligned} \tag{58}$$

but $C_m|0\rangle \neq 0$. The expectation value of the accelerated number operator in the ground state $|0\rangle$ is equal to (not sum over i)

$$N^i(\lambda m) \equiv \langle 0|C^i_m C^{i+}_m|0\rangle = N(\lambda m) = \sum_{n=1}^{\infty} |B_{mn}|^2 \tag{59}$$

which is i-independent. The expectation value of the accelerated M'^2 operator is equal to

$$M'^2 = 24 \sum_1^{\infty} nN(n) - 1 \tag{60}$$

where $N(n)$ is given by equation (59) with $\lambda_n = (2\pi/L_\varepsilon)n$. If $f(x'_+ + l) = e^{\alpha(x'_+ + l)}$, then

$$\begin{aligned} B_{mn}(n \gg 1) &= \mathscr{C}(\lambda m)\frac{n^{-i\lambda m/\alpha}}{\sqrt{n}} + \frac{b}{\sqrt{n}} + O\left(\frac{1}{n^{3/2}}\right) \\ \mathscr{C} &= \frac{i}{2\pi}\frac{e^{-\pi\lambda m/2\alpha}}{\sqrt{m}}\sqrt{\left(1 + i\frac{\lambda m}{\alpha}\right)} \\ b &= -\frac{i}{2\pi}\left(\frac{2\pi}{\sqrt{m}}\right)^{i\lambda m/\alpha} \end{aligned} \tag{61}$$

and

$$\begin{aligned} N(\lambda m) &= \frac{1}{(e^{2\pi\lambda m/\alpha} - 1)} + \frac{\alpha}{2\pi\lambda m} \\ M'^2 &= 24\left[\sum_{n=1}^{\infty} \frac{n}{(e^{n/T_0} - 1)} + T_0\right] - 1 \end{aligned} \tag{62}$$

$$M'^2_{\varepsilon \ll 1} = \frac{1}{2\pi}\ln\frac{2\pi}{\varepsilon} \tag{63}$$

[Recall that in field theory, $B_{\lambda k}\ (k \gg 1) = \mathscr{C}(\lambda)(k/\sqrt{k})^{-i\lambda/\alpha}$, with $\lambda \in R$.] $N(\lambda_n)$ [equation (62)] is equal to a Planckian spectrum at the Hawking-Unruh temperature

$$T_s = \frac{\alpha}{2\pi} \tag{64}$$

In $N(n)$, the temperature is

$$T_0 = \left(\frac{L_\varepsilon}{2\pi}\right) T_s = \frac{a}{2\pi} \underset{\varepsilon \ll 1}{=} \frac{1}{(2\pi)^2} \ln\left(\frac{2\pi}{\varepsilon}\right) \tag{65}$$

The expectation value of M'^2 in the ground state $|0\rangle$ is *positive.* The tachyon level of this state is filled up with accelerated modes following a thermal distribution. For any other mapping f different from the exponential one [equation (41)], B_{nm} $(n \gg 1) \to e^{-n}$ and $N(\lambda_n)$ and $N(n)$ are nonthermal.

It can be noted that the acceleration as well as the temperature T_s appearing in $N(n)$ are rescaled by the factor $(L_\varepsilon/2\pi)$, i.e., the ratio of the string period in the inertial and in the accelerated frame. Thus, the dimensionless temperature T_0 does not depend on the acceleration parameter α of Rindler's space, but on the parameter

$$a = \left(\frac{L_\varepsilon}{2\pi}\right) \alpha = \frac{1}{2\pi} \ln\left(\frac{2\pi}{\varepsilon} + 1\right) \underset{\varepsilon \ll 1}{=} -\frac{l_\alpha}{2\pi} \tag{66}$$

We see that the parameter ε $[l \equiv l(\varepsilon)]$ introduced in the mappings on the world sheet, plays a fundamental role in the string context. ε acts as a regulator or cutoff to avoid the presence of an event horizon in the world sheet and to get a finite string period. Its magnitude is of the order of the Planck length.

6. TRANSFORMATION BETWEEN THE STATES. DERIVATION OF THE PARAMETER ε FROM THE CONSTRAINT EQUATIONS

We see that the transformations (19) are not without consequences but change the ground state and in general, the quantum states, except for mappings belonging to the $O(2, 1)$ group. Such invariance mappings are the Möbius or bilinear transformations. Under the transformations (19) and (27), the states transform as

$$|\ \rangle \to |'\rangle = e^{i\hat{G}}|\ \rangle \tag{67}$$

where $a_n|\ \rangle = 0$, $C_n|'\rangle = 0$, $\forall_n$ and

$$\hat{G} = \sum_n \theta_n (C_n C_n - C_n^+ C_n^+)$$

[This is an operatorial representation for the Bogoliubov transformation, equation (55), with Bogoliubov coefficients $\cosh \theta_n$ and $\sinh \theta_n$.]

The vacuum expectation value of $T_{\mu\nu}$ transforms as

$$\langle T_{\mu\nu}\rangle = \langle T'_{\mu\nu}\rangle + \Theta_{\mu\nu} + P_{\mu\nu} \tag{68}$$

$P_{\mu\nu}$ is any conserved traceless tensor taking into account the dependence of $\langle T_{\mu\nu}\rangle$ on the quantum state. It represents the nonlocal part of $T_{\mu\nu}$. $\Theta_{\mu\nu}$ depends on the mapping and represents the local part. In the conformal gauge we have

$$P_{--} = \mathcal{U}_-(x'_-), \qquad P_{++} = \mathcal{U}_+(x'_+), \qquad P_{+-} = P_{-+} = 0 \tag{69}$$

$$\begin{aligned} \Theta_{--} &= -\frac{1}{12\pi}\sqrt{f'}\, d^2_{x'_-}\left(\frac{1}{\sqrt{f'}}\right) \\ \Theta_{++} &= -\frac{1}{12\pi}\sqrt{g'}\, d^2_{x'_+}\left(\frac{1}{\sqrt{g'}}\right) \\ \Theta_{+-} &= \Theta_{-+} = 0 \end{aligned} \tag{70}$$

$\mathcal{U}_-$ and $\mathcal{U}_+$ are arbitrary functions of the indicated variables. Equations (69) derive from the conditions $\nabla^\nu P_{\mu\nu} = 0$ and $P^\nu_\nu = 0$. Therefore, the constraints

$$\langle T_{\mu\nu}\rangle = 0, \qquad \langle T'_{\mu\nu}\rangle = 0$$

yield to the equations [5]

$$\begin{aligned} \sqrt{f'}\, d^2_{x'_-}\left(\frac{1}{\sqrt{f'}}\right) - 12\pi\mathcal{U}_-(x'_-) &= 0 \\ \sqrt{g'}\, d^2_{x'_+}\left(\frac{1}{\sqrt{g'}}\right) - 12\pi\mathcal{U}_+(x'_+) &= 0 \end{aligned} \tag{71}$$

These are zero-energy Schrödinger equations

$$\frac{d^2}{dx'_\pm}\psi_\pm - 12\pi\mathcal{U}_\pm(x'_\pm)\psi_\pm = 0 \tag{72}$$

for the mappings f and g. By giving the potentials $U_\pm$, equations (72) determine the wave functions

$$\psi_- = 1/\sqrt{f'}, \qquad \psi_+ = 1/\sqrt{g'} \tag{73}$$

Because $\mathcal{U}_\pm$ are arbitrary functions [compatible with the boundary conditions (37)], equations (72) do not yield additional constraints on the mappings f and g, but a way of connecting the mapping to a potential problem.

The first term of equation (71) is the Schwarzian derivative of f:

$$D[f] = \frac{f'''}{f'} - \frac{3}{2}\left(\frac{f''}{f'}\right)^2 \tag{74}$$

which is invariant under the Möbius or bilinear transformations. Under these transformations, f becomes a new function but $D[f]$ is invariant determining the same ground state of the string. In particular, $\mathcal{U}_+ = \mathcal{U}_- = 0$ determine $f(g)$ as

$$f = \frac{\alpha x'_- + \beta}{\gamma x'_- + \delta}, \qquad (\alpha\delta - \beta\gamma) = 1 \tag{75}$$

The ground state defined by this mapping can be considered as a reference or "minimal" state at zero temperature with respect to which other ground states corresponding to nonzero potentials $\mathcal{U}_\pm$ appear as excited or thermal ones. If $\mathcal{U}_+ = \mathcal{U}_- = \mathcal{U}_0 = \text{const} > 0$, then

$$\psi_\pm = A\, e^{-Kx'_\pm}, \qquad f = \frac{1}{\alpha A^2}\, e^{\alpha x'_+} - \varepsilon \tag{76}$$

where A is a normalizing constant (we choose $A^2 = e^{-\alpha l}/\alpha$) and K is the zero-energy transmission coefficient

$$K \equiv \alpha/2 = (12\pi\mathcal{U}_0)^{1/2}$$

For $\varepsilon \to 0$, the mapping (76) defines an event horizon at $x_\pm = 0$ ($x'_\pm = -\infty$) and carries an intrinsic temperature $T_s = \alpha/2\pi$, as can be seen by putting $t = i\tau$ ($x_+ = x + i\tau$) and then $0 \le \tau' \le 2\pi/\alpha$. The temperature appears related to the height of the potential, namely,

$$T_s = (12\mathcal{U}_0/\pi)^{1/2}$$

and the parameter ε arises naturally as an integration constant. This temperature T_s characterizes the spectrum $N(\lambda_n)$ [equation (62)].

Similar equations to (71) also appear in the so-called "back reaction problem" in two dimensions [6] [as a consequence of the ($\pm\,\pm$) components of the semiclassical Einstein equations, the ($+\,-$) component giving rise to the Liouville equation for the geometry because of the conformal anomaly]. Equations (71) can also be derived in the context of conformal field theories on higher genus Riemann surfaces (the potential playing the role of the zero-point energy) in connection with the approach of Ref. 7.

We have shown that with appropriate boundary conditions, the holomorphic mappings of reparametrization invariance of string theory can be interpreted as a change of coordinate frame in the space-time in which strings are embedded. These mappings change the ground state in the quantized theory except for transformations belonging to the $O(2, 1)$ group. This allowed us to discuss in a systematic way the Hawking–Unruh effect in string theory. The transformations describing the world sheet of an accelerated string need the introduction of an additional parameter ε with respect to those describing the trajectories of accelerated point particles.

The results found here apply also to curved space-time. For the most important metrics in general relativity, the presence of isometry groups allows the maximal analytic extension of the (D-dimensional) manifold to be performed through the extension of a relevant two-dimensional manifold containing the time axis and a suitable spatial coordinate. This maximal analytic extension is performed by mappings like equation (27), where X_1, X_0 are Kruskal (maximal) type coordinates and X_1', X_0' are of the Schwarzschild type. (For the role played by these mappings in the context of QFT, see Refs. 8 and 9.)

7. STRINGS NEAR BLACK HOLES

Our investigation of strings in Rindler space-time can be applied to the case of strings in a black-hole background. Black-hole solutions of Einstein equations exist in D-space-time dimensions ($D \geq 4$) [10]. These solutions are asymptotically flat and generalize the Schwarzschild space-time of four dimensions; they have the metric

$$dS^2 = -\left(1 - \frac{C}{R^{D-3}}\right) dT^2 + \frac{dR^2}{1-(C/R^{D-3})} + R^2\, d\Omega_{D-2}^2 \qquad (77)$$

R is the radial coordinate, $d\Omega_D^2$ is the line element on the unit D-sphere, and the constant C is >0. The surface

$$R = C^{1/D-3} \equiv R_S$$

is an event horizon (there are both past and future event horizons) and $R = 0$ is a spacelike singularity. The horizon radius R_S is related to the black-hole mass M by

$$C = 16\pi G \frac{M}{(D-2)A_{D-2}}$$

where

$$A_{D-2} = \frac{2\pi^{(D-1)/2}}{\Gamma((D-1)/2)}$$

is the area of a unit $(D-2)$ sphere and G has dimensions of (length^{D-2}).

The mass and the surface gravity K of the black hole are related by

$$K = \left(\frac{D-3}{2R_S}\right) = \frac{(D-3)}{2}\left[\frac{(D-2)A_{D-2}}{16\pi GM}\right]^{1/(D-3)} \qquad (78)$$

For $D = 4$ this yields the standard relations $R_S = 2GM$ and $K = 1/(4GM)$.

The Kruskal extension of this Schwarzschild manifold is given by the mapping

$$r_K \pm t_K = e^{K(R^* \pm T)} \qquad (79)$$

where

$$R^* = R + R_S^{D-3} \int \frac{dR}{(R^{D-3} - R_S^{D-3})}$$

$$-\infty \le R^*, \qquad T \le +\infty$$

This is the same exponential mapping as the Rindler transform with K instead of α. The Rindler coordinates are similar to the Schwarzschild (R^*, T) ones and the Minkowskian coordinates are similar to the Kruskal (global) coordinates (r_K, t_K). The event horizon $R = R_S$ corresponds to $R^* = -\infty$. As discussed above, a large cutoff $(H \gtrsim 1/K)$ is needed in the negative Schwarzschild coordinate R^*. This shifting of the horizon is equivalent to considering a shifting ε in the mapping

$$r_K \pm t_K + \varepsilon = e^{K(R^* \pm T)} \tag{80}$$

with

$$\varepsilon = e^{-KH} \sim l_p^2$$

and thus

$$-H \le R^* \le +\infty$$

reflecting the fact that a classical description of the geometry is no longer valid at distances of order of the Planck length. We will take

$$\varepsilon = \frac{\lambda_C}{R_S}$$

where

$$\lambda_C = 1/M$$

is the Compton length of the black hole (here $\hbar = c = 1$). Thus the shifting of the horizon is $H = (1/K) \ln \varepsilon$, with

$$\varepsilon = \pi^{1/2} \left[\frac{(D-2)}{8\Gamma((D-1)/2)} \right]^{1/(D-3)} \left(\frac{M_{Pl}}{M} \right)^{(D-2)/(D-3)} \tag{81}$$

and

$$K = \frac{\pi^{1/2}}{2} (D-3) \left[\frac{(D-2)}{8\Gamma((D-1)/2)} \right] \left(\frac{M_{Pl}}{M} \right)^{1/(D-3)} M_{Pl} \tag{82}$$

Following on the same lines of argument discussed above, for the choice of gauge and parametrizations of the string world sheet and considering only left movers, we have

$$\hat{v} \equiv r_K + t_K = p_+ x_+ \tag{83}$$

$$V \equiv R^* + T = x'_+ + \frac{1}{K} \log p_+ \tag{84}$$

and

$$x_+ + \varepsilon = e^{Kx'_+}$$

(Here the longitudinal direction of the string is in the radial direction.)

Positive frequency modes, $\tilde{\varphi}_n$ with respect to the Kruskal time t_K and $\tilde{\phi}_n$ with respect to the Schwarzschild time T can be defined. The Schwarzschild frequency is equal to

$$\lambda n = \frac{2\pi}{L_\varepsilon} n, \qquad n = 1, 2, \ldots \tag{85}$$

where

$$L_\varepsilon = \frac{1}{K} \ln\left(\frac{2\pi}{\varepsilon} + 1\right)$$

For $\varepsilon \ll 1$, the frequency spacing tends to zero reflecting the stretching effect of the string near the horizon as seen by a Schwarzschild external observer. Associated to the modes $\tilde{\varphi}_n$ and $\tilde{\phi}_n$ we will have Kruskal and Schwarzschild operators a_n and C_n, respectively, and a vacuum state defined by

$$a_n | O_K\rangle = 0, \qquad \forall_n > 1$$

On the other hand, in order to have a smooth Euclidean manifold from a black hole space-time with topology $R^2 \times S^{D-2}$, the Schwarzschild imaginary time iT must be identified with a period

$$\beta = 2\pi/K$$

Then the same periodicity in the imaginary time appears in the correlation functions of string coordinates, indicating that the string is in equilibrium with a heat bath at the Hawking temperature

$$T_S = K/2\pi \tag{86}$$

The same temperature T is recovered in the function $N(\lambda_m)$, i.e.,

$$N(\lambda_n) = \langle O_K | C^+_{\lambda n} C_{\lambda n} | O_K\rangle$$

which gives a Planckian distribution for the Schwarzschild modes but with a "filter" $|g(\lambda_n)|^2$ equal to the absorption cross section of the black hole. In the spectrum $N(n)$ in which frequency is measured in dimensionless units $1, 2, \ldots,$ the temperature of the Planckian distribution is equal to

$$T_0 = \frac{1}{4\pi^2} \ln\left(\frac{2\pi}{\varepsilon}\right) \gg 1 \tag{87}$$

One can consider different higher-dimensional black-hole space-times, namely, a 26-dimensional or a four-dimensional black hole with the extra 22 dimensions compactified in a torus [11]. Intermediary situations can also be envisaged, but it must be noticed that the qualitative properties of the string quantization will be the same since they depend upon the horizon structure in the two variables R, T (or X^0, X^1, for Rindler space). We hope to come back to this problem elsewhere.

It can be noticed that the Hagedorn temperature (T_m) in this context is

$$T_m = \frac{1}{\sqrt{\alpha'}} \sim M_{\rm Pl}$$

Then,

$$\frac{T_S}{T_M} \sim \left(\frac{M_{\rm Pl}}{M}\right)^{1/(D-3)}$$

and we have

$$T_S \lesssim T_M$$

since the basic requirement of the present semiclassical treatment is $M \gtrsim M_{\rm Pl}$.

8. NEW APPROACH TO STRING QUANTIZATION IN CURVED SPACE-TIMES

The main feature of strings propagating in a curved space-time is that the equations of motion [equation (16)] are nonlinear in X_A, so right and left movers interact with each other and also with themselves. It must be noticed that purely left modes (or right modes) are exact solutions of equation (16), namely,

$$X_A = q_A(\sigma + \tau) \quad \text{or} \quad X_A = q_A(\sigma - \tau) \tag{88}$$

When G_{AB} is the metric of a symmetric space, the equations of motion possess an associated linear system and exact solutions can be constructed, by using the inverse scattering method. In Ref. 12, we propose a general perturbative scheme to solve the equations of motion and constraints both classically and quantum mechanically. We start from an exact given solution of equations (16) and develop in perturbations around. A possible starting point is a solution for the center-of-mass motion of the string $q_A(\tau)$ where

$$\ddot{q}^A(\tau) + \Gamma^A_{BC}(q)\dot{q}^B(\tau)\dot{q}^C(\tau) = 0 \tag{89}$$

The world-sheet time variable τ is identified here with the proper time of the center-of-mass trajectory. Another possibility is to take pure left (right) mover solutions. Then we set

$$X_A(\sigma, \tau) = q_A(\sigma, \tau) + \eta_A(\sigma, \tau) + \xi_A(\sigma, \tau) + \cdots \tag{90}$$

Here $q_A(\sigma, \tau)$ is an exact solution of equation (16) and η_A is a solution of the linearized perturbation around q_A. That is,

$$\partial^2\eta^A + \Gamma^A_{BC}(q)(\partial_- q^B\partial_+ + \partial_+ q^B\partial_-)\eta^C + \partial_C\Gamma^A_{BD}(q)\partial_+ q^B\partial_- q^D\eta^C = 0 \quad (91)$$

$\xi_A(\sigma, \tau)$ fulfill the second-order perturbation equation around q_A:

$$\begin{aligned}
&\partial^2\xi^A + \Gamma^A_{BC}\partial_+\eta^B\partial_-\eta^C + \tfrac{1}{2}\eta^D\eta^E(\partial^\tau_{DE}\Gamma^A_{BC})\partial_+ q^B\partial_- q^C \\
&\quad + \eta^D(\partial_D\Gamma^A_{BC})(\partial_+ q^B\partial_- + \partial_- q^B\partial_+)\eta^C \\
&\quad + \Gamma^A_{BC}(\partial_+ q^B\partial_- + \partial_- q^B\partial_+)\xi^C + (\partial_D\Gamma^A_{BC})\partial_+ q^B\partial_- q^C\xi^D = 0 \quad (92)
\end{aligned}$$

One can consider the higher-order perturbations, but in this chapter we will restrict ourselves to first and second orders.

It must be noticed that we are treating the space-time metric *exactly* and taking the string oscillations around its center of mass $q_A(\tau)$ as perturbation. So, our expansion corresponds to low excitations of the string as compared with the energy scales of the metric G_{AB}. For example, this method is exact in flat space-time. In the Schwarzschild geometry, it will correspond to an expansion in ω/M where ω is the frequency mode and M the black-hole mass. Thus, our approximation applies to black-hole masses larger than the string energy. In other words, this corresponds to a strong gravitational field expansion. This can be equivalently considered as an expansion in powers of $\sqrt{\alpha'}$. Actually, since $\alpha' = l^2_{\text{Pl}}$ (l_{Pl} is the Planck length), the expansion parameter turns out to be the dimensionless constant

$$g \equiv \sqrt{\pi}\,\frac{l_p}{R_c} = \frac{1}{l_p M} \simeq \frac{\omega}{M}$$

where the length R_c characterizes the curvature radius of the space-time under consideration and M its associated mass (the black-hole mass in the Schwarzschild geometry, the mass of the universe in a cosmological model). In most of the interesting situations, one clearly has $g \ll 1$.

It must be noticed that even for small g, the metric and its derivatives may be very large in some regions of the space-time. This shows that our method has a *larger domain of applicability* than the background field method where one must have everywhere

$$|\sqrt{\alpha'}\partial_c G_{AB}(x)| \ll 1$$

The first-order equation describes the interaction between the string modes and the curved space-time geometry. The interactions between the string modes themselves start to appear from the second-order perturbation (ξ^A) equation.

The constraint equations must also be expanded in perturbations. We find up to terms higher than the second order:

$$T_{\pm\pm} = G_{AB}(q)\partial_{\pm}q^A\partial_{\pm}q^B + 2G_{AB}\partial_{\pm}q^A\partial_{\pm}\eta^B$$

$$+ \eta^C\partial_C G_{AB}(q)\partial_{\pm}q^A\partial_{\pm}q^B + \tfrac{1}{2}\eta^C\eta^D\partial_{CD}G_{AB}(q)\partial_{\pm}q^A\partial_{\pm}q^B$$

$$+ 2\eta^C\partial_C G_{AB}(q)\partial_{\pm}q^A\partial_{\pm}\eta^B + \xi^C\partial_C G_{AB}(q)\partial_{\pm}q^A\partial_{\pm}q^B$$

$$+ 2G_{AB}(q)\partial_{\pm}q^A\partial_{\pm}\xi^B + G_{AB}(q)\partial_{\pm}\eta^A\partial_{\pm}\eta^B \simeq 0$$

See Ref. 12, where we have applied our method to the case in which the exact solution q_A describes the center-of-mass motion, i.e., it fulfills

$$m^2 = G_{AB}(q)\dot{q}^A(\tau)\dot{q}^B(\tau)$$

This defines the (mass)^2 of the string. We have computed the mass spectrum and vertex operator in the de Sitter space. The lower mass states are the same as in flat space up to corrections of order g^2 [here $g = (\pi\alpha')^{1/2}/R$], whereas heavy states *deviate* significantly from the linear Regge trajectories. We find a *maximum* (very large) value of order $1/g^2$ for the quantum number N and spin J of particles. Application of this method to general cosmological models and to the Schwarzschild geometry will be reported elsewhere [13].

The critical dimension for bosonic strings is found to be *25* in de Sitter space time and *9* for the supersymmetric string.

The results obtained here for the de Sitter metric show that our perturbative method is well suited to investigate the quantum dynamics of strings in curved space-time fully taking into account strong curvature effects.

REFERENCES

1. See, for example, N. D. Birrell and P. C. W. Davies, *Quantum Fields in Curved Space*, Cambridge University Press, Cambridge, England, 1982, and references therein.
2. N. Sánchez, *Phys. Rev. D24*, 2100 (1981); N. Sánchez and B. F. Whiting, *Phys. Rev. D34*, 1056 (1986).
3. H. J. de Vega and N. Sánchez, CERN preprint TH. 4681 (1987); *Nucl. Phys. B 229*, 818 (1988).
4. P. Goddard, J. Goldstone, C. Rebbi, and C. B. Thorn, *Nucl. Phys. B 56*, 109 (1973).
5. N. Sánchez, CERN preprint TH. 4733 (1987); *Phys. Lett. 195B*, 160 (1987).
6. N. Sánchez, *Nucl. Phys. B 266*, 487 (1986).
7. M. Martellini and N. Sánchez, CERN preprint TH. 4680 (1987); *Phys. Lett. B 192B*(3, 4), 361 (1987).
8. N. Sánchez, in *Proceedings of the Second Marcel Grossmann Meeting* (R. Ruffini, ed.), North-Holland, Amsterdam, 1982, p. 501.

9. N. Sánchez and B. F. Whiting, in *Quantum Concepts in Space and Time* (C. Isham and R. Penrose, eds.), Oxford University Press, Oxford, 1985, pp. 319–324.
10. R. C. Myers and M. J. Perry, *Ann. Phys.* (*N.Y.*) *172*, 304 (1986).
11. R. C. Myers, *Phys. Rev. D35*, 455 (1987).
12. H. J. de Vega and N. Sánchez, CERN preprint TH.4784 (1987). *Phys. Lett. 197B*, 320 (1987).
13. H. J. de Vega and N. Sánchez, *Nucl. Phys. B 309*, 552 (1988); *309*, 577 (1988).

Chapter 14

Symmetries and Anomalies in Fermionic Theories

Fidel A. Schaposnik

1. INTRODUCTION

The path-integral approach to quantum field theory presents many advantages for the analysis and resolution of relevant questions in which symmetries (gauge and chiral symmetries, conformal symmetry, BRST symmetry) play a central role.*

Concerning chiral symmetry, after Fujikawa's [2] observation on the noninvariance of the fermionic measure under γ_5 rotations (whenever gauge invariance of the measure is imposed) many relevant two-dimensional models (like QCD_2, chiral Gross–Neveu, etc.) were solved using a kind of path-integral approach to bosonization.† This aspect will be discussed in the second part of this chapter, where I analyze the conformal invariant behavior of the chiral Gross–Neveu model at a particular value of the coupling constant.

I will begin by discussing another question, also connected to chiral symmetry, in which the path-integral approach has been shown to lead to

* For a review of the issues of symmetries in the path-integral framework, see, for example, Ref. 1.

† For a review on bosonization in the path-integral approach, see Ref. 3.

FIDEL A. SCHAPOSNIK • Departamento de Física, Universidad Nacional de La Plata, 1900 La Plata, Argentina.

interesting issues: the quantization of potentially anomalous gauge theories.*

2. QUANTIZATION OF GAUGE THEORIES WITH WEYL FERMIONS (OR: ARE ANOMALIES ANOMALOUS?)

In the quantization of theories with Weyl fermions (say left-handed for definiteness) coupled to gauge fields A_μ (in the Lie algebra of some gauge group G), it turns out that the main issue is the definition of the fermionic effective action $S_{\text{eff}}[A]$.

Since the fermionic Lagrangian L_F

$$L_F = \bar{\psi}(i\not{\partial} + \not{A})\frac{(1-\gamma_5)\psi}{2} \equiv \bar{\psi}D[A]\psi \tag{1}$$

is quadratic in the Fermi fields, one expects $S_{\text{eff}}[A]$ to be related to the Dirac operator $D[A]$ determinant in the usual way:

$$S_{\text{eff}}[A] \overset{?}{=} -\log \det D[A] \tag{2}$$

This expression suffers from two problems. First, $D[A]$ maps negative chirality spinors into positive chirality spinors and hence it does not have a well-defined eigenvalue problem. Second, $D[A]$ is an unbounded operator, and some regularization prescription has to be adopted in order to get a finite answer for det $D[A]$.

One usually rectifies the chirality flip problem by considering, instead of (1), the Lagrangian [5, 6]

$$L_F \to L_F = \bar{\psi}\left[i\not{\partial} + \not{A}\frac{(1-\gamma_5)}{2}\right]\psi \equiv \bar{\psi}\hat{D}[A]\psi \tag{3}$$

(i.e., one adds right-handed free fermions hoping that this amounts just to a normalization change in the generating functional).

Then, one adopts the definition

$$\det D[A] \equiv \det \hat{D}[A]|_{\text{reg}} \tag{4}$$

where $\hat{D}$ acts on Dirac fermions and then does define an eigenvalue problem; "reg" means that some regularization has to be adopted in order to make sense from the product of eigenvalues defining the determinant.

The solution of this problem has created a new one: under a gauge transformation $A_\mu \to A_\mu^g$

$$A_\mu^g = g^{-1}A_\mu g - ig^{-1}\partial_\mu g, \qquad g \in G \tag{5}$$

* For a review on anomalous theories, see Ref. 4.

$\hat{D}[A]$ does not transform like a covariant derivative:

$$\hat{D}[A^g] \neq g^{-1}\hat{D}[A]g \tag{6}$$

and, hence, one has in general

$$\det D[A] \neq \det D[A^g] \tag{7}$$

i.e., the Weyl fermion determinant is gauge dependent and so it is the effective action if defined as in (2). Consequently, the gauge current

$$J^a_\mu(x) = \frac{\delta S_{\text{eff}[A]}}{\delta A^a_\mu(x)} \tag{8}$$

is in general anomalous:

$$D_\mu J^\mu = \mathcal{A}[A] \neq 0 \tag{9}$$

Nevertheless, there are several proposals of consistently quantizing these anomalous theories [7, 8]. In fact, it has been recently shown [9] within the path-integral approach that it is possible to define an effective action that is gauge-invariant and leads to a nonanomalous theory. This can be done as follows: consider the generating functional defined as an integral over all fields:

$$\begin{aligned} Z &= \int DA_\mu D\bar{\psi}D\psi \exp\left[-\left(\tfrac{1}{4}\operatorname{tr}\int F^2_{\mu\nu}\,dx + S_F\right)\right] \\ &= \int DA_\mu D\bar{\psi}D\psi \exp(-S[A, \bar{\psi}, \psi]) \end{aligned} \tag{10}$$

If fermions are Dirac fermions, there is no harm in looking at Z as an integral over orbit space. Indeed, using the standard Fadeev–Popov procedure [10, 11] (see Ref. 12 for a geometrical framework), one ends with a Dirac fermion generating functional of the form

$$Z_{\text{Dirac}} = N \int DA_\mu \Delta[A]\delta(F[A]) D\bar{\psi}_D D\psi_D \exp(-S[A, \bar{\psi}, \psi]) \tag{11}$$

Here $F[A^*] = 0$ is the gauge condition selecting one representative over each orbit (a section choice) and $\Delta[A^*]$ is the corresponding Fadeev–Popov determinant, related to the natural metric over orbit space Γ and the scale of the orbit $\rho[A^*]$ through the formula [12]

$$\Delta[A^*] = \rho[A^*](\det\Gamma[A^*])^{1/2} \tag{12}$$

An integration over the volume element on the group of gauge transformations has been factored out and absorbed in the normalization constant N using gauge invariance of DA_μ, $D\bar{\psi}_D D\psi_D$, Δ, and S.

When Weyl fermions are present, this factorization does not take place. In fact, as is evident from (7), the fermionic determinant (i.e., the fermionic path integral) recognizes different points of any orbit (see Fig. 1).

Indeed, instead of (11) one has in the Weyl fermionic case

$$Z_{\text{Weyl}} = \int DA_\mu \Delta[A]\delta(F[A])DgD\bar{\psi}D\psi \exp(-S[A^g, \bar{\psi}, \psi]) \tag{13}$$

If one eliminates the g dependence from S by performing a fermionic change of variables,

$$\begin{aligned} \psi &\to \psi' = g^{-1}\psi \\ \bar{\psi} &\to \bar{\psi}' = \bar{\psi}g \end{aligned} \tag{14}$$

one has to take into account the Fujikawa Jacobian $J(g, A)$:

$$J(g, A) \exp(-S[A, \bar{\psi}', \psi'])D\bar{\psi}'D\psi' = \exp(-S[A^g, \bar{\psi}, \psi)D\bar{\psi}D\psi \tag{15}$$

where, owing to (7),

$$J(g, A) = \left.\frac{\det D[A^g]}{\det D[A]}\right|_{\text{reg}} \neq 1 \tag{16}$$

We then see that now the Dg integration does not factor out:

$$Z_{\text{Weyl}} = \int DA_\mu \Delta[A]\delta(F[A])J(g, A)DgD\bar{\psi}D\psi \exp(-S[A, \bar{\psi}, \psi]) \tag{17}$$

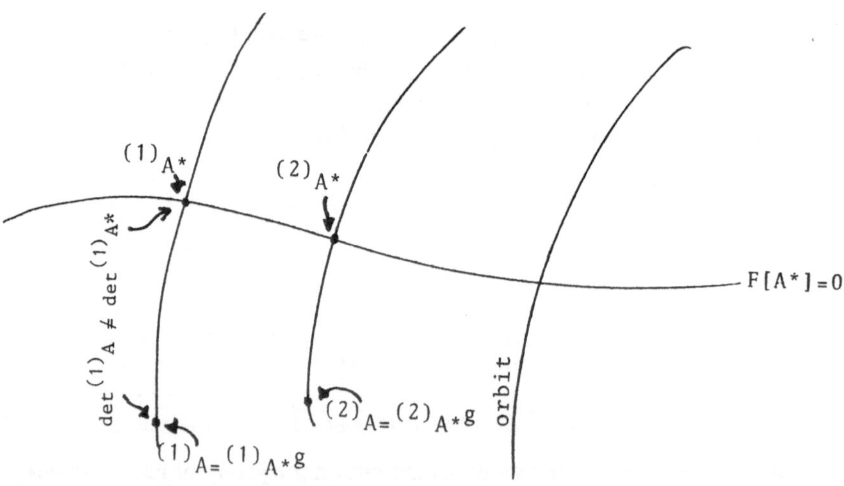

Figure 1. The fermionic path integral recognizes different points of any orbit when Weyl fermions are present.

Note that this expression makes contact with the proposal of Faddeev and Shatashvili [8]. These authors suggested the addition of a group-valued field, with a Wess–Zumino action in order to cure the anomalous behavior of the theory. In the present formulation the group-valued field was already present [it is the gauge-group field $g(x)$!] and the Wess–Zumino term is created by the Jacobian $J(g, A)$:

$$J(g, A) = \exp[-\mathrm{tr} \int_0^1 dt\, dx\, \mathcal{A}[A^{g(t)}]\theta \tag{18}$$

with $g(t)$ such that $g(0) = 1$ and $(1) = \exp[i\theta]$.

It is then natural to define the effective action in the form

$$S_{\text{eff}}[A] = \int DgD\bar{\psi}D\psi J(g, A) \exp\left(-\int dx\, \bar{\psi}D[A]\psi\right) \tag{19}$$

or, owing to (16),

$$S_{\text{eff}}[A] = \int Dg \det D[A^g] \tag{20}$$

Obviously, $S_{\text{eff}}[A]$ is gauge invariant:

$$S_{\text{eff}}[A^h] = S_{\text{eff}}[A], \qquad h \in G \tag{21}$$

and leads to a conserved current through (8).

It is interesting to note that in two-dimensional models this approach leads to a consistent, unitary, Lorentz invariant theory [13]. For example, in the case of chiral QCD_2 one has for the effective action:

$$\begin{aligned} S_{\text{eff}}[A] = \int Dg \exp\Bigg[&-(1+a)\frac{e^2}{16\pi}\int d^2x\, \mathrm{tr}\, A\!\!\!/^2 \\ &-\frac{e^2}{4\pi}\mathrm{tr}\int d^2x \int_0^1 dt \bigg(\gamma_5 \phi A\!\!\!/_t A\!\!\!/_t \\ &+ i\gamma_5\theta A\!\!\!/_t A\!\!\!/_t - \frac{1}{e}\gamma_5 \partial\!\!\!/ 0 A\!\!\!/_t\bigg)\Bigg] \end{aligned} \tag{22}$$

with [3, 13]

$$A\!\!\!/_t = \frac{i}{e}\exp(-\gamma_5\phi t)\partial\!\!\!/ \exp(\gamma_5\phi t) \tag{23}$$

Here a is a parameter related to regularization ambiguities [7].

From the analysis of (22), one concludes that the theory corresponds to $N^2 - 1$ massive scalars ϕ^a (with mass $m^2 = (e^2/16\pi)[a^2/(a-1)] \times (N+1)^{1/2}$) and the same number of massless excitations. One can also study current commutators which are nonanomalous. In summary, the model is consistent provided that $a > 1$.

Of course, a confirmation of this and other results using the Hamiltonian formalism would be welcomed (some results have in fact been presented by different authors [14]). Another approach that can be used to quantize chiral models is the stochastic quantization method.* It has been recently shown [16] within this framework that a very simple treatment of the chirality flip problem of $D[A]$ can be implemented.

Indeed, the Langevin equations governing the (fictitious time τ) evolution of fermion fields in the presence of a Gaussian noise θ

$$\begin{aligned}\frac{\partial\psi}{\partial\tau}(x,\tau) &= -\int dy\, K(x,y)\frac{\delta S}{\delta\bar{\psi}(y,\tau)} + \theta(x,\tau)\\ \frac{\partial\bar{\psi}}{\partial\tau}(x,\tau) &= -\int dy \frac{\delta S}{\delta\psi(x,\tau)} K(x,y) + \bar{\theta}(x,\tau)\end{aligned} \tag{24}$$

allow the presence of a Hermitian kernel K, which can be precisely used to solve the problem posed by Weyl fermions.

The natural choice for K when Dirac fermions are present is

$$\begin{aligned}K &= \not{D}[A]\delta(x-y)\\ \bar{K} &= \overleftarrow{\not{D}}[A]\delta(x-y)\end{aligned} \tag{25}$$

Now, if Weyl fermions are present, one can instead use [16]

$$K_{\text{Weyl}} = \not{D}_a \frac{(1+\gamma_5)}{2}\delta(x-y) \tag{26}$$

where the subscript a indicates the possibility of arbitrary regularization parameters in the operator $\not{D}_a$. Now, the right-handed projector in (26) ensures that the Langevin equations are consistent [in the sense that all terms in (24) have the same chirality]. The only requirement in $\not{D}_a$ is that it has to ensure convergence as $\tau \to \infty$. In Ref. 16 we have shown how this proposal leads to a consistent theory discussing a two-dimensional example. It is interesting to note that the present solution of the chirality-flip problem is related to the one presented by Singer [17].

I will finally mention a recent proposal [18] for defining chiral determinants using the ζ-function method. It also makes use of an auxiliary operator D_a and it is given by the following definition:

$$\det\nolimits_{D_a} D[A] = \frac{\det D[A]D_a}{\det D_a} \tag{27}$$

with "det" in the right-hand side taken as

$$\log\det L = -\left.\frac{d\zeta(L,s)}{ds}\right|_{s=0} \tag{28}$$

* For a review, see Ref. 15.

where $\zeta(L, s)$ is the meromorphic continuation of $\sum_i \lambda_i^{-s}$ with λ_i the eigenvalues of L:

$$\zeta(L, s) = \text{tr} \int dx\, K_{-s}(L; x, x) \tag{29}$$

with $K_{-s}(L; x, x)$ the continuation of the evaluation of the Kernel of L^{-s} on the diagonal $x = y$. Of course, if $\det DD_a = \det D$, then the $\det D_a$ expression (27) coincides with the usual ζ-function definition. However, it has been shown [18] that for chiral fermions this is no more the case and then definition (27) provides a more general way of regularizing determinants, allowing, in particular, the introduction of arbitrary regularization parameters. In this way, the Schwinger model determinant can be easily reobtained in the form [7, 18]

$$\log \frac{\det_{D_a} D}{\det i\not{\partial}} = \frac{e^2}{8\pi} \int d^2x\, A_\mu \left[a\delta_{\mu\nu} + (\delta_{\mu\alpha} + i\varepsilon_{\mu\alpha}) \frac{\partial_\alpha \partial_\nu}{\Delta} \right] A_\nu \tag{30}$$

by choosing

$$D_a = D[A] + \frac{ea}{2} \not{A} \frac{(1 + \gamma_5)}{2} \tag{31}$$

3. CONFORMAL INVARIANCE OF THE CHIRAL GROSS–NEVEU MODEL

This model has a Lagrangian (in two-dimensional Minkowski space-time) given by

$$L = \bar{\psi}^i i\not{\partial} \psi^i + \tfrac{1}{2} g_s^2 j_\mu j^\mu - g_N^2 j^{\mu a} j_\mu^a \tag{32}$$

with $g_s^2 = g_N^2/2N$ [if this relation does not hold it is known as the $SU(N)$ Thirring model]. Here j and j^a are $U(1)$ and $SU(N)$ currents:

$$j_\mu = \bar{\psi}^i \gamma_\mu \psi^i, \qquad j_\mu^a = \bar{\psi}^i \gamma_\mu t_{ij}^a \psi^j, \qquad i, j = 1, 2, \ldots, N \tag{33}$$

with fermions in the fundamental representation of $SU(N)$ and

$$\begin{gathered} [t^a, t^b] = if^{abc} t^c \\ \text{tr}\, t^a t^b = \frac{\delta^{ab}}{2} \\ f^{abc} f^{dbc} = C(G) \delta^{ad} \end{gathered} \tag{34}$$

The model is invariant under global $U(1) \times U(1)$ and global $SU(N)$ rotations [the non-Abelian interaction breaks chiral $SU(N)$ invariance of the free model].

Using trivial identities of the form

$$\exp\left(-i\frac{g_N^2}{2}\int dx\, j^{\mu a} j_\mu^a\right) = \int DA_\mu \exp\left[\frac{i}{2}\int dx\,(A_\mu^a A^{\mu a} - 2g_N A_\mu^a j^{\mu a})\right] \tag{35}$$

with $A_\mu = A_\mu^a t^a$ a vector field and a similar identity for the $U(1)$ current-current interaction, one obtains, within the path-integral approach to quantization, an effective Lagrangian of the form

$$L_{\text{eff}} = \bar{\psi}(i\partial\!\!\!/ - g_N A\!\!\!/ - g_s B\!\!\!/)\psi + \tfrac{1}{2}B_\mu B^\mu + \tfrac{1}{2}A_\mu^a A^{\mu a} \tag{36}$$

with B^μ the Abelian vector field associated to the $U(1)$ current. Now, with an appropriate chiral rotation, the Abelian vector field can be factorized:

$$\begin{aligned} \psi &= \exp(\gamma_5\phi + i\eta)\chi \\ \bar{\psi} &= \bar{\chi}\exp(\gamma_5\phi - i\eta) \\ B_\mu &= -\frac{i}{g_s}(\varepsilon_{\mu\nu}\partial^\nu\phi - \partial_\mu\eta) \end{aligned} \tag{37}$$

We skip details and simply write the answer one easily obtains, due to this factorization, for the two-point function (see Ref. 21 for details):

$$\langle\psi(x)\bar{\psi}(0)\rangle = \mu|x|^{\kappa(g_s)}\langle\chi(x)\bar{\chi}(0)\rangle_{SU(N)} \tag{38}$$

with

$$\kappa(g_s) = \frac{g_s^2}{2\pi}\left(1 - \frac{1}{1 + \dfrac{g_s^2/\pi}{1 + ag_s^2/\pi}}\right)\frac{1}{1 + ag_s^2/\pi} \geq 0 \tag{39}$$

We then see that in the infrared region an almost long-range order in the manner of Kosterlitz–Thouless occurs, as conjectured by Witten using the $1/N$ expansion [22].

Now, it has been shown [19] within the $1/N$ expansion that fermions have a dynamically generated mass. This seems somehow contradictory with the result of Ref. 20 concerning conformal invariance of the model for a particular value of the $SU(N)$ coupling constant g_N. Of course, both results occur for different regions, but it is interesting to see how conformal invariance appears for a certain value of g_N, g_N^*, using the path-integral approach.

One begins by evaluating the fermion determinant associated to Lagrangian (36) once the $U(1)$ sector has been decoupled. The answer is [24]

$$\det(i\partial\!\!\!/ - g_N A\!\!\!/) = \exp\left[-iW(gh^{-1}) - i\alpha\int g^{-1}\partial_+ g h^{-1}\partial_- h\,dx\right] \tag{40}$$

where $W(g)$ is the Wess–Zumino Witten action in Minkowski space. We have written

$$A_+ = \frac{-i}{g_N} g^{-1}\partial_+ g$$
$$A_- = \frac{-i}{g_N} h^{-1}\partial_- h \tag{41}$$

with $g, h \in SU(N)$. The α-dependent term reflects again the regularization ambiguities that we have discussed before.

Now, it remains to express DA_+DA_- in terms of $DgDh$ in order to have an effective bosonic theory. We use the identity [24–26]

$$J_\pm = \det D_\pm^{\text{Adj}} = [\det D_\pm^{\text{Fund}}]^{C(G)} \tag{42}$$

for the Jacobians defined through the relation

$$DA_+DA_- = J_+J_-DgDh \tag{43}$$

with

$$D_\pm^{\text{Adj}} = \partial_\pm - g_N[A_\pm, \ ]$$
$$D_\pm^{\text{Fund}} = \partial_\pm - g_N A_\pm \tag{44}$$

and $C(G)$ the Casimir in the adjoint representation, $C(G) = N$.

Then, we have for the $SU(N)$ sector

$$Z_{SU(N)} = N \int DgDh \exp\left\{-i(N+1)W(gh^{-1}) - i\left[\frac{1}{g_N^2} + \alpha(1+N)\right]\right.$$
$$\left. \times \operatorname{tr} \int dx\, g^{-1}\partial_+ g h^{-1}\partial_- h\right\} \det i\partial\!\!\!/ \tag{45}$$

For arbitrary α we now have a global $SU(N) \times (SU(N)$ invariance, $g \to \Omega g$, $h \to \Lambda h$, and $\Omega, \Lambda \in SU(N)$. Moreover, if for fixed α we choose

$$g_N = g_N^*$$
$$g_N^* = \frac{-1}{\alpha(1+N)} \tag{46}$$

we see that $Z_{SU(N)}$ becomes

$$Z_{SU(N)} = N \int DgDh \exp[-i(1+N)W(gh^{-1})] \det i\partial\!\!\!/$$
$$= \bar{N} \int Dg \exp[-i(1+N)W(g)] \det i\partial\!\!\!/ \tag{47}$$

and then for this particular value of the $SU(N)$ coupling constant one has a local invariance under the transformation

$$g \to \Omega(x_+) g \Lambda(x_-) \tag{48}$$

i.e., for a particular value of g_N one has an associated Kac–Moody algebra with level $k = N + 1$ (not a free one). Conformal invariance is then an invariance of the resulting bosonized Lagrangian (47) for $g_N = g_N^*$ [$W(g)$ is the Wess–Zumino–Witten action at the infrared stable fixed point]. When $\alpha = -1/4\pi$ (the regularization choice which leads to the usual boson-fermion equivalence in the free theory), relation (46) coincides with the result of Ref. 20. (But, as we stressed, we have gotten a $k = N + 1$ theory and not a $k = 1$ one as in that work.)

In summary, the original model had a current–current interaction term which breaks $SU(N) \times SU(N)$ symmetry of the free fermionic Lagrangian. There is, however, a value $g_N^* \neq 0$ of the coupling constant at which chiral symmetry is exhibited and a Kac–Moody algebra emerges.

We end by noting that recently it has been shown [28] in two-dimensional models that there exists a function $c(g)$ related to the β-function which, at fixed points g^* [$\beta(g^*) = 0$], coincides with the central charge of the Virasoro algebra of the corresponding conformal field theory. The chiral Gross–Neveu model exhibits, as we have seen, such a behavior and it is important to stress that at the fixed point g_N^* it is not a free model ($k \neq 1$).

ACKNOWLEDGMENTS. I would like to thank C. Teitelboim and J. Zanelli for the splendid days I spent during the Santiago meeting. I am partially supported by C.I.C., Buenos Aires, Argentina.

REFERENCES

1. K. Fujikawa, *Proceedings of the Nato Advanced Research Workshop on Super Field Theories*, Vancouver, 1986, World Scientific, Singapore; P. van Niewenhuizen, *Proceedings of the Trieste Spring School on Supersymmetry, Supergravity and Superstrings* (B. de Witt *et al.*, eds.), World Scientific, Singapore, 1986.
2. K. Fujikawa, *Phys. Rev. Lett. 42*, 1195 (1979); and references in Ref. 1.
3. F. A. Schaposnik, *Lectures at the IV Brazilian School of Physics Jorge André Swieca "Particle and Fields,"* 1987, World Scientific, Singapore (in press), and references therein.
4. R. Jackiw, *Update on Anomalous Theories*, in *Quantum Mechanics of Fundamental Systems 1* (C. Teitelboim, ed.), Plenum Press, New York, 1988 and references therein.
5. D. Gross and R. Jackiw, *Phys. Rev. D 6*, 477 (1972).
6. L. Alvarez-Gaumé and P. Ginsparg, *Nucl. Phys. B 243*, 449 (1984).
7. R. Jackiw and R. Rajaraman, *Phys. Rev. Lett. 54*, 1219, 2060(E); *55*, 2224 (1985).
8. L. D. Faddeev and S. L. Shatashvili, *Phys. Lett. 167B*, 223 (1986).
9. O. Babelon, F. A. Schaposnik, and C. M. Viallet, *Phys. Lett. 177B*, 385 (1986); K. Harada and I. Tsutsui, *Phys. Lett. 83B*, 311 (1987); V. Kulikov, Serpukhov Report 86-83 (unpublished).

10. L. D. Faddeev and V. N. Popov, *Phys. Lett. 25B*, 29 (1967).
11. L. D. Faddeev, *Theor. Math. Phys. 1*, 3 (1969).
12. O. Babelon and C. M. Viallet, *Phys. Lett. 85B*, 246 (1979).
13. M. V. Manías, F. A. Schaposnik, and M. Trobo, *Phys. Lett. 195B*, 209 (1987).
14. K. Harada and I. Tsutsui, TIT-HEP reports No. 102, 112, 113 (unpublished); N. K. Falck and G. Krammer, *Phys. Lett. 193B*, 257 (1987); *Ann. Phys (N.Y.) 176*, 330 (1987); E. Abdalla and K. D. Rothe, *Phys. Rev. D 36*, 3190 (1987); R. D. Ball, *Phys. Lett. 183B*, 315 (1987); K. Harada, TIT-HEP report No. 118 (unpublished).
15. P. Damgaard and H. Huffel, *Phys. Rep. 152*, 227 (1987), and references therein.
16. H. Montani and F. A. Schaposnik, *Ann. Phys. (N.Y.) 181*, 161 (1988).
17. I. Singer, Asterisque, hors série, p. 223 (1985).
18. R. E. Gamboa Saraví, M. A. Muschietti, F. A. Schaposnik, and J. E. Solomin, La Plata University report, *Lett. Math. Phys. 16*, 325 (1988).
19. D. Gross and A. Neveu, *Phys. Rev. D 10*, 3235 (1974).
20. R. Dashen and Y. Frishman, *Phys. Lett. 46B*, 469 (1973); *Phys. Rev. D 11*, 2781 (1974).
21. K. Furuya, R. E. Gamboa Saraví, and F. A. Schaposnik, *Nucl. Phys. B 208*, 159 (1982).
22. E. Witten, *Nucl. Phys. B 145*, 110 (1978).
23. E. Moreno and F. A. Schaposnik, La Plata University report (unpublished).
24. A. M. Polyakov and P. B. Wiegmann, *Phys. Lett. 131B*, 121 (1983); *141B*, 223 (1984).
25. A. N. Redlich and H. Schnitzer, *Phys. Lett. 193B*, 471 (1987).
26. A. Ceresole, A. Lerda, P. Pizzochero, and P. van Niewenhuizen, ITP report No. 86/96 (unpublished).
27. E. Fradkin, C. M. Naón, and F. A. Schaposnik, *Phys. Lett. 200B*, 95 (1988); *Phys. Rev. D 36*, 3809 (1987).
28. A. B. Zamolodchikov, *JETP Lett. 43*, 731 (1986).
29. A. W. W. Ludwig and J. Cardy, University of California report No. 83/86 (unpublished).

Chapter 15

Differential Algebras in Field Theory

R. Stora

1. INTRODUCTION

Differential algebras have been known for a long time to be mathematical tools appropriate to field theory (cf., e.g., Ref. 1). Classically, the de Rham algebra of differential forms on the space-time manifold is favored. More exotic differential algebras made their appearance some fifteen years ago, in the context of perturbatively quantized gauge theories and by now belong to the standard equipment of field theorists as either necessary or only convenient tools, to deal with gauge theories of rather general types. It is easier to talk about these things now than ten or fifteen years ago because many people are not so allergic anymore to the mathematics of the 1950s as they used to be. Preparing these notes served as an opportunity to recollect some memories and dig out some documents that were not widely circulated. I have not attempted any completeness, not only because it is very difficult, possibly impossible, given the abundance of literature on the subject, but also because of personal prejudices, which I shall try to justify.

The exotic example at the start of this industry has to do with the Slavnov [3] symmetry of the Faddeev–Popov [2] gauge fixed Yang–Mills

R. Stora • LAPP, F-74019 Annecy-le-Vieux, France and Theory Division, CERN, CH-1211 Geneva 23, Switzerland.

theory, whose main property is to manifestly maintain locality throughout the renormalization procedure. This example has served as a prototype for a number of extensions. Among the best developed and most remarkable is the analysis of constrained Hamiltonian systems developed by the Lebedev group [4]. I will not touch this body of knowledge, because, in spite of its recent popularity in string theory, it fails to accommodate simultaneously the locality of field theory and geometry, which the original Lagrangian version allows. It has, however, the advantage of allowing for a Hilbert space framework. Actually the operator Q, as well as $\bar{Q}$, popular in string theory [5] and related to the conformal structure of the theory, and which bear some formal similarity to that of the Lebedev construction [4] in the particular case where the constraints generate a Lie algebra, is nothing else than the coboundary operator for the cohomology of the Virasoro algebra via a formula due to J. L. Koszul [6]. At the Lagrangian level, solving specific examples shows that a general solution is missing for what may be called the auxiliary field problem.* Unfortunately most known examples result in more or less trivial extensions of the cohomology algebra of some (gauge) Lie algebra. Even there, the auxiliary field problem has been solved only in particular cases, e.g., in $N = 1$ supergravities where the solution is not unique [28]. The standard way out [10, 14, 29] is to introduce external fields, but the correct definition of physical observables is even more conjectural. Apart from these classes of examples two others are instructing by themselves: the Torino group manifold approach [7], which grounds the construction of geometrical theories (e.g., supergravities) on generalized Weyl algebras; the other prototype example is that of quantized p-differential forms [8]. Much of the present activity is concerned with applications to string theory, first and second quantized. This area is covered elsewhere. I will therefore limit myself to a discussion of the principles that may be abstracted from the examples I know best.

Section 2 recalls the situation in the Yang–Mills theories and expounds a parallel treatment of the first-quantized bosonic string.

Section 3 is devoted to a sketchy account of a nice by-product: the algebraic description of anomalies.

Section 4 poses a general class of problems suggested by the examples of Section 2.

2. YANG–MILLS THEORIES AND FIRST QUANTIZED STRINGS

2.1. Yang–Mills Theories

Let us recall that the Faddeev–Popov gauge fixed action, in a Feynman-

* That is, can a nonintegrable situation be viewed as a restriction of an integrable one?

like gauge, reads

$$S(a, \omega, \bar{\omega}) = S_{\text{inv}}(a) + S_{\text{gf}}, \qquad S_{\text{gf}} = \int_{M^4} [\tfrac{1}{2}\{g, g\} + \{\bar{\omega}, M\omega\}]\, d^4x \tag{1}$$

where a is the Yang–Mills field

$$a = \sum_{\substack{\mu=0\ldots3 \\ \alpha}} a_\mu^\alpha(x)\, dx^\mu e_\alpha \tag{2}$$

e_α is a basis of Lie G, the Lie algebra of the compact group G, with commutation relations

$$[e_\alpha, e_\beta] = f_{\alpha\beta}^\gamma e_\gamma$$

$$F(a) = da + \tfrac{1}{2}[a, a] = \tfrac{1}{2} \sum_{\mu,\nu} F_{\mu\nu}\, dx^\mu \wedge dx^\nu$$

$$= \tfrac{1}{2} \sum_{\mu,\nu,\alpha} F_{\mu\nu}^\alpha\, dx^\mu \wedge dx^\nu e_\alpha$$

$$F_{\mu\nu}^\alpha = \partial_\mu a_\nu^\alpha - \partial_\nu a_\mu^\alpha + f_{\beta\gamma}^\alpha a_\mu^\beta a_\nu^\gamma \tag{3}$$

$$S_{\text{inv}}(a) = -\tfrac{1}{4} \int d^4x\, (F_{\mu\nu}, F^{\mu\nu}) \tag{4}$$

where (,) is a G invariant bilinear form on Lie G, g is a Lie G-valued* local function of a (and its derivatives), { , } another quadratic form on Lie G [e.g., $g^\alpha(x) = \partial_\mu a^{\mu\alpha}(x) + \cdots$]

$$\omega = \sum_\alpha \omega^\alpha e_\alpha \tag{5}$$

is the Faddeev–Popov geometrical ghost which serves as a generator of the Grassmann algebra of the dual of Lie $\mathcal{G}$, the Lie algebra of the gauge group $\mathcal{G}$:

$$\mathcal{G} = \{\text{Maps } M_4 \to G\} \tag{6}$$

with the obvious pointwise composition rules. If $\Lambda \in$ Lie $\mathcal{G}$ is represented by

$$\Lambda(x) = \sum_\alpha \Lambda^\alpha(x) e_\alpha \tag{7}$$

$$\langle \omega^\alpha(x), \Lambda \rangle = \Lambda^\alpha(x) \tag{8}$$

$M\omega$ is the variation of g under an operation of Lie $\mathcal{G}$, of parameter ω, induced by

$$\delta_\Lambda a = d\Lambda + [a, \Lambda] \tag{9}$$

$\bar{\omega}$, the Faddeev–Popov antighost, also a Grassmann-type field is a Lagrange multiplier which, in the present instance generates another copy of Λ Lie

* This is a convenient labeling of the set of gauge functions whose number is dim G, but it is also misleading: the geometrical nature of g need not in principle have anything to do with Lie G. This will become apparent, e.g., in the string case (Section 2).

$\mathcal{G}$.* This is a peculiarity of this example, as we shall see that, in general $\bar{\omega}$ has geometrical properties that depend on those of the gauge function.

The full action S is found to be invariant under the Slavnov symmetry [3]:

$$\begin{aligned} sa &= -d\omega - [a, \omega] \\ s\omega &= -\tfrac{1}{2}[\omega, \omega] \\ s\bar{\omega} &= -g \end{aligned} \tag{10}$$

This was derived in two steps: using a beautiful trick† due to A. A. Slavnov [3] one first derives the Slavnov identity for, say, the connected Green's functional [9]‡; then one recognizes that the Slavnov identity expresses the invariance of the action under the Slavnov symmetry equation (10).§

In the initial version, the transformation contained an anticommuting infinitesimal parameter that was soon gotten rid of [11], as above, considering s as a graded derivation on the algebra of functionals of a, ω, $\bar{\omega}$—the total grading being given by $\#\omega - \#\bar{\omega}$ (+ form degree, if one anticipates Section 3).

Also note that in the form given in equation (10) one does not have $s^2 = 0$ because

$$s^2\bar{\omega} = -M\omega \tag{11}$$

i.e., it is only nilpotent modulo the ghost equation of motion. This phenomenon, which creates mathematical inconveniences, will be repeatedly met. Here it is easily eliminated by the introduction of a multiplier field b, turning S_{gf} into

$$S_{\text{gf}}(a, \omega, \bar{\omega}b) = \int b_\alpha(x) g^\alpha(x)\, dx + \int \bar{\omega}_\alpha M^\alpha_\beta \omega^\beta\, dx + F(b) \tag{1'}$$

invariant under

$$s\omega_\alpha = -b_\alpha, \qquad sb_\alpha = 0, \qquad s^2 = 0 \tag{12}$$

* See, however, the remark in the previous footnote.

† Write the Ward identity that characterizes the breaking of gauge invariance of the total action, only gauge transforming the gauge field, and use as an infinitesimal gauge parameter $\Lambda = M^{-1}\omega$.

‡ The computational appendix on pp. 61–63 of Ref. 9 acknowledges C. Itzykson for generous help. This memory goes with another one: the classical paper of I. A. Batalin and G. A. Vilkovisky refers to my "advice to verify the supersymmetry of the theory." Actually, when the first paper by E. S. Fradkin and G. A. Vilkovisky came out, Claude Itzykson and I conjectured over the telephone that their action was invariant under a Slavnov symmetry. The next day, Claude Itzykson called me back to give the formula, which I immediately forgot.

§ This remark was made by C. Becchi and A. Rouet in January 1974. The hard work of pushing the Slavnov symmetry through renormalization, proving gauge independence and unitarity, resulted in the works in Ref. 10. The symmetry was found independently by I. V. Tyutin, Lebedev Institute preprint No. 39 (1975), who did not publish it because some of our work was already published.

so that $\mathcal{F}(b)$ can be chosen as a quadratic functional reproducing the initial action upon elimination of the b-field—whereby nilpotency of s is lost. I have always called b the Stueckelberg multiplier because it was very similar to the ghost field introduced by Stueckelberg to remedy the nonrenormalizability of massive QED by passing to the Stueckelberg gauge [12], but this multiplier field is commonly called the Nakanishi–Lautrup field.

This algorithm can be generalized to allow the metric $\{\ ,\ \}$ in equation (1) to be field-a-dependent at the expense of introducing an extra ghost [13].

In order to write down properly the Slavnov identity, it is convenient to add to the action a source term

$$S_{\text{source}} = \int \mathcal{A}sa + \Omega s\omega \tag{13}$$

so that

$$S_{\text{tot}} = S_{\text{inv}} + S_{\text{gf}} + S_{\text{source}} \tag{14}$$

fulfills

$$\int \left(\frac{\delta S_{\text{tot}}}{\delta \mathcal{A}} \frac{\delta S_{\text{tot}}}{\delta a} + \frac{\delta S_{\text{tot}}}{\delta \Omega} \frac{\delta S_{\text{tot}}}{\delta \omega} - b \frac{\delta S_{\text{tot}}}{\delta \bar{\omega}} \right) = 0 \tag{15}$$

which "easily" passes through renormalization—modulo anomalies, cf. section 3—by substituting the vertex functional Γ against S_{tot}.

Legendre transforming with respect to b^* and calling $\bar{\Omega}$ the Legendre transform yields for the Legendre transform $\tilde{S}$ of S the more symmetrical identity [14]*

$$\int \frac{\delta \tilde{S}}{\delta \mathcal{A}} \frac{\delta \tilde{S}}{\delta a} + \frac{\delta \tilde{S}}{\delta \Omega} \frac{\delta \tilde{S}}{\delta \omega} + \frac{\delta \tilde{S}}{\delta \bar{\Omega}} \frac{\delta \tilde{S}}{\delta \bar{\omega}} = 0 \tag{16}$$

Whereas b has to be considered as a quantum field, $\bar{\Omega}$ has to be considered as a classical field serving as a source of the gauge function (S_{gf} now gets a term $\frac{1}{2}(g - \bar{\Omega}, g - \bar{\Omega}) + \bar{\omega}M\omega$). The Slavnov identity encodes two symmetries of the action:

$$\begin{aligned} s\bar{\omega} &= 0 \\ s\bar{\Omega} &= M\omega = sg \end{aligned} \tag{17}$$

* Another way to construct $\tilde{S}$ is to argue that the symmetry of $S_{\text{inv}} + S_{\text{gf}} + S_{\text{source}}$ is best expressed by adding a term $\int \Omega g$, so that the resulting action, Σ fulfills

$$\int \frac{\delta \Sigma}{\delta \mathcal{A}} \frac{\delta \Sigma}{\delta a} + \frac{\delta \Sigma}{\delta \Omega} \frac{\delta \Sigma}{\delta \omega} = \bar{\Omega} \frac{\delta \Sigma}{\delta \bar{\omega}}, \qquad \text{then } \tilde{S} = \Sigma - \frac{\bar{\Omega}^2}{2}$$

which is nilpotent,

$$s\bar{\omega} = -(g - \bar{\Omega})$$
$$s\bar{\Omega} = 0 \tag{18}$$

which is not.

The Slavnov operator is defined as

$$\mathscr{S} = \int \frac{\delta\tilde{S}}{\delta\mathscr{A}}\frac{\delta}{\delta a} + \frac{\delta\tilde{S}}{\delta a}\frac{\delta}{\delta\mathscr{A}} + \frac{\delta\tilde{S}}{\delta\Omega}\frac{\delta}{\delta\omega} + \frac{\delta\tilde{S}}{\delta\omega}\frac{\delta}{\delta\Omega} + \frac{\delta\tilde{S}}{\delta\bar{\Omega}}\frac{\delta}{\delta\bar{\omega}} + \frac{\delta\tilde{S}}{\delta\bar{\omega}}\frac{\delta}{\delta\bar{\Omega}} \tag{19}$$

and $\mathscr{S}^2 = 0$ follows from equation (16). This is the important operator in perturbation theory. The discussion of anomalies, i.e., the appearance of a right-hand side in equation (16), when S is replaced by Γ goes through the discussion of local functionals $\int \mathscr{a}$, which fulfill the Wess-Zumino [15] consistency condition—actually a generalization thereof:

$$\mathscr{S} \int_{M^4} \mathscr{a} = 0 \tag{20}$$

modulo trivial anomalies of the form

$$\int \mathscr{a} = \mathscr{S}\Gamma^{\text{loc}} \tag{21}$$

where Γ^{loc} is a local functional.

The nice thing* is that the only nontrivial part of this cohomology involves α's which are local in a, ω and fulfill

$$s \int \mathscr{a}(\omega, a) = 0 \qquad \int \alpha \text{ defined modulo } s\Gamma^{\text{loc}}(a, \omega) \tag{22}$$

which takes us back to the conventional Wess-Zumino consistency conditions [15]. Given the Slavnov identity, one can prove gauge independence and unitarity whenever an asymptotic theory exists [10]. The definition of asymptotic physical states has been nicely algebraized by T. Kugo and I. Ojima [17], using the b field. This goes through the construction of the Noether current associated with the Slavnov symmetry and the definition of the corresponding asymptotic charge Q which is nilpotent, and commutes with the S operator if there is no anomaly. Physical states are then defined by the cohomology of Q for zero ghost number. The Noether current algebra

* See Refs. 11 and 16. The preprint corresponding to Ref. 16 is dated 1975; we finally decided to publish it because of the theorem in Appendix D.

is interesting in itself [18], although it is not unique because of the occurrence of derivatives in the transformation laws.

2.2. The Free Bosonic String

The free bosonic string can be treated in a way parallel to Yang–Mills theories with a few distinctions. The invariant action is

$$S_{\text{inv}} = \int_{\Sigma} d^2x\, g^{\alpha\beta}\sqrt{g}\,\partial_\alpha X \cdot \partial_\beta X \tag{23}$$

$x = (\sigma, \tau) \in \Sigma$ parametrizing a world sheet, $X \in \mathbb{R}^D$, . the Euclidean metric in $\mathbb{R}^D$,* g a metric on Σ.

One may take two attitudes. Either one has S_{inv} depend on g [19]. Then S_{inv} is invariant under $\text{Diff}\,\Sigma \times \text{Weyl}$, where Weyl transformations locally scale g. One has to pick three gauge functions which one may choose as the three components of g and the full Slavnov symmetry results as a current algebra Ward identity for the stress energy tensor coupled to g, the multiplier field being trivially eliminated in a Landau-type gauge [$F(b) = 0$ in equation (1′)].

Or one may have S_{inv} dependent only on the conformal class of g [20] and choose the gauge fixing in such a way that the conformal structure of the theory is manifestly exhibited [5].

Now, there is a nice parametrization of conformal classes of metrics, in terms of Beltrami differentials: Picking a reference complex structure with local complex analytic coordinates $z, \bar{z}$, the conformal class is characterized by

$$ds^2 \propto |dz + \mu\, d\bar{z}|^2, \qquad |\mu| < 1 \tag{24}$$

Then, in terms of μ, μ, S_{inv} reads

$$S_{\text{inv}} = \int \frac{dz \wedge d\bar{z}}{2i} \frac{1}{1 - \mu\bar{\mu}} (\partial_z - \bar{\mu}\partial_{\bar{z}} X, \partial_{\bar{z}} - \mu\partial_z X) \tag{25}$$

C. Becchi has found that the *unique* choice of gauge functions, which exhibits the theory as a conformal theory, is the pair $\mu - \mu^0, \bar{\mu} - \bar{\mu}^0$.†

* We use the Euclidean version throughout, although locality does not care.

† The easiest way is to proceed via the elimination of the b field as well as of the Weyl ghost and antighost. One may also start directly with conformal classes of metrics, eliminate the b field in the manner of Kato Ogawa [20, 5], introduce a source $\mu, \bar{\mu}$ coupled to the s variation of the antighost, and recover a Slavnov identity in the manner of Zinn Justin (C. Becchi, private communication).

Eliminating the multiplier field yields the gauge fixed action:

$$S = S_{\text{inv}} + S_{\text{gf}}$$

$$S = S_{\text{inv}} + S_{\text{gf}}$$

$$S_{\text{gf}} = \int \frac{dz \wedge d\bar{z}}{2i} b_{zz}(\partial_{\bar z} - \mu\partial_z + \partial_z\mu)C^z + \text{c.c.} \tag{26)*$$

Here the ghost C^z is related to the diffeomorphism ghost $\xi = (\xi^z, \xi^{\bar z})$ through

$$C^z = \xi^z + \mu\xi^{\bar z} \ \& \ \text{c.c.} \tag{27}$$

and the Slavnov symmetry associated with diffeomorphisms reads

$$\begin{aligned} sX &= (\xi \cdot \partial)X = (\xi^z\partial_z + \xi^{\bar z}\partial_{\bar z})X \\ sC^z &= C^z\partial_z C^z \ \& \ \text{c.c.} \end{aligned}$$

or

$$\begin{aligned} s\xi &= (\xi \cdot \partial)\xi \\ s\mu &= \partial_{\bar z}C^z + C^z\partial_z\mu - \mu\partial_z C^z \\ sb_{zz} &= 0 \end{aligned} \tag{28}$$

The corresponding Slavnov identity defines the current algebra for the two components Θ_{zz}, $\Theta_{\bar z\bar z}$ of the stress energy tensor, whose correlation functions are obtained through functional differentiation with respect to μ, $\bar\mu$.

The conformal properties of the system are furthermore characterized by a set of Ward identities which assign conformal weights 0, −1, +2 to X, C^z, b_{zz}, namely, the action is invariant under the following action of diffeomorphisms†:

$$\begin{aligned} \delta_\lambda X &= (\lambda \cdot \partial)X \\ \delta_\lambda \mu &= \partial_{\bar z}\Lambda + \Lambda\partial_z\mu - \mu\partial_z\Lambda \\ (\Lambda &\overset{\text{def}}{=} \lambda^z + \mu\lambda^{\bar z}) \\ \delta_\lambda C^z &= (\lambda \cdot \partial)C^z - (\partial_z\lambda^z + \mu\partial_z\lambda^{\bar z})C^z \\ \delta_\lambda b_{zz} &= (\lambda \cdot \partial)b_{zz} + 2(\partial_z\lambda^z + \mu\partial_z\lambda^{\bar z})C^z \end{aligned} \tag{29}$$

The action defines Feynman rules which entails factorization of the Green's functions into factors involving z indices alone, and factors involving $\bar z$

* Note that the geometrical nature of the antighosts b_{zz}, $b_{\bar z\bar z}$ is related to that of the gauge function as mentioned in Section 2.1.

† L. Baulieu and M. Bellon [20, 21] have found another action of diffeomorphisms which leaves S invariant but does not apply to more general conformal models involving Thirring-like couplings.

indices alone. The anomalies are easily computed and of course found to be proportional to $D-26$.*

As a side remark, transformations (29) apply to any conformal model, i.e., Lagrangian model, which one can couple to a conformal class of metrics—or complex structure—represented by μ: For a field $\varphi_{q\bar{q}}$ with conformal weights $q, \bar{q}$ one finds

$$\delta_\lambda \varphi_{qq} = [(\lambda \cdot \partial) + q(\partial_z \lambda^z + \mu \partial_z \lambda^{\bar{z}}) + \bar{q}(\partial_{\bar{z}} \lambda^{\bar{z}} + \bar{\mu} \partial_{\bar{z}} \lambda^z)] \varphi_{qq} \tag{30}$$

which checks with equation (29) for $(q, \bar{q}) = (0,0), (-1,0), (+2,0)$, + c.c. for X, C^z, b_{zz}.

The whole scheme can be extended to include the Noether current algebra of the Slavnov symmetry [21]. Just as in the Yang–Mills case, the anomaly is rigidly tied up with the "holomorphy anomaly" ($\propto D$-26) and yields a local analog of the Kato–Ogawa nilpotency anomaly for the charge.

Another similarity with the Yang–Mills case is the form of the Slavnov identity: Introducing

$$S_{\text{source}} = \int \mathscr{X} sX + \mathscr{C} sC^z + \text{c.c.} \tag{31}$$

we have for

$$S_{\text{tot}} = S_{\text{inv}} + S_{\text{gf}} + S_{\text{source}}$$

$$\int \frac{dz \wedge d\bar{z}}{2i} \left(\frac{\delta S}{\delta \mathscr{X}} \frac{\delta S}{\delta X} + \frac{\delta S}{\delta \mathscr{C}} \frac{\delta S}{\delta C^z} + \frac{\delta S}{\delta b_{zz}} \frac{\delta S}{\delta \mu} + \text{c.c.} \right) = 0 \tag{32}$$

The last term may be interpreted in two ways: the invariance of S under (28), which is nilpotent, or under

$$\begin{aligned} s'\mu &= 0 \\ s'b_{zz} &= \frac{\delta S}{\delta \mu} = \Theta_{zz} \ \&\ \text{c.c.} \end{aligned} \tag{33}$$

which is only nilpotent modulo the ghost equations of motion [5]. The existence of the Slavnov identity can, however, be most simply obtained via the existence of the nilpotent transformation (28).

* Their algebraic form reads

$$\int \frac{dz \wedge d\bar{z}}{2i} [\Lambda^z \partial_z^3 \mu - \mu \partial_z^3 \Lambda^z + R(\Lambda^z \partial_z \mu - \mu \partial_z \Lambda^z)] + \text{c.c.}$$

where R is a projective connection [30]. I thank L. Bonora and C. Reina for suggesting the use of a projective connection in order to get a form of the anomaly valid on an arbitrary Riemann surface.

A last remark is in order: The conformal structure of the model is obtained at the expense of having chosen a gauge function that is not in general a good gauge function—unless there is only one orbit of μ's under diffeomorphisms. For instance, if Σ is a compact Riemann surface of genus $g > 1$, there are $3g - g$ zero modes for b_{zz} to be gauge fixed (the analog of the Gribov horizon in the YM case). We shall return to this problem in Section 3.

3. THE ALGEBRAIC STRUCTURE OF LOCAL ANOMALIES

We have mentioned in Section 2 the possible occurrence at the quantum level of local anomalies fulfilling the proper Wess–Zumino consistency condition, and mentioned that only the dual Lie algebra part of the Slavnov symmetry was involved. In the Yang–Mills case, the $\mathscr{S}$ operation when restricted to a and ω reduces [22] to the coboundary operator defining the local cohomology of the gauge Lie algebra with values in the algebra of local functionals of a.

This makes it possible to give compact formulas for the Adler–Bardeen anomalies derived, e.g., in Ref. 16, and this turned into a small industry a few years ago [23]. This is a subject of its own for which we have indicated [23] a few references that allow one to trace back most of the literature. The temperature has now decreased on this topic, but there is still one interesting mathematical problem to be settled.

There are in fact two theories of anomalies. There is an algebraic one, the "local" theory sketched out here, which reduces the problem in each model to computing a set of numerical coefficients, but in particular does not explain the rationality of various coefficients at the one-loop level, a property that is stable thanks to the nonrenormalizability theorems in the manner of Bardeen. Nor does it give any clue as to why it occurs only when chiral fields are involved. But this explains $\pi^0 \to 2\gamma$.

On the other hand, there is the topological theory [23, 24] due to Atiyah, Bott, Singer, Quillen, and others, which is purely Euclidean as it deals with the Index of families of elliptic operators and relates anomalies to de Rham cohomology classes of, e.g., gauge groups. The puzzle is that the local anomaly responsible for $\pi^0 \to 2\gamma$ corresponds to a trivial de Rham cohomology class, a fact which was known to physicists [25]. This phenomenon presumably also happens in the case of gravitational anomalies. This situation raises an interesting mathematical problem: can one construct an Index theory in terms of local cohomology classes "on the fibers"?

Besides, on the local side, there is still much to be done. Work is still in progress [26] and progress is definitely needed.

4. LESSONS TO BE DRAWN

The examples of Section 2 pose the problem of quantizing a degenerate classical action. This means that $S(\varphi)$, $\varphi = \{\varphi^i\}$, is invariant under some transformations

$$\delta_\omega \varphi = P(\varphi, \omega) \tag{34}$$

where $P(\varphi, \omega)$ is linear in $\omega = \{\omega^\alpha\}$. We are mostly concerned with local actions and local transformations. Note that any action is invariant under uninteresting transformations

$$\delta\varphi^i = M^{ij} \frac{\delta S}{\delta \varphi^j} \tag{35}$$

where M^{ij} is a (graded) symmetric kernel. If the transformations (34) are integrable, i.e., express differentiation along the leaves of a foliation in field space, they give rise to a nilpotent Slavnov symmetry s, which is nothing but differentiation along the leaves. By a formal Frobenius property, if the operations δ_ω are in involution, namely,

$$\left[\int \delta_{\omega_1}\varphi \frac{\delta}{\delta\varphi}, \int \delta_{\omega_2}\varphi \frac{\delta}{\delta\varphi}\right] = \int \delta_{f(\omega_1, \omega_2, \varphi)}\varphi \frac{\delta}{\delta\varphi} \tag{36}$$

with f antisymmetric in ω, ω_2, there will be a foliation, locally, and therefore an s operation*:

$$\begin{aligned} s\varphi &= P(\varphi, \omega) \\ s\omega &= -\tfrac{1}{2} f(\omega, \omega, \varphi) \end{aligned} \tag{37}$$

$s^2 = 0$ is a consequence of the Jacobi identity.

This situation occurs not only when a Lie algebra is involved but for instance genuinely to define σ models [27] where one has to express that splitting the field into a background and fluctuations is invariant under an infinitesimal change of the background and an adequately corresponding change of the fluctuating part.

A more general situation, which occurs frequently, and which we have already met, is when closure of the algebra—i.e., nilpotency of the s operation—is fulfilled only modulo the equations of motion, i.e., the "uninteresting" operations of equation (35). This occurs in supergravity where closure can be achieved in a variety of ways via the introduction of either of several possible sets of auxiliary fields [28]. Once closure is achieved gauge fixing can be performed by introducing multiplier fields and antighosts, after choosing a gauge function.

* By abuse of notation we denote by ω the $\phi\pi$ ghost corresponding to the infinitesimal transformations parametrized by ω.

A large amount of work has been performed on the general problem [29].

My present understanding is limited and not quite in agreement with the published literature (mostly the articles by I. A. Batalin and G. A. Vilkovisky). It relies on unpublished work by M. Tonin, and generalizes the Yang–Mills case treated à la Zinn Justin [14] [equation (16)] and the case of the string as treated by C. Becchi [20] [equation (32)]. One assumes s given by equation (37) with $s^2 = 0$ only modulo the irrelevant transformations equation (35). One introduces sources Φ, Ω coupled to $s\varphi$, $s\omega$. One may first look for an action $S(\varphi, \omega; \Phi, \Omega)$ fulfilling

$$\tfrac{1}{2}[S, S] \equiv \int \frac{\delta S}{\delta \varphi}\frac{\delta S}{\delta \Phi} + \frac{\delta S}{\delta \omega}\frac{\delta S}{\delta \Omega} = 0 \tag{38}$$

Or one may directly look for $S(\varphi, \omega, \omega; \Phi\Omega\Omega)$ such that

$$\tfrac{1}{2}[S, S] \equiv \int \frac{\delta S}{\delta \varphi}\frac{\delta S}{\delta \Phi} + \frac{\delta S}{\delta \omega}\frac{\delta S}{\delta \Omega} + \frac{\delta S}{\delta \bar{\Omega}}\frac{\delta S}{\delta \bar{\omega}} = 0 \tag{39}$$

S has zero ghost number ($= \#\omega - \#\omega - \#\Phi - 2\#\Omega$). The notation $[S, S]$ refers to the graded Lie bracket of the vector fields

$$\mathscr{S} \equiv \int \frac{\delta S}{\delta \varphi}\frac{\delta}{\delta \Phi} + \frac{\delta S}{\delta \Phi}\frac{\delta}{\delta \varphi} + \frac{\delta S}{\delta \omega}\frac{\delta}{\delta \Omega} + \frac{\delta S}{\delta \Omega}\frac{\delta S}{\delta \omega} + \frac{\delta S}{\delta \bar{\Omega}}\frac{\delta}{\delta \bar{\omega}} + \frac{\delta S}{\delta \bar{\omega}}\frac{\delta}{\delta \bar{\Omega}} \tag{40}$$

One solves equation (39) for local functionals with the boundary conditions imposed by the data $S(\varphi)$, $s\varphi$, $s\omega$. This fixes the Feynman rules for the quantum fields $\varphi, \omega, \bar{\omega}$, in presence of the classical fields $\Phi, \Omega, \bar{\Omega}$. One then has to study the quantum extensions Γ fulfilling (39), or the occurrence of anomalies. Note that the defining equation for S is invariant under the restricted canonical transformations—which do not alter the quantized fields—

$$\begin{aligned} &(\varphi, \omega, \bar{\omega}) \to (\varphi, \omega, \bar{\omega}) \\ &(\Phi, \Omega, \bar{\Omega}) \to \left(\Phi + \frac{\delta \chi}{\delta \varphi}, \Omega + \frac{\delta \chi}{\delta \omega}, \bar{\Omega} + \frac{\delta \chi}{\delta \bar{\omega}}\right) \end{aligned} \tag{41}$$

where χ is an odd generating function which depends only on $(\varphi, \omega, \bar{\omega})$. This reflects the freedom in the choice of gauge functions. Besides locality, power counting restricts this construction. Since the b-field is absent from this scheme, there does not appear to be a transparent algebraization of the asymptotic theory, when it exists, but one may try to construct the current algebra for a Noether current. Observables may be defined in the zero ghost number sector either by the local cohomology of $\mathscr{S}$ or by the local cohomology of $\mathscr{S}$ modulo d (d being the exterior derivative). In the Yang–Mills case, the first local cohomology is nonempty and defines "gauge invariant operators." In the string case—presumably in all cases where diffeomorphisms are involved—the physics is presumably concentrated in the local cohomology of $\mathcal{S}$ mod d, which yields integrated observables. This has to be studied in each case and is really the most important point.

There is one last amusing example with which this section will be concluded: in the treatment of the bosonic string in the conformal gauge, I mentioned the global zero mode problem which occurs because of the bad choice of the gauge function. The ghost action

$$S_{\rm gh} = \int \frac{dz \wedge d\bar{z}}{2i} b_{zz} s\mu = \int b_{zz}(\partial_z c^z + c^z \partial_z \mu - \mu \partial_z c^z) \frac{dz \wedge d\bar{z}}{2i} \tag{42}$$

is invariant under

$$b_{zz} \to b_{zz} + \delta b_{zz},\ \delta b_{zz} = \sum_{i=1}^{3g-3} c_i \phi^i \tag{43}$$

where the ϕ^i's are solutions of

$$(-\partial_{\bar{z}} + \mu \partial_z + 2\partial_z \mu)\phi^i = 0 \tag{44}$$

i.e., the ϕ^i's correspond to holomorphic quadratic differentials for the complex structure defined by μ. The question is to extend the s operation equation (28) in order to take these zero modes into account. One may try

$$sb_{zz} = \sum c_i \phi^i \tag{45}$$

but ϕ^i depends on μ and there is no way to achieve $s^2 b_{zz} = 0$. A nilpotent s operation can be constructed by noticing that there are also fermionic zero modes $s\phi^i$. The following s operation is nilpotent:

$$\begin{aligned} sb_{zz} &= \sum c_i \phi^i + d_i s\phi^i \\ sd_i &= -c_i, \qquad sc_i = 0! \end{aligned} \tag{46}$$

It is not quite clear whether it is reasonable to use (46) as a basis for gauge fixing.

To conclude, I would say that not much is known about situations where a noninvolutive distribution is involved, e.g., can it be viewed as the restriction of an involutive one? (The auxiliary field problem.) Further knowledge is clearly needed!

ACKNOWLEDGMENTS. I wish to thank C. Becchi for detailed information on his work on strings; M. Tonin for communicating his interpretation of the Batalin Vilkovisky generalization and allowing me to report on it here; L. Baulieu and M. Henneaux for stimulating my interest in this subject; L. Bonora and C. Reina for pointing out the use of projective connections in gluing anomalies on Riemann surfaces. I also wish to thank the organizers of the meeting that this volume arose from for their kind invitation and warmest welcome.

REFERENCES

1. E. Cartan, *Leçons sur les invariants intégraux*, Hermann, Paris, 1971 (first edition, 1922), *Les systèmes différentiels extérieurs et leurs applications géométriques*, Hermann, Paris, 1971 (first edition 1945, Act. Sci. Ind. No. 994).
2. L. D. Faddeev and V. N. Popov, *Phys. Lett. 25B*, 29 (1967).
3. A. A. Slavnov, *Theor. Math. Phys. 10*, 153 (1972).
4. E. S. Fradkin and G. A. Vilkovisky, *Phys. Lett. 55B*, 224 (1975), CERN report No. TH.2332 (1977); I. A. Batalin and G. A. Vilkovisky, *Phys. Lett. 69B*, 309 (1977); M. Henneaux, *Phys. Rep. 126*, 1–66 (1985); B. Kostant and S. Sternberg, *Ann. Phys.* (*N.Y.*) *176*, 49 (1987); M. Dubois-Violette, *Ann. Inst. Fourier, 37*, 45 (1987).
5. M. B. Green, J. H. Schwarz, and E. Witten, *Superstring Theory*, Vol. 1, Cambridge University Press, Cambridge, 1987; D. Friedan, E. Martinec, and S. Shenker, *Nucl. Phys. B 271*, 93 (1986).
6. J. L. Koszul, *Bull Soc. Math. France 78*, 65 (1950).
7. T. Regge, in *Relativity Groups Topology II*, Les Houches 1983, session XL (B. S. de Witt and R. Stora, eds.), North-Holland, Amsterdam, 1984; L. Castellani, R. d'Auria, and P. Fré, *Supergravity, A Geometric Perspective*, World Scientific, Singapore, to appear; S. Boukraa, The BRS algebra of a free minimal differential algebra, *Nucl. Phys. B 303*, 237 (1988).
8. L. Baulieu and J. Thierry-Mieg, *Nucl. Phys. B 288*, 259 (1983); L. Baulieu, *Phys. Lett. 126B*, 455 (1983); L. Baulieu and J. Thierry-Mieg, *ibid. 145B*, 53 (1984); L. Baulieu and M. Bellon, *ibid. 161B*, 96 (1985).
9. A. Rouet and R. Stora, Models with renormalizable Lagrangians: Perturbative approach to symmetry breaking, Enseignement du 3ème cycle de la Physique en Suisse romande, pp. 47–50, 61–63, June 1973.
10. C. Becchi, A. Rouet, and R. Stora, *Phys. Lett. 52B*, 344 (1974); *Commun. Math. Phys. 42*, 127 (1975), *Ann. Phys. 98*, 287 (1976); C. Becchi, *Commun. Math. Phys. 39*, 329 (1975); G. Bandelloni, C. Becchi, A. Blasi, and R. Collina, *Ann. IHP XXVIII*, 225 (1978), 255 (1978).
11. C. Becchi, A. Rouet, and R. Stora, in *Renormalization Theory*, Erice 1976 (G. Velo and A. S. Wightman, eds.), Reidel, NATO ASI Series C, Vol. 23, 1976.
12. J. H. Lowenstein and B. Schroer, *Phys. Rev. D 6*, 1533 (1972); K. Symanzik, Lectures in Lagrangian Quantum Field Theory (Islamabad 1968), DESY report T-71/1 (1971).
13. N. K. Nielsen, *Phys. Lett. 103B*, 197 (1981); R. E. Kallosh, *Nucl. Phys. B 141*, 141 (1978); F. R. Ore and P. van Nieuwenhuizen, *Phys. Lett. 112B*, 364 (1982).
14. J. Zinn Justin, in *Trends in Elementary Particle Theory*, Bonn, 1974 (H. Rollnick and K. Dietz, eds.), Lecture Notes in Physics, Vol. 37, Springer-Verlag, Berlin, 1975.
15. J. Wess and B. Zumino, *Phys. Lett. 37B*, 95 (1971).
16. C. Becchi, A. Rouet, and R. Stora, in *Field Theory Quantization and Statistical Physics* (E. Tirapegui, ed.), Reidel, Dordrecht, 1981.
17. T. Kugo and I. Ojima, *Prog. Theor. Phys. 66*, Suppl. 1979.
18. F. R. Ore and P. van Nieuwenhuizen, *Nucl. Phys. B 204*, 317 (1982); L. Baulieu, B. Grossmann and R. Stora, *Phys. Lett. 180B*, 55 (1986).
19. L. Brink, P. di Vecchia, and P. S. Howe, *Phys. Lett. 65B*, 471 (1976); S. Deser and B. Zumino, *ibid. 65B*, 396 (1976); A. M. Polyakov, *ibid. 103B*, 207 (1981); L. Baulieu, C. Becchi, and R. Stora, *ibid. 180B*, 55 (1986).

20. M. Kato and K. Ogawa, *Nucl. Phys. B 212*, 443 (1983); R. Marnelius, *ibid. 211*, 14 (1983); S. Hwang, *Phys. Rev. D 28*, 2614 (1983); C. Becchi, On the covariant quantization of the free string: The conformal structure, *Nucl. Phys. B 304*, 513 (1988); L. Baulieu and M. Bellon, *Phys. Lett. 196B*, 142 (1987); L. Baulieu, M. Bellon, and R. Grimm, *ibid. 198B*, 343 (1987).
21. L. Baulieu and M. Bellon, *Phys. Lett. 196B*, 142 (1987); J. P. Ader and J. C. Wallet, *ibid. 192B*, 103 (1987); *192B*, 103 (1987); *200B*, 285 (1988); M. Abud and J. P. Ader, *ibid. 212B*, 320 (1988); J. P. Ader, A. Bonda, and J. C. Wallet, *ibid. 215B*, 111 (1988).
22. D. Sullivan, IHES Publication No. 47, p. 269 (1977).
23. *Symposium on Anomalies, Geometry, Topology*, Argonne Chicago 1985 (W. A. Bardeen and A. R. White, eds.), World Scientific, Singapore, 1985; M. Dubois Violette, M. Talon, and C. M. Viallet, *Commun. Math. Phys. 102*, 105 (1985); J. Mañes, R. Stora, and B. Zumino, *Commun. Math. Phys. 102*, 157 (1985); L. Alvarez Gaumé and P. Ginsparg, *Ann. Phys. 161*, 423 (1985); L. Alvarez Gaumé, in *Erice 1985* (G. Velo and A. S. Wightman, eds.), Reidel, Dordrecht, 1986. L. Bonora, P. Cotta Ramusino, M. Rinaldi, and J. Stasheff, *Commun. Math. Phys. 112*, 237 (1987); *114*, 381 (1988).
24. I. M. Singer, in Colloque Elie Cartan, Lyon 1984, Asterisque 1985.
25. D. Gross and R. Jackiw, *Phys. Rev. D 6*, 477 (1972).
26. G. Bandelloni, private communication.
27. D. Friedan, *Ann. Phys. 163*, 318 (1985); C. Becchi, talk at the *Ringberg Workshop 1987* (P. Breitentohner, D. Maison, and K. Sibold, eds.), to appear.
28. K. Stelle and P. C. West, *Phys. Lett. 74B*, 330 (1978); *Nucl. Phys. B 140*, 285 (1978); S. Ferrara, M. T. Grisaru, and P. van Nieuwenhuizen, *ibid. 138*, 430 (1978); L. Baulieu and M. Bellon, *Phys. Lett. 161B*, 96 (1985); *Nucl. Phys. B 266*, 75 (1986); *Phys. Lett. 169B*, 59 (1986); R. Grimm, *Nucl. Phys. B 294*, 279 (1987).
29. B. de Wit and J. van Holten, *Phys. Lett. 79B*, 389 (1979); I. A. Batalin and G. A. Vilkovisky, *ibid. 102B*, 27 (1981); I. A. Batalin, *Nucl. Phys. B 234*, 106 (1984); *J. Math. Phys. 26*, 172 (1985); *Phys. Rev. D 28*, 2567 (1983); I. V. Tyutin and Sh. M. Shwartsman, *Phys. Lett. 169B*, 225 (1986).
30. R. C. Gunning, *Lectures on Riemann Surfaces*, Princeton Mathematical Notes, Princeton University Press, Princeton, New Jersey, 1966.

Chapter 16

Supermembranes, Superstrings, and Supergravity

P. van Nieuwenhuizen

1. INTRODUCTION

A fundamental problem in particle physics concerns the structure of elementary particles: whether they are pointlike or extended objects. Lorentz [1] already considered relativistic models for an extended electron, but his description was not Lorentz invariant, and contrary to what is done in modern string theory, the boundaries of his objects did not propagate with the speed of light. Dirac [2] considered a shell-like electron with a surface charge whose action is the sum of the Maxwell action outside the shell, plus a Nambu–Goto area term for the shell. The repulsive electric forces counterbalance the attractive surface tension, and one can compute the quantum mechanical energy levels, choosing the strength of the surface tension such that the lowest state is the electron mass. Of course, the size of this electron is far too large. His action was Lorentz invariant, but it did not contain the notion of spin (it did, however, predict a muon as an excited state of the electron, with a mass of $53m_e$).

At the conference, the author presented two talks, one on the questions of massless modes for supermembranes and one on nonlinear σ models with nonvanishing Nijenhuis tensors. See the "Note to the reader" at the end of this chapter.

P. VAN NIEUWENHUIZEN • Institute for Theoretical Physics, State University of New York at Stony Brook, Stony Brook, New York 11794.

Strings are the first really successful relativistic models for extended objects. Membranes are the next in the series

Points, strings, membranes, . . .

More generally, one can consider p-dimensional objects called "p-branes" propagating in a (flat or curved) d-dimensional space-time. It is widely believed that supersymmetry is necessary for the consistency of strings. Indeed, the bosonic string, which is obviously not supersymmetric, has a tachyon (negative mass-squared) and the $O(16) \times O(16)$ heterotic string [3] which has no massless gravitinos in its spectrum and thus leads to a nonsupersymmetric theory in space-time, generates at the one-loop level an unwanted cosmological constant and is therefore presumably not finite. Supersymmetry seems necessary for quantum-finiteness, and we shall extend this belief in supersymmetry to p-branes and consider only super p-branes. Super p-branes were discovered in the beginning of 1987 by Bergshoeff, Sezgin, and Townsend [4], building on the work of Polchinsky *et al.* [5].

Superstrings correspond to the case $p = 1$ and $d = 10$, while, as we shall see, supermembranes correspond to $p = 2$ and $d = 11$. In $d = 10$ one has also objects with $p = 5$. From the mathematical point of view, a necessary condition for a super p-brane to exist in flat d-dimensional space-time is that the Dirac matrices satisfy the identity [6]

$$(\Gamma^{m_1})_{(\alpha\beta}(\Gamma_{m_1 \dots m_p})_{\gamma\delta)} = 0 \tag{1}$$

where $(\alpha\beta, \gamma\delta)$ means totally symmetric in $\alpha\beta\gamma\delta$. The relation (1) is satisfied for the cases shown in Fig. 1.* In all cases, super p-branes indeed exist. As is well known, the $d = 10$ string case is consistent while the $d = 11$ membrane case may be consistent as well, but the other models are believed to be inconsistent [7].

Supergravity models exist in dimensions up to $d = 11$. The $d = 10$ supergravity models are the 2a, 2b, 1a, and 1b models. The 2a and 2b models have two local supersymmetries and two gravitinos (spin-3/2 particles) with opposite or equal chiralities, respectively, and are the low-energy limit of the IIA and IIB closed string, whereas the 1a supergravity model (the model with an antisymmetric tensor $A_{\mu\nu}$) corresponds to the closed type I string. There is also the 1b model, obtained by dualizing the curl $\partial_{[\mu} A_{\nu\rho]}$ into a curl $\partial_{[\mu_1} A_{\mu_2 \cdots \mu_7]}$ of a six-index antisymmetric tensor. Finally in $d = 10$ there is also the super Yang–Mills model, which is the low-energy limit of the open string.

It was until a year ago an open question whether the $n = 1$ $d = 11$ and $n = 1\text{b}$ $d = 10$ supergravities did correspond to an extended object. The

* In certain cases one has symplectic Majorana spinors or Majorana–Weyl spinors instead of ordinary Majorana spinors. In these cases, (1) is slightly modified.

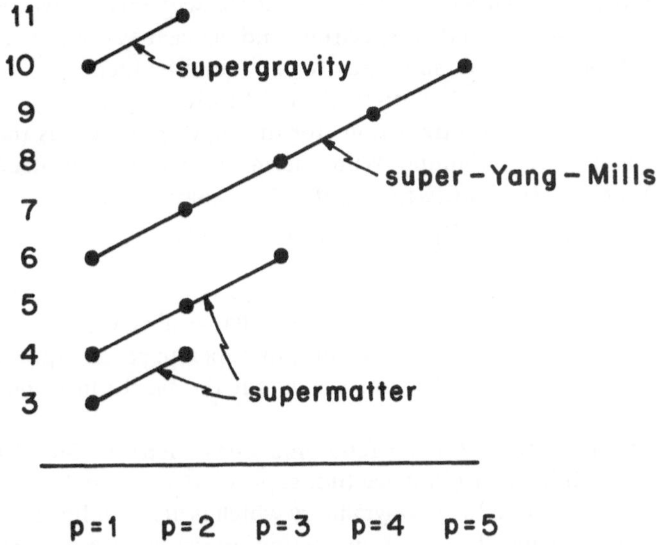

Figure 1. *p* branes in *d*-dimensional Minkowski space-time.

answer is now known: they correspond to the $p = 2$ $d = 11$ and the $p = 5$ $d = 10$ superbrane, respectively.

To be more precise, in the $d = 11$ $p = 2$ case it has been shown that if one couples the supermembrane to external supergravity superfields, then these superfields must satisfy the $d = 11$ superspace supergravity constraints (and hence the $d - 11$ supergravity field equations) in order that there be world-volume supersymmetry (induced by local κ symmetry; see below) [4].

[One remaining mystery concerns a possible $n = 1$b $d = 11$ supergravity theory. So far attempts to construct it have failed (because the action contains also bare $A_{\mu\nu}$ fields and hence obvious dualization is not possible). Also there is no super *p*-brane in Fig. 1 to which it could correspond. Yet, less obvious dualization methods have been useful in constructing new models (new $d = 7$ supergravity models), and some physicists, including the author, keep wondering about a possible $n = 1$b $d = 11$ model. We will not pursue this subject any further.]

For the lower three branches in Fig. 1, the *p*-brane theories seem inconsistent according to the following argument due to Bars, Pope, and Sezgin. Coupling the *p*-brane to background supergravity fields, one expects that for consistency the massless modes of the free completely collapsed *p*-brane must be the same as the massless modes of the supergravity background model. However, the lower three sequences have no supergravity massless modes in their spectrum, and hence they seem inconsistent. Thus, finally $n = 1$ $d = 11$ supergravity has found its niche.

In what follows we shall discuss the following aspects:

(1) The Green–Schwarz actions for the super p-branes as the sum of a volume and a Wess–Zumino–Witten term. We will see the need for the $d = 11$ Dirac matrix identity for the $d = 11$ membrane [6]

$$(d\bar{\theta}\Gamma^{mn}\, d\theta)(\bar{\theta}\Gamma_m\, d\theta) + (\bar{\theta}\Gamma^{mn}\, d\theta)(d\bar{\theta}\Gamma_m\, d\theta) = 0 \tag{2}$$

where $d\theta$ is a commuting fermionic 1-form. [Taking the exterior derivative of (2) reproduces (1).] We will consider a flat ordinary spacetime rather than putting the string in a background of superspace supergravity fields because this is much simpler and brings out the basics in a more transparent way.

(2) The need for the local fermionic κ-symmetry of Siegel [8]. The supermembranes have a rigid space-time supersymmetry which is *nonlinearly* realized ($\delta\theta = \varepsilon$) and a local κ-symmetry which is also nonlinearly realized [$\delta\theta = (1+\Gamma)\kappa$] where $1+\Gamma$ is a projection operator. In a suitable gauge, they combine into a rigid world-sheet supersymmetry, which is linearly realized {so that δ(boson) = fermion, δ(fermion) = boson [9]}. Hence super p-branes *with* κ symmetry *must* have an equal number of bosonic and fermionic modes on the world volume. Indeed, they all do. The question arises whether there are super p-branes without local κ-symmetry where the number of bosons and fermions is not equal. This question has recently been considered, and the answer is negative. For example, for the string in $d = 10$ one has 8 X-modes (10 coordinates X^μ minus two world-sheet general coordinate transformations; choose for example the gauge $X^0 = \tau$ and $X^9 = \sigma$) and also 8 θ-modes (because there are two Majorana–Weyl spinors θ^A for the superstring, having 2×16 components, but local κ-symmetry cancels half of them, while the Dirac equation in two dimensions tells one that the modes are either left- or right-moving). For the membrane one has $11 - 3 = 8$ bosonic modes, but now one has only one (32-component) spinor θ^a. The κ-symmetry leaves 16 fermionic modes, while half of the θ's are conjugate momenta and the other half coordinates, so that there are again 8 fermionic modes.

(3) For the $d = 10$ string the massless modes in its spectrum correspond in one-to-one fashion to the massless background fields to which one can couple the string (these are the $d = 10$ supergravity fields). Consistency presumably requires that the same be true for the $d = 11$ supermembrane. The $n = 11$ supergravity fields are the graviton $g_{\mu\nu}$ (44 states), the antisymmetric tensor $A_{\mu\nu\rho}$ (84 states), and the gravitino ψ_μ (128 states). The question thus poses itself: are there massless modes in the spectrum of the supermembrane, and if so, do they correspond to $g_{\mu\nu}$, $A_{\mu\nu\rho}$, and ψ_μ?

We shall begin, however, in Section 2 with two properties of membranes that are potentially disastrous. Perhaps, however, a certain benevolence towards membranes is temporarily justified.

Comment 1. Each of the four series in Fig. 1 can be obtained from the model with highest d by "double-dimensional reduction" (torus-compactification): putting $\xi^{p+1} = X^{p+1}$ and letting all $(d-1)$-dimensional fields become independent of ξ^{p+1}. The Nambu-Goto action goes then over into the action for one dimension lower [15].

Comment 2. There are indications that all super p-branes except the $d = 10$ string and $d = 11$ membranes in the light-cone gauge seem to have Lorentz anomalies. As to local κ anomalies, a classification of possible anomalies (cohomologies) has been given for the string, but whether these anomalies are really present is not known.

Comment 3. For the superstring, a candidate counterterm containing extrinsic curvature [starting with $(\Box X)^2$] has recently been given [10]. One could try to do the same for the supermembrane.

2. NONLINEARITY, ABSENCE OF WEYL INVARIANCE, AND NONRENORMALIZABILITY

There are two fundamental differences between (super)membranes and superstrings. In the so-called light-cone gauge, (super)strings in a flat background (linear σ-models) are described by two-dimensional free field theories, and, consequently, one can exactly solve their spectrum. The reason that this is possible is that a graviton has three field components in two dimensions, while the action has also three local symmetries: two general coordinate invariances and local Weyl invariance. (Under local Weyl invariance, $\delta X^\mu = 0$, $\delta\lambda^\mu = -\frac{1}{2}\Lambda\lambda^\mu$ or $\delta\theta = 0$ and $\delta g_{\mu\nu} = \Lambda g_{\mu\nu}$. For a particle this works, too: The one-component metric can be removed by the one-degree general coordinate invariance.) For membranes the metric of the three-dimensional world volume has six components, but there are only three general coordinate invariances and *no local Weyl symmetry.* Consequently, (super) membranes are not described by three-dimensional free field theories in *any* gauge, and thus it seems from the outset a hopeless task to try to repeat all impressive accomplishments of string theory. This, of course, is not a defect of membranes but rather of our present abilities, but it may well prevent real progress in membrane theory.*

The second defect of (super)membranes is their nonrenormalizability. Let us compare the situation with strings. On the two-dimensional world

* One can write down a membrane model with Weyl invariance. It starts with $\partial x\, \partial x\, \partial x\, \partial x$, but the model is still not free.

sheet of strings, the gravitational coupling constant κ is dimensionless. One can consider two kinds of nonlinear σ-models.

(I) Those in which the scalar (and spin-$\frac{1}{2}$) fields couple only to external massless fields, namely, to the graviton, antisymmetric tensor $A_{\mu\nu}$, and dilaton ϕ. (One can also consider a massless external gravitino. That is best studied by using external superfields with the Green–Schwarz formalism.) These models are renormalizable in a generalized sense.

(II) Models in which one couples to all, massless *and* massive, external fields. (Each external field corresponds one-to-one to a mode of the string.) One can require that β-functions for these models vanish, which fixes some of the coefficients in front of these extra terms but not all, so that the theory is not finite. The action by itself is not conformally invariant, but (or, rather, because) "the theory" (Green's functions) are. The fact that the models in I are a consistent truncation of the models in II may not be so important. Perhaps we should be doing string theory only with the more general class of models in II, even though they are nonrenormalizable. (Note that the coupling constants for the higher-derivative terms in the action have the wrong dimension.)

For membranes, the gravitational coupling constant is dimensionful and it has the wrong kind of dimension. (To see this, note that the metric g can be decomposed as $\eta_{\mu\nu} + \kappa h_{\mu\nu}$, where $h_{\mu\nu}$ has the standard type of action $I = \int d^dx(\partial h)^2$. Hence in $d = 3$, the dimension of $h_{\mu\nu}$ equals $[h_{\mu\nu}] = \frac{1}{2}$, and thus $[\kappa] = -\frac{1}{2}$. In $d = 4, [h_{\mu\nu}] = 1$ and $[\kappa] = -1$.) Hence, the linear σ-model in three dimensions is not a free theory, and hence membranes are not renormalizable. One might wonder why one bothers to study $d = 3$ models, instead of directly tackling the $d = 4$ models, since both classes of models have the same disease. One reason might be that anomalies cancel only in the $d = 3$ case. Another (rather weak) argument might be that divergences in $d = 3$ are less severe than in $d = 4$. If one adopts the point of view, however, that supermembranes should be treated like the type II σ models in two dimensions, membranes are not worse than strings. In both cases one is dealing with a nonrenormalizable quantum field theory on the world sheet or world volume.

It is not ruled out that for particular (AdS) backgrounds in a particular gauge ($X^0 = \tau, X^1 = \rho, X^2 = \sigma, \Gamma^+\theta = 0$) the models become free in a particular limit (radius $R \to \infty$) [11]. This is the best we can say about this problematic situation.

3. CLASSICAL GS SUPERSTRINGS AND CLASSICAL GS SUPERMEMBRANES

A supersymmetric string can be formulated either as a Ramond–Neveu–Schwarz (RNS) string or as a Green–Schwarz string. In the first case, world-sheet spinors are space-time vectors (written as λ^μ with $\mu =$

$0, \ldots, d-1$), while in the latter case they are space-time spinors. In the RNS formulation one has a supergravity theory in two dimensions. There is $N = 1$ local worldsheet SUSY. (The $N = 2$ and $N = 4$ models have critical dimension 2 and -2, respectively, and seem not useful.) To obtain a supersymmetric spectrum one needs Gliozzi–Scherk–Olive (GSO) projection operators. The critical dimension of space-time (the dimension where the *quantum* theory is consistent because Weyl anomalies cancel) is $d = 10$, although classically the RNS strings can exist in any d. An action for the classical GS string can only be written down in $d = 10, 6, 4, 3$. It has (one or two) rigid space-time supersymmetries and a new local fermionic symmetry, called κ-symmetry, due to Siegel [8]. At the quantum level, one has only been able to quantize the GS string in the light-cone gauge, and then Lorentz invariance is only preserved in the $d = 10$ model.

For the supermembrane, also a RNS membrane and a GS membrane exist. The RNS membrane is now a supergravity theory in three dimensions. More precisely: it is a scalar multiplet coupled to supergravity fields. There seems no way to write this action only in terms of the scalar and spinor fields X^μ and λ^μ because one needs for that purpose a cosmological constant, and in supergravity theories a cosmological constant requires the presence of the Einstein action, so that the graviton field equation becomes propagating. This excludes the possibility to eliminate the metric from its own nonpropagating field equation. [A recent possible alternative is to consider instead of the Polyakov-type action $(\partial_\mu X)^2$ an action like $(\partial X)^4$ in which case no cosmological constant is needed to recover the Nambu-Goto forms. A propagator could only arise by expansion around a background.] In addition, the GSO projection operators, if existing, are unknown. Given these problems, it is not clear how to define the critical dimension for a RNS membrane, but note that the membrane action is not conformally invariant to begin with, so that conformal anomalies cannot even occur. One might perhaps get global gravitational anomalies ("modular anomalies").

The GS membrane, on the other hand, fares better: it exists classically in $d = 11, 7, 5, 4$, has one ($N = 1$) rigid space-time SUSY, and local κ supersymmetry. It is not clear whether any of these models are consistent at the quantum level, although in analogy with the string one might conjecture that only the $d = 11$ model is consistent, if at all.

These properties of the RNS/GS models for string/membranes are summarized in Fig. 1. We now give details of the construction of actions.

The action for a bosonic p-brane is

$$I = -T \int d^{p+1}\xi(-\det \partial_\alpha X^\mu \, \partial_\beta X_\mu)^{1/2}, \qquad \alpha, \beta = 1, \ldots, p+1 \qquad (3)$$

For the bosonic membrane this action was written down in 1962 by Dirac, while it was proposed as the starting point for Lagrangian string theory by

Nambu and Goto in 1969. It is proportional to the world volume, and the constant T is the p-brane tension with dimension

$$[T] = [m][l]^{1-p}[t]^{-1} \tag{4}$$

The action can be obtained from a first-order action with cosmological constant

$$\mathscr{L} = T(-\tfrac{1}{2}(-g)^{1/2}g^{\alpha\beta}\,\partial_\alpha X^\mu\,\partial_\beta X_\mu + \tfrac{1}{2}(p-1)(-g)^{1/2}) \tag{5}$$

by eliminating the independent $(p+1)$-dimensional metric $g_{\alpha\beta}$ from its algebraic field equation. One finds then that $g_{\alpha\beta}$ is the induced metric

$$g_{\alpha\beta} = \partial_\alpha X^\mu\,\partial_\beta X_\mu \tag{6}$$

In what follows we shall use the Nambu–Goto formulation of the action. For the string, $p = 1$, and the cosmological constant is absent; as a result the string, and only the string, has an extra local symmetry, namely, local Weyl invariance.

The simplest way to supersymmetrize the action of the bosonic p-brane is to replace it by the coupling of a scalar multiplet (whose bosonic components include X^μ) to $(p+1)$-dimensional supergravity. In this case one must start from the first term in (5). However, in order still to obtain (3), one has to also get a cosmological constant in the action [the second term in (5)], and in supergravity models this means that one must add a supercosmological constant, proportional to

$$S(-g)^{1/2} + \bar{\psi}_\mu\gamma^{\mu\nu}\psi_\nu(-g)^{1/2} \tag{7}$$

In order to eliminate the field S while keeping the cosmological constant, one adds the supergravity gauge action, which itself contains not only the Einstein action but also a term $(-g)^{1/2}S^2$, and by eliminating S one indeed gets the cosmological constant $(-g)^{1/2}$. However, owing to the presence of the Einstein action, one can no longer solve for $g_{\mu\nu}$ algebraically. Hence as supersymmetric extensions of the Nambu–Goto action, spinning p-branes are excluded. Instead, one must use GS p-branes. [Replacing $(\partial X)^2$ by a $(\partial X)^4$ type of action does not require S; see remarks before.]

Let us remark that for the string with $p = 1$, one has no cosmological constant and one can use a spinning string as well as a superstring formulation. In the light-cone gauge, the spinors in the former case are $SO(8)$ vectors, while in the latter case they are $SO(8)$ spinors. One can then prove the equivalence of both formulations by using "triality." [One uses $\lambda^\mu = \chi^{\dot{a}}(\Gamma^\mu)_{\dot{a}a}\lambda^a$, where $\chi^{\dot{a}}$ is a commuting nowhere vanishing spinor. Such spinors globally exist on any two-dimensional surface, see Ref. 12.]

The complete spectrum of the spinning string coincides with that of the superstring if one takes both periodic and antiperiodic solutions in both cases [13]. In that case, both formulations have a tachyon. One can also prove that the GSO-projected spectrum of the spinning string coincides with the spectrum of the superstring with periodic boundary conditions. In that case neither has a tachyon.

To construct a super p-brane, we first of all replace dX^μ by its supersymmetric-invariant extension

$$\pi^\mu = dX^\mu + \bar{\theta}\Gamma^\mu\, d\theta \tag{8}$$

(It is supersymmetric under $\delta\theta \sim \varepsilon$ provided δX^μ is such that $\delta\pi^\mu = 0$.) Next one adds a Wess-Zumino-Witten term. The general procedure is to first find a closed $(p+2)$-form ω_{p+2}, and then to show that $\omega_{p+2} = d\omega_{p+1}$. In other words ω_{p+2} is exact. Since cohomology in superspace is usually trivial, one can start with a closed $p+2$ form and expect that $d\omega_{p+2} = 0$ implies that $\omega_{p+2} = d\omega_{p+1}$. This is how we shall construct ω_{p+1}. The WZW term is then the $(p+1)$-dimensional integral of ω_{p+1}. (The corresponding cohomology in *local* superspace is also trivial. This may be related to the fact that the bosonic topological charges, such as the Euler number, do not seem to get fermionic corrections.)

$d = 10$ GS superstrings. (See also Ref. 14.) A closed 3-form is in this case

$$\omega_3 = (d\bar{\theta}^1\, \Gamma^m\, d\theta^1 - d\bar{\theta}^2\, \Gamma^m\, d\theta^2)\pi_m \tag{9}$$

The fermionic 1-forms $d\theta^A = d\xi^j \partial_j \theta^A$ are commuting (because both θ^A and $d\xi^j$ are anticommuting) and π^m is the bosonic supersymmetric 1-form

$$\pi^m = dX^m + \bar{\theta}^1\Gamma^m\, d\theta^1 + \bar{\theta}^2\Gamma^m\, d\theta^2 \tag{10}$$

The closure of ω_3 follows from the identity

$$(d\bar{\theta}\,\Gamma^m\, d\theta)(d\bar{\theta}\,\Gamma_m\, d\theta) = 0 \qquad \text{in } d = 10 \tag{11}$$

We claim that the 2-form ω_2 is given by (we will fix α presently)

$$\omega_2 = (\bar{\theta}^1\Gamma^m\, d\theta^1 - \bar{\theta}^2\Gamma^m\, d\theta^2)\pi_m + \alpha(\bar{\theta}^1\Gamma^m\, d\theta^1)(\bar{\theta}^2\Gamma_m\, d\theta^2) \tag{12}$$

To show that $d\omega_2 = \omega_3$ we will use the identity

$$\boxed{\Gamma^m\, d\theta\, d\bar{\theta}\,\Gamma_m\, d\theta = 0 \qquad \text{in } d = 10} \tag{13}$$

which is equivalent to (11). Then $d\omega_2$ indeed equals ω_3 provided $\alpha = -1$. The GS superstring action is thus (we will fix a presently)

$$I_{\text{GS}}^{\text{string}} = I_{\text{bos}} + I_{\text{WZW}}$$

$$I_{\text{bos}} = -T\int d^2\xi[-\det(\Pi_\alpha^\mu\, \Pi_{\beta\mu})]^{1/2}$$

$$\begin{aligned} I_{\text{WZW}} = a\int d^2\xi\, \varepsilon^{\alpha\beta}[&(\bar{\theta}^1\Gamma_m\partial_\alpha\theta^1 - \bar{\theta}^2\Gamma_m\partial_\alpha\theta^2)\Pi_\beta^m \\ &- (\bar{\theta}^1\Gamma^m\partial_\alpha\theta^1)(\bar{\theta}^2\Gamma_m\partial_\beta\theta^2)] \end{aligned} \tag{14}$$

In order that this action be supersymmetric under

$$\delta\theta^A = \varepsilon^A, \qquad \delta X^\mu = \sum_{A=1}^{2} \bar{\theta}^A \Gamma^m \delta\theta^A \qquad (\text{hence } \delta\pi^\mu_\alpha = 0) \tag{15}$$

we rewrite I_{WZW} in a simpler form, using $(\bar{\theta}^1\Gamma^m\, d\theta^1)(\bar{\theta}^1\Gamma_m\, d\theta^1) = 0$

$$\begin{aligned} I_{\text{WZW}} = a \int [(&\bar{\theta}^1\Gamma_m\, d\theta^1 - \bar{\theta}^2\Gamma_m\, d\theta^2)\, dX^m \\ &+ (\bar{\theta}^1\Gamma^m\, d\theta^1)(\bar{\theta}^2\Gamma_m\, d\theta^2)] \end{aligned} \tag{16}$$

The variation δI_{WZW} due to $\delta\theta^1 = \varepsilon^1$ contains terms involving only θ^1, cross-terms involving both θ^1 and θ^2 and terms with dX^m. The last type of terms cancels, being a total derivative. This is the lesson to be learned from the WZW term: A Lagrangian built from geometrical objects need not be invariant, but may still transform into a total derivative (a well-known fact from general relativity and supersymmetry, but relatively late realized for nonlinear σ-models). The cross-terms cancel at once, and the terms involving only θ^1 cancel using (13). [These terms read $(\bar{\theta}^1\Gamma_m\, d\theta^1)(d\bar{\theta}^1\Gamma^m\varepsilon)$ and are totally symmetric in θ^1, $d\theta^1$, and $d\bar{\theta}^1$ if one is allowed partial integration.]

The constant a in front of I_{WZW} is still left free by rigid SUSY, but local κ symmetry will fix it. Under local κ symmetry $\delta\theta = (1+\Gamma)\kappa$, $\delta X^\mu = -\bar{\theta}\Gamma^\mu\,\delta\theta$, with $\Gamma^2 = +1$ (see below), and I_{bos} will give a variation proportional to $(1+\Gamma)\kappa$ while I_{WZW} will give $-a(1+\Gamma)\kappa$. Together one obtains

$$\delta I(\text{total}) = (1+\Gamma)(1-a\Gamma)\kappa(\cdots) \tag{17}$$

and hence for local κ invariance one needs $a = 1$.

For the IIB string, θ^1 and θ^2 have the same chirality, while for the IIA string they have opposite chirality. For the open string, boundary terms in the SUSY (and κ-symmetry) variations cancel provided $\theta^1(\sigma) = \theta^2(\sigma)$ at $\sigma = 0$ and $\sigma = \pi$, so this action has only $N = 1$ SUSY. One can drop θ_2 in the discussion above, but then the action becomes inconsistent at the quantum level. (Anomalies do not cancel.) To make it consistent one must add heterotic fermions and consider closed strings. To obtain the $N = 1$ closed string one must put $\theta(\sigma) = \theta(\pi - \sigma)$.

$d = 11$ GS supermembranes. The action is now $I_{\text{bos}} + \int \omega_3$, where $d\omega_3 = \omega_4$ and ω_4 is a closed 4-form. Since we are in $d = 11$, there are only Majorana spinors, and, as discussed before, we need one spinor θ^a ($a = 1, 32$) in order to get equal numbers of bosons and fermions. (The equality $11 - 3 = \frac{32}{4}$). In order that ω_4 be supersymmetrically invariant, it must be constructed

from π^m and $d\theta$, and it must be a 4-form. This uniquely leads to

$$\omega_4 = d\bar{\theta}\,\Gamma^{mn}\,d\theta\,\pi_m\pi_n \tag{18}$$

This 4-form is closed due to the $d = 11$ identity [6]

$$d\bar{\theta}\,\Gamma^{mn}\,d\theta\,d\bar{\theta}\,\Gamma_m\,d\theta = 0 \tag{19}$$

To find ω_3 with $d\omega_3 = \omega_4$ we make the ansatz

$$\omega_3 = \bar{\theta}\Gamma^{mn}\,d\theta[\pi_m\pi_n + \alpha\bar{\theta}\Gamma_m\,d\theta\,\pi_n + \beta\bar{\theta}\Gamma_m\,d\theta\bar{\theta}\Gamma_n\,d\theta] \tag{20}$$

with α and β arbitrary constants. Using the Dirac matrix identity

$$\boxed{(d\bar{\theta}\Gamma^{mn}\,d\theta)(\bar{\theta}\Gamma_m\,d\theta) + (\bar{\theta}\Gamma^{mn}\,d\theta)(d\bar{\theta}\Gamma_m\,d\theta) = 0 \qquad \text{in } d = 11} \tag{21}$$

which is equivalent to (19) we see that $d\omega_3 = \omega_4$ provided $\alpha = -1$ and $\beta = -\frac{1}{3}$. Hence the supermembrane action reads

$$\begin{aligned} I = -T \int d^3\xi\{[(-\det \pi_i^\mu \pi_j^\nu \eta_{\mu\nu})]^{1/2} - \frac{a}{2}\varepsilon^{ijk}(\bar{\theta}\Gamma_{\mu\nu}\partial_i\theta) \\ \times [\partial_j X^\mu \pi_k^\nu - \tfrac{1}{3}(\bar{\theta}\Gamma^\mu\,\partial_j\theta)(\bar{\theta}\Gamma^\nu\,\partial_k\theta)]\} \end{aligned} \tag{22}$$

(Our way of writing the action is simpler than other forms in the literature because it contains one term less.) The action is supersymmetric under

$$\delta\theta = \varepsilon, \qquad \delta X^\mu = \bar{\theta}\Gamma^\mu\varepsilon \qquad (\text{hence } \delta\pi^\mu = 0 \text{ with } \pi^\mu = dX^\mu + \bar{\theta}\Gamma^\mu\,d\theta) \tag{23}$$

For example, the variations with $\partial_j X^m$ give

$$a\varepsilon^{ijk}(\bar{\varepsilon}\Gamma_{mn}\partial_i\theta)(\partial_j X^m\bar{\theta}\Gamma^n\partial_k\theta) + a\varepsilon^{ijk}\bar{\theta}\Gamma_{mn}\partial_i\theta(\partial_j\bar{\theta}\Gamma^m\varepsilon\partial_k X^n) = 0 \tag{24}$$

which cancels due to (21). [To see this, note that they are equal to

$$\begin{aligned} -\tfrac{1}{3}\Gamma_{mn}(\partial_i\theta)(\bar{\theta}\Gamma^n\partial_j\theta) + \tfrac{1}{3}\Gamma_{mn}\partial_j\theta(\bar{\theta}\Gamma^n\partial_i\theta) + \tfrac{1}{3}\Gamma_{mn}\theta\partial_i\bar{\theta}\Gamma^n\partial_j\theta \\ -\tfrac{1}{3}(\Gamma^n\partial_i\theta)(\bar{\theta}\Gamma_{mn}\partial_j\theta) + \tfrac{1}{3}\Gamma^n\partial_j\theta\bar{\theta}\Gamma_{mn}\partial_i\theta + \tfrac{1}{3}\Gamma^n\theta\partial_i\bar{\theta}\Gamma_{mn}\partial_j\theta \end{aligned} \tag{25}$$

which vanishes due to the identity in (21).] The variations without ∂X^m but with only five θ's vanish too because they combine again into the identity (21).

4. LOCAL κ SYMMETRY

The action for the $d = 11$ supermembrane reads, as we have seen,

$$\begin{aligned} I = -T \int d^3\xi\{(-\det \pi_i^\mu \pi_{j\mu})^{1/2} - \frac{a}{2}\varepsilon^{ijk}(\bar{\theta}\Gamma_{\mu\nu}\partial_i\theta) \\ \times [(\partial_j X^\mu)\pi_k^\nu - \tfrac{1}{3}(\bar{\theta}\Gamma^\mu\partial_j\theta)(\bar{\theta}\Gamma^\nu\partial_k\theta)]\} \end{aligned} \tag{26}$$

and is supersymmetric, even though it still contains the free parameter a. Local κ-symmetry will fix a to the value

$$a = 1 \tag{27}$$

as we now show.

Any variation of the action can be written as

$$\delta I = -T \int d^3\xi (B_\mu \delta X^\mu + \delta\bar{\theta} F) \tag{28}$$

We therefore first obtain the field equations. For the field equation of X^μ we find directly

$$B^\mu = -\partial_i[(-g)^{1/2} g^{ij} \pi_j^\mu + \frac{a}{2} \varepsilon^{ijk} (\bar{\theta}\Gamma^{\mu\nu}\partial_j\theta)(\pi_{k\nu} + \partial_k X_\nu)] = 0 \tag{29}$$

where g^{ij} is the inverse of $g_{ij} \equiv \pi_i^\mu \pi_{j\mu}$. (The variation of $\partial_j X^\mu$ gives the $\pi_{k\nu}$ term and the variation of the X in π gives the $\partial_k X_\nu$ term.) By using the identity

$$\varepsilon^{ijk}(\partial_i\bar{\theta}\Gamma^{\mu\nu}\partial_j\theta)(\bar{\theta}\Gamma_\nu\partial_k\theta) = \varepsilon^{ijk}(\bar{\theta}\Gamma^{\mu\nu}\partial_j\theta)(\partial_i\bar{\theta}\Gamma_\nu\partial_k\theta) = 0 \tag{30}$$

[which is the same identity as (21) if one replaces ε^{ijk} by $dx^i \wedge dx^j \wedge dx^k$], we can rewrite the bosonic field equation as

$$B^\mu = -\partial_i[(-g)^{1/2} g^{ij} \pi_j^\mu] - a\varepsilon^{ijk}(\partial_i\bar{\theta}\Gamma^{\mu\nu}\partial_j\theta)\pi_{k\nu} = 0 \tag{31}$$

In this form it is manifestly supersymmetric.

The field equation for θ is obtained by varying the explicit θ's as well as the θ's contained in the π's. One obtains after a straightforward but tedious computation

$$F = 2(-g)^{1/2} g^{ij} \pi_j^\mu \Gamma_\mu \partial_i\theta - B^\mu \Gamma_\mu \theta - a\varepsilon^{ijk}(\Gamma_{\mu\nu}\partial_i\theta)\pi_j^\mu \pi_k^\nu = 0 \tag{32}$$

Using that

$$\begin{aligned} \varepsilon^{ijk}\Gamma_{\mu\nu}\pi_j^\mu \pi_k^\nu &= 2(-g)^{1/2}\Gamma\Gamma_\mu \pi^{i\mu} \\ \Gamma &\equiv \tfrac{1}{6}\varepsilon^{ijk}\Gamma_{\mu\nu\rho}\pi_i^\mu \pi_j^\nu \pi_k^\rho \end{aligned} \tag{33}$$

one can (I thank Dr. Sezgin for pointing this out to me) rewrite the fermionic field equation as

$$F = 2(-g)^{1/2}(1 - a\Gamma)(\pi^{i\mu}\Gamma_\mu)(\partial_i\theta) - B^\mu \Gamma_\mu \theta = 0 \tag{34}$$

This follows easily from the identity

$$\pi_i^m \pi_j^n \pi_\kappa^p \pi_l^q \Gamma_{mnp} \Gamma_q = 3\pi_i^m \pi_j^n g_{kl} \Gamma_{mn} \tag{35}$$

which is due to the fact that i, j, k and l run only over three values, so that total antisymmetrization in all four of them yields zero (the Schouten "theorem" of the early days of supergravity). Notice that F is supersymmetric, except for the term $-\not{B}\theta$, which varies into $-\not{B}\varepsilon$. The origin of the terms in F is clear: The variation of the θ's in the Nambu-Goto action gives

$$\delta[(\bar{\theta}\Gamma^{\mu}\partial_i\theta)(\cdots)] = 2\delta\bar{\theta}\Gamma^{\mu}(\partial_i\theta)(\cdots) + (\delta\bar{\theta}\Gamma^{\mu}\theta)\partial_i(\cdots) + \text{a total derivative} \tag{36}$$

where $(\cdots) = (-g)^{1/2}g^{ij}\pi_j^{\mu}$ and $\partial_i(\cdots)$ is completed to $-B^{\mu}$. The variation of both θ's in $\bar{\theta}\Gamma^{\mu\nu}\partial_j\theta$ gives twice as much as varying only $\bar{\theta}$, except for terms with one or no π factor. These terms cancel separately, as one might either directly demonstate, or indirectly prove by showing that the action is supersymmetric with this F.

To show that the action is supersymmetric by using (32), we evaluate δI, using

$$\delta X^{\mu} = \bar{\theta}\Gamma^{\mu}\delta\theta, \qquad \delta\theta = \varepsilon \tag{37}$$

These are the correct rules (in particular the correct signs) since they leave π_i^{μ} invariant

$$\delta\pi_i^{\mu} = \partial_i\bar{\theta}\Gamma^{\mu}\varepsilon + \bar{\varepsilon}\Gamma^{\mu}\partial_i\theta = 0 \tag{38}$$

We thus obtain for the supersymmetry variation

$$\delta(\varepsilon)I = \int [\bar{\theta}\not{B}\varepsilon + 2\bar{\varepsilon}(-g)^{1/2}g^{ij}\not{\pi}_j\partial_i\theta - \bar{\varepsilon}\not{B}\theta - a\varepsilon^{ijk}(\varepsilon\bar{\Gamma}_{\mu\nu}\partial_i\theta)\pi_j^{\mu}\pi_k^{\nu}]\, d^3\xi \tag{39}$$

Partially integrating the second term, we obtain

$$-2(\bar{\varepsilon}\Gamma_{\mu}\theta)\partial_i[(-g)^{1/2}g^{ij}\pi_j^{\mu}] \tag{40}$$

which cancels $-2\bar{\varepsilon}\not{B}\theta$ except for a term

$$2a(\bar{\varepsilon}\Gamma_{\mu}\theta)\varepsilon^{ijk}(\partial_i\bar{\theta}\Gamma^{\mu\nu}\partial_j\theta)\pi_{k\nu} \tag{41}$$

The last term in $\delta(\bar{\varepsilon})I$ can also be partially integrated, and it yields

$$2a\varepsilon^{ijk}(\bar{\varepsilon}\Gamma_{\mu\nu}\theta)(\partial_i\pi_j^{\mu})\pi_k^{\nu} \tag{42}$$

Both terms are now totally antisymmetric in the spinors θ, $\partial_i\theta$, and $\partial_j\theta$, and their sum cancels according to the identity in (21).

To prove the local κ-invariance of the action, we begin with

$$\delta X^{\mu} = -\bar{\theta}\Gamma^{\mu}\delta\theta \text{ (note } -\text{sign)} \tag{43}$$

Owing to the minus sign, the B terms in δI in (28) cancel straightaway. (This fixes δX^{μ}.) We are left with

$$\delta(\kappa)I = -T\int d^3\xi\delta\bar{\theta}\{2(-g)^{1/2}(1-a\Gamma)g^{ij}\not{\pi}_i\partial_j\theta\} \tag{44}$$

The basic reason for κ-supersymmetry is now the observation that, as we shall show in a moment,

$$\Gamma^2 = 1 \tag{45}$$

Therefore, if we take

$$\delta\theta = (1+\Gamma)\kappa \tag{46}$$

then we are left with

$$\delta(\kappa)I = -T\int d^3\xi\,\bar{\kappa}(1+\Gamma)(1-a\Gamma)[2(-g)^{1/2}g^{ij}\not{\pi}_i\partial_j\theta] \tag{47}$$

and taking $a = 1$, we find $\delta(\kappa)I = 0$. Thus, local κ-invariance is due to two reasons: (i) the B-terms in the varied action cancel by taking the sign in $\delta(\kappa)X^\mu$ opposite to that in $\delta(\varepsilon)X^\mu$; (ii) the F-terms in the varied action are proportional to

$$\bar{\kappa}(1+\Gamma) \quad \times \quad (1-a\Gamma) \quad \times \quad (\cdots) = 0 \tag{48}$$

$\swarrow$ from $\delta\theta$ $\quad\swarrow$ from $\mathscr{L}(\text{bos})$ $\quad\searrow$ from $\mathscr{L}(\text{WZW})$

The proof that $\Gamma^2 = 1$ is straightforward. In the contractions of the product of $\Gamma_{\mu\nu\rho}$ with $\Gamma_{\mu'\nu'\rho'}$ we only need the terms with no Dirac matrices, because the terms with 6 and 4 Dirac matrices vanish since $i, j, \ldots$ run only over three values, while the terms with two Dirac matrices can only give a result proportional to $\Gamma_{mn}\pi_i^m\pi_j^n g^{ij}$, which also vanishes. There are six contractions of the Γ-matrices. In this way one finds

$$\begin{aligned}\Gamma^2 &= \frac{1}{36}\varepsilon^{ijk}\frac{\varepsilon^{i'j'k'}}{(-g)}\pi_i^\mu\cdots\pi_{k'}^{\rho'}(-6\eta_{\mu\mu'}\eta_{\nu\nu'}\eta_{\rho\rho'})\\ &= \frac{1}{6g}\varepsilon^{ijk}\varepsilon^{i'j'k'}g_{ii'}g_{jj'}g_{kk'} = \frac{1}{g}\varepsilon^{ijk}g_{1i}g_{2j}g_{3k} = 1\end{aligned} \tag{49}$$

5. RIGID WORLD-SHEET SUSY [9]

As we have seen, the actions of GS super p-branes are invariant under two distinct fermionic symmetries:

(1) Rigid space-time supersymmetry: Since this symmetry is *nonlinearly* realized ($\delta\theta = \varepsilon$), we cannot conclude on the basis of this symmetry alone that the number of bosonic and fermionic modes on the world volume must be equal.

(2) Local κ-symmetry. This is also a nonlinearly realized space-time symmetry: $\delta\theta = (1+\Gamma)\kappa$ (where κ is a space-time spinor), instead of $\delta\theta \sim \kappa X$.

Yet, in a particular gauge, called "the physical gauge" by its inventors [9], both fermionic symmetries fuse into:

A linearly-realized rigid supersymmetry on the world sheet, which maps bosons onto fermions and vice versa. World-sheet SUSY is desirable, because it is probably needed to get massless fermions in the space-time spectrum.

Since the number of modes should not depend on the gauge chosen, it follows that local κ-symmetry leads to equal numbers of bosonic and fermionic modes. For the string one could have reversed the argument as follows. From the fact that the spinning string has a linearly realized (local world-sheet) SUSY, it must have equal numbers of bosons and fermions, and from the equivalence of the spinning and GS string in the light-cone gauge, it follows that the GS string must have equal numbers of bosons and fermions in any gauge. This then, explains to some extent why the GS string has local κ-symmetry, although one cannot rule out the possibility of achieving world-sheet SUSY by other means than local κ-symmetry.

Models with rigid space-time SUSY and local κ-symmetry need not always have equal numbers of bosonic and fermionic modes. For example, if one drops one of the two spinors θ^A in the closed GS superstring, the model has still $N = 1$ rigid SUSY and local κ-symmetry, yet there are eight bosonic modes and four fermionic modes [10]. All one can say is that this model is not equivalent to a spinning string model, while it is also not a consistent model (anomalies do not cancel). It seems that consistent GS models always need local κ-symmetry.

The physical gauge is defined by the gauge condition

$$X^i = \xi^i \qquad (i = 9, 10, 0)$$

$$(1 + {}^*\Gamma)\theta^a = 0 \qquad (a = 1, 32), \qquad {}^*\Gamma = \Gamma^1 \cdots \Gamma^8 \tag{50}$$

Since $({}^*\Gamma)^2 = +1$, the operator $(1 + {}^*\Gamma)$ is a projection operator, and the condition $(1 + {}^*\Gamma)\theta = 0$ is expected to eliminate 16 components of θ, just as many as the number of κ components which are effectively present in $\delta\theta = (1 + \Gamma)\kappa$. Indeed, also $(1 + \Gamma)$ is a projection operator and also κ has 32 components. So, that gauge $(1 + {}^*\Gamma)\theta = 0$ seems at first sight a good gauge to fix the local κ symmetry. The reason the gauge $(1 + {}^*\Gamma)\theta = 0$ is chosen, rather than $(1 + \Gamma)\theta = 0$, is that one can very easily solve the former whereas the latter cannot be solved in closed form at all.

As we shall explicitly show,

$$\Gamma = {}^*\Gamma + \text{field-dependent terms} \tag{51}$$

Hence, if we put all fields to zero, the condition $(1 + {}^*\Gamma)\theta^a = 0$ fixes the gauge completely, because then $\delta\theta^a = 0$ [since in this limit $\delta(1 + {}^*\Gamma)\theta^a$ equals $(1 + {}^*\Gamma)\kappa$ and hence fixes all effective gauge parameters $(1 + \Gamma)\kappa$ for vanishing fields]. The usual kind of argument can then suggest that also in the presence of fields the gauge condition $(1 + {}^*\Gamma)\theta = 0$ completely fixes the gauge:

Claim: If $(1 + {}^*\Gamma)\theta = 0$, then $\delta\theta = 0$.

[Note that κ is not fixed, only its effective part $(1 + \Gamma)\kappa$.] It may be instructive to see how this happens, and for this reason we will explicitly show that the above claim is valid to second order in X-fields. Since we will need for this purpose details of Dirac matrices, we postpone this demonstration to the end of this subsection.

We return to the issue of rigid world-sheet SUSY. The set of all transformations of the fields $X^\mu(\rho, \sigma, \tau)$ with $\mu = 0, 1, \ldots, 10$ reads

$$\delta X^\mu = \eta^i \partial_i X^\mu + l_\nu{}^\mu X^\nu + a^\mu + \bar{\theta}\Gamma^\mu\varepsilon - \bar{\theta}\Gamma^\mu(1 + \Gamma)\kappa \tag{52}$$

Hence, in order that $\delta X^i = 0$ for $i = 9, 10$, and 0, we must accompany any space-time Poincaré transformation or any fermionic symmetry transformation by a compensating general world volume coordinate transformation, whose parameter η^i is given by

$$\eta^i = -l^i_\nu X^\nu - a^i - \bar{\theta}\Gamma^i\varepsilon + \bar{\theta}\Gamma^i(1 + \Gamma)\kappa \tag{53}$$

The remaining eight transversal fields X^I $(I = 1, 8)$ then transform as follows:

$$\begin{aligned} \delta X^I = &-(l^i_j\xi^j)\partial_i X^I - (l^i_J X^J)\partial_i X^I - a^i\partial_i X^\mu - \bar{\theta}\Gamma^i\varepsilon\partial_i X^I \\ &+ \bar{\theta}\Gamma^i(1 + \Gamma)\kappa\partial_i X^I + l^I_J X^J + l^I_i\xi^i + a^I + \bar{\theta}\Gamma^I\varepsilon - \bar{\theta}\Gamma^I(1 + \Gamma)\kappa \end{aligned} \tag{54}$$

The first term on the right-hand side shows that X^I has *become* a world-volume scalar under rigid world-volume Lorentz transformations because a scalar field S transforms under rigid Lorentz transformations with precisely this orbital term. The l^I_J term corresponds to the $SO(8)$ subgroup of $SO(10, 1)$, of which the X^I form a linear representation.

In Kaluza-Klein theory one has a similar but not exactly the same situation: There one starts with a fixed X^μ depending on all 11 coordinates, and by reduction on the 8-torus, one gets in $d = 3$ Minkowski space-time a theory with a linearly realized group $SO(2, 1) \times SO(8)$, whereas the off-diagonal part of the original symmetry group $SO(10, 1)$ is fixed by putting the off-diagonal part of the vielbein equal to zero: $e^m_I = 0$. The original $d = 11$ general coordinate invariance is left unfixed, but by requiring that all fields depend only on ρ, σ, τ, we get in $d = 3$ dimensions the following local and global symmetries: general coordinate invariance [with $\eta(\rho, \sigma, \tau)$], Yang-Mills invariance [with $\eta^I(\rho, \sigma, \tau)$], and global $SO(8)$ invariance (with $\eta^I = M^I_J y^J$). *Here* we fix the $d = 3$ general coordinate invariance (by

$X^i = \xi^i$) and leave Lorentz invariance unfixed (because it is a rigid symmetry). Consequently, we still have a rigid symmetry with Lorentz parameter l_i^I, but it is nonlinearly realized:

$$\delta X^I = l_i{}^I \xi^i - l^i_J X^J \partial_i X^I \tag{55}$$

In the language of the theory of nonlinear realizations, the X^I form a "spectator-representation" and the $l_i{}^I$ are the constant parameters of the generators of broken symmetry, which correspond in one-to-one fashion to the Nambu–Goldstone fields (which are not present in this model).

Let us repeat the analysis of general coordinate invariance and Lorentz symmetry, but now for the case of κ and ε symmetry. The fields $\theta^a(\rho, \sigma, \tau)$ transform before any gauge fixing as follows:

$$\delta\theta = \eta^i \partial_i \theta + \tfrac{1}{4} l^{\mu\nu} \Gamma_{\mu\nu} \theta + \varepsilon + (1 + \Gamma)\kappa \tag{56}$$

We already know that $\eta^i = -\bar{\theta}\Gamma^i \varepsilon + \bar{\theta}\Gamma^i (1 + \Gamma)\kappa$, and hence it produces in $\delta\theta$ a term quadratic in fields, which we disregard since we work only to second order in X-fields. We now use the following facts, derived later:

$$^*\Gamma = \begin{pmatrix} 1 & & & \\ & -1 & & \\ & & 1 & \\ & & & -1 \end{pmatrix}, \qquad \Gamma = {}^*\Gamma + {}^*\Gamma(M^{(1)} + M^{(2)} + \cdots) \tag{57}$$

$$M^{(1)} = \Gamma^i(\partial_i X^I)\Gamma_I = \begin{pmatrix} 0 & * & 0 & * \\ * & 0 & * & 0 \\ 0 & * & 0 & * \\ * & 0 & * & 0 \end{pmatrix} \tag{58}$$

Thus the constraint $(1 + {}^*\Gamma)\theta = 0$ leads to $\theta^T = (0, \theta_2^T, 0, \theta_4^T)$, where θ_2 and θ_4 are eight-component spinors, and the compensating κ-transformation, which must accompany a rigid SUSY transformation in order that $(1 + {}^*\Gamma)\delta\theta = 0$, is given by

$$(1 + {}^*\Gamma)[\varepsilon + (1 + {}^*\Gamma)\kappa + {}^*\Gamma M^{(1)}\kappa + \cdots] = 0 \tag{59}$$

Since $\delta\theta_1$ and $\delta\theta_3$ must vanish, this will lead to a relation for κ in terms of ε.

The vanishing of $\delta\theta_1$ and $\delta\theta_3$ requires

$$\begin{aligned} 2\varepsilon_1 + 4\kappa_1 + 2(M^{(1)}\kappa)_1 + \cdots &= 0 \\ 2\varepsilon_3 + 4\kappa_3 + 2(M^{(1)}\kappa)_3 + \cdots &= 0 \end{aligned} \tag{60}$$

For the transformation rules of θ_2 and θ_4, we then find from

$$\delta\theta = \varepsilon + (1 + {}^*\Gamma + {}^*\Gamma M^{(1)} + \cdots)\kappa$$

the following results:

$$\begin{aligned}\delta\theta_2 &= \varepsilon_2 - (M^{(1)})_2{}^4\kappa_1 - (M^{(1)})_2{}^3\kappa_3 + \cdots \\ \delta\theta_4 &= \varepsilon_4 - (M^{(1)})_4{}^1\kappa_1 - (M^{(1)})_4{}^3\kappa_3 + \cdots\end{aligned} \tag{61}$$

Thus, the parameters $(\varepsilon_1, \kappa_1)$ and $(\varepsilon_3, \kappa_3)$ fuse into a rigid linearly realized supersymmetry [with parameter ε_1 and ε_3, since κ_1, κ_3 are expressed in terms of $\varepsilon_1, \varepsilon_3$; see (60)]. The $\varepsilon_2, \varepsilon_4$ correspond to remaining rigid nonlinear symmetries. For the bosons the transformation rule contains already a linear piece

$$\delta X^I = \bar\theta\Gamma^I\varepsilon - \bar\theta\Gamma^I(1 + {}^*\Gamma)\kappa + \cdots \tag{62}$$

Using that

$$\Gamma^0\Gamma^I = \left(\begin{array}{cc|cc} 0 & & I & 0 \\ & & 0 & I \\ \hline -I & 0 & & \\ 0 & -I & & 0 \end{array}\right)\begin{pmatrix} 0 & & & m^I \\ & & \tilde m^I & \\ & m^I & & \\ \tilde m^I & & & 0 \end{pmatrix} = \left(\begin{array}{cc|cc} 0 & m^I & & \\ \tilde m^I & 0 & & \\ \hline & & 0 & -m^I \\ & & -\tilde m^I & 0 \end{array}\right) \tag{63}$$

where m^I are real 8×8 matrices (see below) and " $\sim$ " denotes transposition, we find

$$\delta X^I = \theta_2^T\tilde m^I\varepsilon_1 - \theta_4^T\tilde m^I\varepsilon_3 - \theta_2^T\tilde m^I 2\kappa_1 + \theta_4^T\tilde m^I 2\kappa_3 + \cdots \tag{64}$$

As we have seen, $\varepsilon_1 + 2\kappa_1$ and $\varepsilon_3 + 2\kappa_3$ vanishes to zeroth order in fields [see (60)], hence one finally obtains

$$\begin{aligned}\delta X^I &= 2\theta_2^T\tilde m^I\varepsilon_1 - 2\theta_4^T\tilde m^I\varepsilon_3 + \text{non linear terms} \\ \delta\theta_2 &= \tfrac12(M^{(1)})_2^1\varepsilon_1 + \tfrac12(M^{(1)})_2^3\varepsilon_3 + \text{nonlinear terms} \\ \delta\theta_4 &= \tfrac12(M^{(1)})_4^1\varepsilon_1 + \tfrac12(M^{(1)})_4^3\varepsilon_3 + \text{nonlinear terms}\end{aligned} \tag{65}$$

These are the linear world-volume SUSY transformations with eight two-component SUSY spinors. (The original two eight-component spinors ε_1 and ε_3 of space-time have become eight two-component spinors on the world volume.)

Let us finally study how θ transforms under l^{ij} transformations. Since Γ_{ij} and Γ_{IJ} commute with $(1 + {}^*\Gamma)$, the l^{ij} transformations do not break the gauge $(1 + {}^*\Gamma)\theta = 0$. Hence there are no compensating κ-transformations but only compensating η^i transformations [see (53)]:

$$\delta\theta = (-l^i_j\xi^j)\partial_j\theta + \tfrac14 l^{ij}\Gamma_{ij}\theta + \tfrac14 l^{IJ}\Gamma_{IJ}\theta \tag{66}$$

Now Γ_i acts like the unit matrix on the eight-component spinors θ_2, θ_4, and Γ_{ij} acts like $(-i\tau_2, -\tau_1, \tau_3)$ on the eight doublets (θ_2, θ_4). The matrices Γ_{IJ}, on the other hand, are diagonal in $(\theta_1, \theta_2, \theta_3, \theta_4)$ space:

$$\Gamma_{IJ} = \mathrm{diag}([m_I, \tilde{m}_J], [\tilde{m}_I, m_J], [m_I, \tilde{m}_J], [\tilde{m}_I, m_J]) \tag{67}$$

So we see that the θ's have become world-volume spinors, as well as $SO(8)$ spinors, in the same $SO(8)$ representations. Hence, the spinors are in

$$(\mathbf{2}, \mathbf{8}) \text{ of } SO(2,1) + SO(8) \tag{68}$$

The nonlinearly realized symmetry with l^{Ii}, on the other hand, interchanges θ_2 and θ_4 as well as reshuffles their components.

Finally we come back to our promise to explicitly demonstrate to first nontrivial order in fields that $(1 + {}^*\Gamma)\theta = 0$ indeed fixes local κ-symmetry, and exhibit explicit Dirac matrix representation used above. Under a local κ-symmetry the gauge condition transforms into

$$(1 + {}^*\Gamma)(1 + \Gamma)\kappa = 0$$

$$\Gamma = \tfrac{1}{6}\varepsilon^{ijk}(-g)^{-1/2}\partial_i X^\mu \partial_j X^\nu \partial_k X^\rho \Gamma_{\mu\nu\rho} + \theta\text{-terms} \tag{69}$$

The field-independent terms are due to taking X^μ only equal to X^i. This yields

$$\Gamma = \varepsilon^{9,10,0}\Gamma_9\Gamma_{10}\Gamma_0 = \Gamma^9\Gamma^{10}\Gamma^0 \equiv {}^*\Gamma \tag{70}$$

taking $\varepsilon^{0,9,10} = -1$ and $\Gamma_9 = \Gamma^9$ but $\Gamma_0 = -\Gamma^0$. We choose

$$\Gamma^9 = \begin{pmatrix} 1 & 0 \\ 0 & -1 \end{pmatrix}_{16\times 16} \times \tau_1, \qquad \Gamma^0 = 1_{16\times 16} \times i\tau_2, \qquad \Gamma^{10} = 1_{16\times 16} \times \tau_3$$

$$\Gamma^I = \{(\sigma_2 \times \gamma^i) \times \tau_1 \quad \text{or} \quad (\sigma_1 \times 1_{8\times 8}) \times \tau_1\} \qquad \text{with } i = 1, 7 \tag{71}$$

In this representation

$${}^*\Gamma = \begin{pmatrix} 1 & & & \\ & -1 & & \\ & & 1 & \\ & & & -1 \end{pmatrix}, \qquad \Gamma^I = \begin{pmatrix} & & & m^I \\ & & \tilde{m}^I & \\ & m^I & & \\ \tilde{m}^I & & & \end{pmatrix} \tag{72}$$

The γ^i are the 8×8 Dirac matrics for $d = 7$ Euclidean space and are completely antisymmetric and purely Hermitian satisfying $\gamma^1\gamma^2 \cdots \gamma^6\gamma^7 = i$. Hence $m^I = \{-i\gamma^i, I\}$ are real. The terms in Γ that are linear in the fields X^I are given by

$$({}^*\Gamma)(\Gamma^i\partial_i X^I)\Gamma_I = {}^*\Gamma(\Gamma^9\partial_\rho + \Gamma^{10}\partial_\sigma + \Gamma^0\partial_\tau)\Gamma_I X^I$$

$$= \left(\begin{array}{cc|cc} \partial_\sigma & 0 & \partial_\rho + \partial_\tau & 0 \\ 0 & -\partial_\sigma & 0 & \partial_\rho - \partial_\tau \\ \hline \partial_\rho - \partial_\tau & 0 & -\partial_\sigma & 0 \\ 0 & \partial_\rho + \partial_\tau & 0 & +\partial_\sigma \end{array}\right)\Gamma_I X^I \tag{73}$$

which is of the form (58).

Let us now work out the consequences of $(1+{}^*\Gamma)(1+\Gamma)\kappa=0$ for κ, and then verify whether $\delta\theta=(1+\Gamma)\kappa$ indeed vanishes.

From $(1+{}^*\Gamma)(1+\Gamma)\kappa=0$ we find

$$\begin{aligned} 2\kappa_1+(\partial_\rho+\partial_\tau)X^I(m_I\kappa_2)+\partial_\sigma X^I(m_I\kappa_4)+\cdots&=0\\ 2\kappa_3+(-\partial_\sigma X^I)(m_I\kappa_2)+(\partial_\rho-\partial_\tau)X^I(m_I\kappa_4)+\cdots&=0 \end{aligned} \tag{74}$$

Clearly, taking κ^2 and κ^4 as arbitrary functions of ρ, σ, τ, we can determine κ_1 and κ_3 iteratively such that the gauge conditions are satisfied. We have then still a two-function freedom in κ.

For the variation of θ, we need to consider only $\delta\theta_2$ and $\delta\theta_4$. The two X^I terms in Γ come not only from expanding $\partial_i X^\mu\partial_j X^\nu\partial_k X^\rho$ but also from expanding $(-g)^{-1/2}$. One finds

$$\begin{aligned} \Gamma={}^*\Gamma+{}^*\Gamma[\Gamma^k(\partial_k X^I)\Gamma_I-\tfrac{1}{2}\Gamma^j(\partial_j X^I)\Gamma^k(\partial_k X^J)\Gamma_{IJ}\\ -\tfrac{1}{2}\eta^{ij}\partial_i X^I\partial_j X^J\delta_{IJ}\}+\cdots \end{aligned} \tag{75}$$

In this way we find, combining (74) and (75),

$$\begin{aligned} \delta\theta_2=(\partial_\rho-\partial_\tau)X^I\tilde{m}_I[-(\partial_\rho+\partial_\tau)X^J m_J\kappa_2-\partial_\sigma X^J m_J\kappa_4\cdots]^{1/2}\\ -(\partial_\sigma X^I)\tilde{m}_I[\partial_\sigma X^J m_J\kappa_2-(\partial_\rho-\partial_\tau)X^J m_J\kappa_4\cdots]^{1/2}\\ +\tfrac{1}{2}[(\Gamma^j\partial_j X^I)(\Gamma^k\partial_k X^J)\Gamma_{IJ}]\kappa\\ +\tfrac{1}{2}\eta^{ij}\partial_i X^I\partial_j X_I\kappa+\cdots \end{aligned} \tag{76}$$

In the last two terms only κ_2 and κ_4 appear, because $\Gamma^i\Gamma^j$ only interchanges the components κ_2 and κ_4 but never maps them into the range of κ_1 and κ_3. In fact, direct evaluation reveals that indeed $\delta\theta_2=0$. Similarly one finds that $\delta\theta_4=0$. Thus, the gauge condition $(1+{}^*\Gamma)\theta=0$ indeed fixes $\delta\theta=0$ to second order in X^I.

Comment. In the superstring, the mechanism by which the θ become world-sheet spinors is somewhat different in the light-cone gauge. There one divides θ by $(p+)^{1/2}$, and since $(p+)^{1/2}$ transforms as a world-sheet spinor, one finds that $\theta^a/(p+)^{1/2}\equiv S^a$ is a world-sheet spinor.

Note to the reader. This article presents a pedagogical introduction to the theory of extended objects, in particular, supersymmetric membranes. At the conference, the author presented a complete and exhausting calculation based on path-integrals done with L. Mezincescu and R. Nepomechie, according to which there are no massless states in the spectrum of the supermembrane. This would mean that supermembranes are not consistent. Since then Gandour, Pope, and Stelle have analyzed the role of zero modes in the path-integral quantization. Although these authors have not performed a complete calculation, and some questions remain, they believe that there are massless particles in the spectrum, and that supermembranes are consistent. On the other hand, recent work by de Wit and Nicolai claims that there is a continuous spectrum and, hence, supermembranes would not make sense at the quantum level. Since the question is unsettled at this moment, I decided not to include the details of my work. They are published in *Nucl. Phys. B 309*, 317 (1988).

At the conference, the author also presented a talk on a new class of rigidly $N = 4$ supersymmetric nonlinear σ models with nonvanishing Nijenhuis tensor, and the action and transformation roles of the $N = 4$ locally supersymmetric nonlinear σ model with torsion. These results were obtained with de Wit and can be found in *Nucl. Phys. B 312*, 58 (1989).

REFERENCES

1. H. A. Lorentz, *The Theory of the Electrons*, Dover, New York, 1953.
2. P. A. M. Dirac, *Proc. Roy. Soc. London 268*, 57 (1962).
3. L. Alvarez-Gaumé, P. Ginsparg, G. Moore, and C. Vafa, *Phys. Lett. 171B*, 155 (1986).
4. E. Bergshoeff, E. Sezgin, and P. K. Townsend, *Phys. Lett. 189B*, 75 (1987).
5. J. Hughes, J. Liu, and J. Polchinsky, *Phys. Lett. 180B*, 370 (1986).
6. A. Achúcarro, J. M. Evans, P. K. Townsend, and D. L. Wiltshire, *Phys. Lett. 198B*, 441 (1987); R. D'Auria and P. Fré, *Nucl. Phys. B201*, 101 (1982).
7. I. Bars and C. N. Pope, *Class. Qu. Grav. 5*, 1152 (1988).
8. W. Siegel, *Phys. Lett. 128B*, 397 (1983).
9. E. Bergshoeff, E. Sezgin, and P. K. Townsend, *Ann. Phys. 185*, 330 (1988).
10. T. Curtright and P. van Nieuwenhuizen, *Nucl. Phys. B294*, 125 (1987).
11. E. Bergshoeff, M. J. Duff, C. N. Pope, and E. Sezgin, *Phys. Lett. 199B*, 69 (1978).
12. C. J. Isham, C. N. Pope, and N. P. Warner, *Class. Qu. Grav. 5*, 1297 (1988).
13. R. Nepomechie, *Phys. Lett. 178B*, 207 (1986); *180B*, 423 (1986).
14. M. Henneaux and L. Mezincescu, *Phys. Lett. 152B*, 340 (1985).
15. M. J. Duff, P. S. Howe, T. Inami, and K. S. Stelle, *Phys. Lett. 191B*, 70 (1987); M. J. Duff, T. Inami, C. N. Pope, E. Sezgin, and K. S. Stelle, *Nucl. Phys. B 297*, 515 (1988).

Chapter 17

Thirring Strings: Use of Generalized Nonabelian Bosonization Techniques

Elcio Abdalla

1. INTRODUCTION: CONFORMAL INVARIANT STRING THEORIES IN A COMPACTIFIED SPACE

In the last few years strings have been proved to be extremely important objects for describing fundamental interactions.[1] However, there are many technical difficulties in the description of string dynamics, and, perhaps, one of the most relevant aspects to be fully understood is compactification. It is my aim to relate the string defined on a compactified space time to a fermionic model in such a way that, in the latter, complicated and important operators, such as the vertices,[1,2] turn out to be elementary fields. Therefore, correlators involving vertices, which are nonlinear in the string field variables, will turn out to be linear in terms of fermionic variables.

The action describing a string on a compactified manifold is given by

$$S = \frac{1}{2\pi}\int d\sigma\, d\tau\, \partial_\alpha X^a \partial_\beta X^b(\eta^{\alpha\beta} g_{ab}(x) + \varepsilon^{\alpha\beta} B_{ab}(x)) \tag{1}$$

where the first term is the usual Polyakov string action, and the latter is the Wess–Zumino term, which is needed to maintain conformal invariance in the compactified space,[3,4] because the first action describes a nonlinear sigma model that, in general, is asymptotically free.[5]

Elcio Abdalla • Instituto de Física, Universidade de São Paulo, 20516 São Paulo, Brazil.

The separation between left and right movers is rather subtle in this theory. To substantiate this statement, we elaborate. In conformal theory, left and right movers are equivalent to holomorphic and antiholomorphic fields in the euclidian version of the theory.[6]

We have

$$X^a_\pm = X^a_{0\pm} + P^a_{0\pm} x_\pm + \sum_{n\neq 0} \alpha^a_{n\pm}\, e^{inx_\pm} \tag{2}$$

where the position variable is simply the sum of left and right movers

$$X^a(x) = X^a_+(x_+) + X^a_-(x_-) \tag{3}$$

We also define the dual field $\tilde{X}^a$, given by

$$\tilde{X}^a(x) = X^a_+(x_+) - X^a_-(x_-) + B^{ab}(x)(X^a_+(x_+) + X^a_-(x_-)) \tag{4}$$

motivated by the algebra-valued fields of WZW theory,

$$J^+(x_+) = g^{-1}\partial^+ g \tag{5}$$

$$J^-(x_-) = \partial^-_g g^{-1} \tag{6}$$

where X^a is the Lie-algebra-valued field corresponding to the group-valued g-field described by the action

$$S = \frac{1}{8\pi}\int d^2x \operatorname{tr} \partial_\mu g^{-1}\partial^\mu g + \frac{1}{4\pi}\int_0^1 dr \int d^2x \operatorname{tr} \varepsilon^{\mu\nu}\tilde{g}^{-1}\partial_\mu \tilde{g}\tilde{g}^{-1}\partial_\nu \tilde{g}\tilde{g}^{-1}\partial_r \tilde{g} \tag{7}$$

We can formally identify

$$J^{\pm a}(x_\pm) = \mp P^{a\pm}(x_\pm) = -\frac{1}{\pi}\partial^\pm X^a_\pm \tag{8}$$

If the symmetry group is abelian, such as the torus $[\mathrm{U}(1)]^d_1$ we may assume that all X^a's are independent. Their commutation relations are easily derived from J's Kac–Moody algebra and are given by[7]

$$[X^a_+(x_+), X^b_+(y_+)] = -\frac{i}{4}\pi\delta^{ab}\varepsilon(x_+ - y_+) \tag{9a}$$

$$[P^a_+(x_+), X^b_+(y_+)] = -\frac{i}{2}\delta^{ab}\delta(z_+ - y_+) \tag{9b}$$

$$[J^a_+(x_+), J^b_+(y_+)] = \frac{i}{2\pi}\delta^{ab}\delta'(x_+ - y_+) \tag{9c}$$

which is an abelian Kac–Moody algebra. We will elaborate the abelian theory later. Now we discuss the (nonabelian) WZW theory in more detail.

Starting from (7), we can formally integrate over the r variable and obtain an effective action with an unknown expression $A(g)$[8]:

$$S = \frac{1}{8\pi}\int d^2x \,\mathrm{tr}\, \partial_\mu g^{-1} \partial^\mu g + \frac{1}{4\pi}\int d^2x \,\mathrm{tr}\, A(g)\partial_0 g \tag{10}$$

For canonical quantization, $A(g)$ is not needed. All necessary information is provided by the derivative

$$F_{ij;kl}(g) = \frac{\partial A_{ij}}{\partial g_{lk}} - \frac{\partial A_{kl}}{\partial g_{ji}} \tag{11}$$

where

$$F_{ij;kl} = \partial_1 g^{-1}_{il} g^{-1}_{kj} - g^{-1}_{il} \partial_1 g^{-1}_{kj} \tag{12}$$

The momentum conjugate to g is

$$\pi_{ij} = \frac{1}{4\pi}(\partial_0 g^{-1}_{ji} + A_{ji}(g)) \tag{13}$$

We also define

$$\hat{\pi}_{ij} = \pi_{ij} - \frac{1}{4\pi} A_{ji} \tag{14}$$

Canonical commutation relations have been discussed in the literature; they are[8]:

$$[\pi_{ij}(x), \pi_{kl}(y)] = 0 \tag{15a}$$

$$[g_{ij}(x), \pi_{kl}(y)] = i\delta_{ik}\delta_{jl}\delta(x^1 - y^1) \tag{15b}$$

$$[g_{ij}(x), g_{kl}(y)] = 0 \tag{15c}$$

at equal time; it follows that the current, which can be written in terms of the elementary field as

$$J^a_+ = \frac{-i}{4\pi}\mathrm{tr}\, g^{-1}\partial^+ g \tau^a = \mathrm{tr}\left(i\hat{\pi} g - \frac{i}{4\pi} g^{-1}\partial_1 g\right)\tau^a \tag{16}$$

(there is also an expression for J^a_-), has well-defined commutation relations. Actually since equation (16) is purely left moving, equation (17) is valid for any time:

$$[J^a_+(x), J^b_+(y)] = if^{abc}J^c_+(x)\delta(x-y) + \frac{i}{2\pi}\delta^{ab}\delta'(x-y) \tag{17}$$

The energy momentum tensor may be readily computed, being of the Sugawara form[4]

$$\theta_+(x_+) = \frac{\pi}{c_v + k} :(J^a(x_+))^2: \tag{18}$$

We define the field operator

$$X^a(x) = \int_{-\infty}^{x} J^a(y)\,dy \tag{19}$$

which, in view of (17), obeys the algebra

$$[X^a(x), X^b(y)] = \frac{i}{2} f^{abc}(X^c(x) - X^c(y))\varepsilon(x-y)$$

$$+\frac{i}{4\pi}\,\delta^{ab}\varepsilon(x-y) \tag{20}$$

Our problem is to implement compactification, that is, to realize the identifications

$$X^a \equiv X^a + 2\pi E^a_\mu n^\mu \tag{21}$$

where n^μ are integers and E^a_μ generate a lattice Λ in such a way that $\mathbf{R}^d/\Lambda$ is the target manifold of our sigma model described by the action (1). A string theory defined on a compact manifold has a further symmetry[7] associated with the dual field $\tilde{X}$ defined in (4). This symmetry is related to the dual lattice $\tilde{\Lambda}$ generated by $\tilde{E}^a_\mu$ and defined as

$$E^a_\mu \tilde{E}^{a\nu} = \delta^\nu_\mu \tag{22}$$

In the one-dimensional case $E = R$, and the symmetry is

$$X \equiv X + 2\pi R \tag{23}$$

In order to understand the second symmetry, consider the mode expansion of the closed string field on a compact space

$$X(x) = X_0 + \frac{M}{R}\tau + 2LR\sigma + \text{oscillators} \tag{24}$$

Both zero modes are quantized—the former ($P_\tau = M/R$) because the string is in a compact space of radius $1/R$ and P_τ must be quantized (M is an integer); the other ($P_\sigma = 2LR$, L an integer) because σ is defined up to multiples of π, in which case X can change only by multiples of 2π [equation (23)].

Therefore, left and right movers are

$$X_L = X_{0L} + \frac{1}{2}\left(\frac{M}{R} + 2LR\right)(\tau + \sigma) + \text{oscillators} \tag{25}$$

$$X_R = X_{0R} + \frac{1}{2}\left(\frac{M}{R} - 2LR\right)(\tau - \sigma) + \text{oscillators} \tag{26}$$

implying that

$$\tilde{X} = \tilde{X}_0 + \frac{M}{R}\sigma + 2LR\tau + \text{oscillators} \tag{27}$$

which has a momentum $P_\tau = 2LR$, quantized in units of $2R$; thus the identification

$$\tilde{X} \equiv \tilde{X} + \pi/R \tag{28}$$

is valid.

2. BOSONIZATION AND FERMIONIZATION IN CONFORMALLY INVARIANT TWO-DIMENSIONAL FIELD THEORIES

The principal nonlinear σ model with a Wess–Zumino term written as a functional of a group-valued field $g(x)$ is described by the action

$$S = \frac{1}{4\lambda^2}\,\mathrm{tr}\int d^2x\,\partial_\mu g^{-1}\partial^\mu g + \frac{k}{8\pi}\,\varepsilon^{\mu\nu}\,\mathrm{tr}\int dr\int d^2x\,\tilde{g}^{-1}\dot{\tilde{g}}\tilde{g}^{-1}\partial_\mu\tilde{g}\tilde{g}^{-1}\partial_\nu\tilde{g} \tag{29}$$

This is conformally invariant only if the coupling constant is[4]

$$\lambda^2 = 4\pi/k \tag{30}$$

The constant k is quantized (integer) since the topological term is defined up to redefinitions of the extension $g(x) \to \tilde{g}(r, x)$. Different choices of boundaries differ by multiples of 2π. This system has been related to free fermions by several authors, when $k = 1$. This so-called nonabelian bosonization prescription is realized by the identification

$$g_{ij}(x) = \mu^{-1}N[\Psi^\dagger_{1i}(x)\Psi_{2j}(x)] \tag{31}$$

where N is a normal product prescription and μ is an arbitrary mass parameter. The resulting theory is a multiplet of free fermions.[3,4,9]

Aiming at general values of the central charge k, we study the G-invariant Thirring model (we will specialize $G = \mathrm{SU}(n)$, when writing explicit formulas). The lagrangian is[10]

$$\mathscr{L} = i\bar{\psi}\partial\!\!\!/\psi - \tfrac{1}{2}g'\bar{\psi}\gamma^\mu\psi\bar{\psi}\gamma_1\psi - \tfrac{1}{2}g\bar{\psi}\gamma^\mu\tau^a\psi\bar{\psi}\gamma_\mu\tau^a\psi \tag{32}$$

with $[\tau^a, \tau^b] = if^{abc}\tau^c$.

The formal field equation is

$$i\partial\!\!\!/\psi = gj^a_\mu\gamma^\mu\tau^a\psi + g'j_\mu\gamma^\mu\psi \tag{33}$$

where the currents are formally defined by the expressions

$$j_\mu^a = \bar{\psi}\gamma_\mu \tau^a \psi \tag{34a}$$

$$j_\mu = \bar{\psi}\gamma_\mu \psi \tag{34b}$$

According to the symmetries of the model, we have three fundamental conservation laws:

$$\partial^\mu j_\mu = 0 \tag{35a}$$

$$\partial^\mu j_\mu^a = 0 \tag{35b}$$

$$\varepsilon^{\mu\nu}\partial_\mu j_\nu = 0 \tag{35c}$$

The curl of j_μ^a is nonzero, however, thus obeying a "nonconservation" law

$$\varepsilon^{\mu\nu}\partial_\mu j_\nu^a = g\varepsilon_{\mu\nu} f^{abc} j^{\mu b} j^{\nu c} \tag{36}$$

Dashen and Frishman[10] studied the conditions under which the quantum model displays conformal invariance. They considered the equal time commutators

$$[j_0^a(t,x), j_0^b(t,y)] = if^{abc} j_0^c(t,x)\delta(x-y) \tag{37a}$$

$$[j_0^a(t,x), j_1^b(t,y)] = if^{abc} j_1^c(t,x)\delta(x-y) + \frac{ik}{2\pi}\delta^{ab}\delta'(x-y) \tag{37b}$$

$$[j_1^a(t,x), j_1^b(t,y)] = if^{abc} j_0^c(t,x)\delta(x-y) \tag{37c}$$

If the theory is scale-invariant, it follows that j_μ^a has scale dimension 1. We can prove now that j_μ^a is divergenceless and curl-free. Consider the vector $J_\mu(x)$ with dimension 1, and the two-point function

$$\langle 0|J_+(x_+, x_-)J_+(y_+, y_-)|0\rangle = \frac{C}{(x_+ - y_+ + i\varepsilon)^2} \tag{38a}$$

The right side is fixed because under a Lorentz transform $J_+ = J_0 + J_1$ becomes $1/x_+$.

We may consider analogously

$$\langle 0|J_-(x_+, x_-)J_-(y_+, y_-)|0\rangle = \frac{C}{(x_- - y_- + i\varepsilon)^2} \tag{38b}$$

From these expressions it follows that

$$\partial^\mu J_\mu(x) = 0 \tag{39a}$$

and

$$\varepsilon^{\mu\nu}\partial_\mu J_\nu(x) = 0 \tag{39b}$$

Therefore the nonconservation law (36) transforms, due to quantum fluctuations, to a conservation law in quantum theory. We will verify which conditions are left by the imposition of conformal invariance.

Let us set up the commutation relations:

(i) Singlet currents obey an (abelian) Kac–Moody algebra:

$$[j_\pm(x_\pm), j_\pm(y_\pm)] = 2i\mathbf{C}_0\delta'(x-y) \tag{40a}$$

$$[j_+(x_+), j_-(y_-)] = 0 \tag{40b}$$

(ii) Singlet currents act on fermions in the same way as the abelian Thirring model does.

$$[j_\pm(x_\pm), \psi_{(y)}] = -(a \pm \bar{a}\gamma_5)\psi_{(y)}\delta(x_\pm - y_\pm) \tag{41}$$

(iii) Nonabelian currents satisfy a Kac–Moody algebra

$$\begin{aligned}[j_+^a(x_+), j_+^b(y_+)] &= 2if^{abc}j_+^c(x_+)\delta(x_+ - y_+) \\ &\quad + \frac{ik}{\pi}\delta^{ab}\delta'(x_+ - y_+)\end{aligned} \tag{42a}$$

$$\begin{aligned}[j_-^a(x_-), j_-^b(y_-)] &= 2if^{abc}j_-^c(x_-)\delta(x_- - y_-) \\ &\quad + \frac{ik}{\pi}\delta^{ab}\delta'(x_- - y_-)\end{aligned} \tag{42b}$$

$$[j_+^a(x_+), j_-^b(y_-)] = 0 \tag{42c}$$

(iv) Nonsinglet currents act on fermions as

$$[j_\pm^a(x_\pm), \psi_{(y)}] = -\sigma(1 \pm \delta\gamma_5)\tfrac{1}{2}\lambda^a\psi_{(y)}\delta(x_\pm - y_\pm) \tag{43}$$

where the Jacobi identity requires

$$\sigma = 1 \qquad \delta^2 = 1 \tag{44}$$

The energy momentum tensor is of Sugawara form; fixing the constants so that currents and the energy momentum tensor satisfy the usual form of the Virasoro Kac–Moody algebra, we have

$$\theta_\pm(x_\pm) = \frac{1}{2\mathbf{C}_0}:(j_\pm(x_\pm))^2: + \frac{\pi}{c_v + k}:(j_\pm^a(x_\pm))^2: \tag{45}$$

The constant $\mathbf{C}_0$ defined in (40) is arbitrary and will depend, as we shall see, on the dimension and spin of the fermionic field. The Casimir c_v is given by the relation

$$f^{abc}f^{dbc} = c_v\delta^{ad} \tag{46}$$

For SU(n) we have

$$c_v = n \tag{47}$$

and k is the central charge of the Kac–Moody algebra; thus it is an integer.

The energy momentum tensor satisfies the Virasoro algebra

$$[\theta_\pm(x_\pm), \theta_\pm(y_\pm)] = 2i(\theta_\pm(x_\pm) + \theta_\pm(y_\pm))\delta'(x_\pm - y_\pm) - \frac{i}{6\pi}\delta'''(x_\pm - y_\pm) \tag{48}$$

where the central charge is[11]

$$c = \frac{k \dim G}{c_v + k} = \frac{k(n^2 - 1)}{n + k} \tag{49}$$

Equation (49) has been specialized for $G = \mathrm{SU}(n)$.

Using (41) and (43), we may compute the action of the energy momentum tensor (45) on the fermionic field. On the other hand, we know that it generates translations. Therefore, we may compute equations of motion. They are

$$i\partial^- \psi_1 = \frac{1}{2}\left\{2\pi\frac{1-\delta}{c_v + k}\frac{1}{2}\lambda^b :j_-^b\psi_1: + \frac{a - \bar{a}}{c_0}:j_-\psi_1:\right\} \tag{50}$$

$$i\partial^+ \psi_2 = \frac{1}{2}\left\{2\pi\frac{1-\delta}{c_v + k}\frac{1}{2}\lambda^b :j_+^b\psi_2: + \frac{a - \bar{a}}{c_0}:j_+\psi_2:\right\} \tag{51}$$

$$i\partial^+ \psi_1 = \frac{1}{2}\left\{2\pi\frac{1+\delta}{c_v + k}\frac{1}{2}\lambda^b :j_+^b\psi_1: + \frac{a + \bar{a}}{c_0}:j_+\psi_1:\right\} \tag{52}$$

$$i\partial^- \psi_2 = \frac{1}{2}\left\{2\pi\frac{1+\delta}{c_v + k}\frac{1}{2}\lambda^b :j_-^b\psi_2: + \frac{a + \bar{a}}{c_0}:j_-\psi_2:\right\} \tag{53}$$

Comparing these to the formal field equations, we write

$$g' = \frac{a - \bar{a}}{c_0} \tag{54}$$

$$g = 2\pi\frac{1-\delta}{c_v + k} \tag{55}$$

Notice that $\delta = 1$ corresponds to the abelian Thirring model ($g = 0$). The point $\delta = -1$, or

$$g = \frac{4\pi}{c_v + k} \tag{56}$$

corresponds to a nontrivial zero of the β-function, which has not been seen in other treatments (there have recently been some hints in this direction through the path integral procedure[12]).

Notice also that there is a doubling of the field equations. This is necessary because the formal expression (34) can no longer be used in view of (39), which replaced the formal equation (36). Thus (52) and (53) are interpreted as definitions of $j^a_{\mp}$ by Ref. 10. Moreover, if $j^a_{\pm}, j_{\pm}$ are free fields, as predicted by conformal invariance (see (35), (36)), the equations obeyed by ψ_1 and ψ_2 decouple and the system may be solved. Finally, at the nontrivial coupling ($\delta = -1$) we may adjust the constants a and $\bar{a}$ such that (52) and (53) will be holomorphic (antiholomorphic) conditions to be obeyed by ψ_1 and ψ_2 in the euclidianized version.

From this point we specialize to the case $G = \mathrm{SU}(n)$ and compute the two- and four-point functions. Conformal invariance is enough to compute the two-point functions

$$\langle 0|\psi_1(x)\psi_1^+(y)|0\rangle = [i(x_+ - y_+) + \varepsilon]^{-2s}[-(x-y)^2 + i\varepsilon(x_0 - y_0)]^{s-\gamma} \quad (57)$$

where γ is the dimension and s is the spin.

The spin and dimension may be computed in terms of the previously defined parameters a, $\bar{a}$, c_0, c_v, and k:

$$s = \frac{1}{2}\left[\frac{a\bar{a}}{\pi c_0} - \frac{n^2 - 1}{n(k+n)}\right] \quad (58)$$

$$\gamma = \frac{1}{4\pi}\left[\frac{a^2 + \bar{a}^2}{c_0} + 2\pi\frac{n^2 - 1}{n(n+k)}\right] \quad (59)$$

To compute functions of ψ_2, change x_+, y_+ into x_-, y_-.

To compute the four-point function, we need the field equations (50)–(53). The last piece of information comes from the normal product of the current and the elementary field ψ. We use the commutation relations

$$[j_+^{a(-)}(x_+), \psi_1(y)] = \frac{1+\delta}{2\pi}\frac{1}{2}\lambda^a\psi(y)\frac{1}{i(x_+ - y_+) - \varepsilon} \quad (60)$$

and others arising from (41), (43), and separation from the creation and annihilation operators

$$j^a_{\pm}(x_{\pm}) = \int_0^{\infty} dp\,(a^a_{\pm}(p)\,e^{-ipx_{\pm}} + a^{a+}_{\pm}(p)\,e^{ipx_{\pm}}) \quad (61)$$

Therefore, taking the derivative of the correlator

$$\mathcal{F}_{ii',jj'}(x, x', y, y') = \langle \psi_1^i(x)\psi_1^{i'}(x)\psi_1^{+j}(y)\psi_1^{+j'}(y')\rangle \quad (62)$$

with respect to x_+, we get correlators involving the product $\frac{1}{2}\lambda^a\psi_1^i(x)j_+^{a(-)}(x)$; at this point we use the commutator of $j^{a(-)}$ with the remaining fields to obtain a differential equation for the correlator $\mathscr{F}$. Since this is a technically straightforward but long computation, we refer to Refs. 13 and 14 for details, and simply write down the results. To compare to known results of conformal theory, we write the results for euclidian theory. In terms of the z, $\bar{z}$ variables, which correspond to $e^{\tau \pm i\pi}$, the field equations become

$$i\partial_z \psi_1(z,\bar{z}) = \frac{\pi}{2}\frac{1-\delta}{c_v+k}\lambda^b :J_-^b(z)\psi_j(z,\bar{z}): + \frac{1}{2}\frac{a-\bar{a}}{c_0}:J(z)\psi_1(z,\bar{z}): \tag{63}$$

$$i\partial_{\bar{z}} \psi_1(z,\bar{z}) = \frac{\pi}{2}\frac{1+\delta}{c_v+k}\lambda^b :\bar{J}^b(\bar{z})\psi_1(z,\bar{z}): + \frac{1}{2}\frac{a+\bar{a}}{c_0}:\bar{J}(\bar{z})\psi_1(z,\bar{z}): \tag{64}$$

Formulas for ψ_2 are analogous. The foregoing equations mean that ψ_1 (analogously ψ_2) is a representation of the simplest Verma module[15] of nonabelian theory[16] in terms of fermions, since the constraint is

$$\left(J_{-1}^a \tau^a + \frac{1}{2\pi}(n+k)L_{-1}\right)\psi = 0 \tag{65}$$

The ansatz for the four-point function is now[10,13]

$$\langle \psi_1^a(z_1)\psi_1^{b+}(z_2)\psi_1^{c+}(z_3)\psi_1^d(z_4)\rangle = [(z_1-z_4)(z_2-z_3)]^{-2\Delta}\{\delta^{ab}\delta^{cd}A_1(x) + \delta^{ac}\delta^{bd}A_2(x)\} \tag{66}$$

where

$$x = \frac{(z_1-z_2)(z_3-z_4)}{(z_1-z_4)(z_3-z_4)} \tag{67}$$

is invariant under modular transformations (generated by L_0, $L_{\pm 1}$). As a result of the fermionic field equations together with (65) [specialized to $G = SU(n)$], the functions $A_{1,2}(x)$ obey

$$x(x-1)\frac{dA_1}{dx} = \left[(x-1)\frac{1/n-n}{2(n+k)} + \frac{x}{2n(n+k)}\right]A_j - (x-1)\frac{1}{2(n+k)}A_2 \tag{68a}$$

$$x(x-1)\frac{dA_2}{dx} = \left[x\frac{1/n-n}{2(n+k)} + \frac{x-1}{2n(n+k)}\right]A_2 - \frac{x}{2(n+k)}A_1 \tag{68b}$$

The solution is given in terms of hypergeometric functions

$$A_1(x) = \mathscr{F}_1^{(0)}(x) + h\mathscr{F}_1^{(1)}(x) \tag{69a}$$

$$A_2(x) = \mathscr{F}_2^{(0)}(x) + h\mathscr{F}_2^{(1)}(x) \tag{69b}$$

where

$$\mathscr{F}_1^{(0)}(x) = x^{-2\Delta}(1-x)^{\Delta_1-2\Delta}F\left(-\frac{1}{\lambda}, \frac{1}{\lambda}, 1+\frac{n}{\lambda}; x\right) \tag{70a}$$

$$\mathscr{F}_2^{(0)}(x) = -\frac{x^{1-2\Delta}}{\lambda+n}(1-x)^{\Delta_1-2\Delta}F\left(1-\frac{1}{\lambda}, 1+\frac{1}{\lambda}, 2+\frac{n}{\lambda}; x\right) \tag{70b}$$

$$\mathscr{F}_1^{(0)}(x) = x^{\Delta_1-2\Delta}(1-x)^{\Delta_1-2\Delta}F\left(-\frac{n-1}{\lambda}, -\frac{n+1}{\lambda}, 1-\frac{n}{\lambda}; x\right) \tag{70c}$$

$$\mathscr{F}_2^{(1)}(x) = -nx^{\Delta_1-2\Delta}(1-x)^{\Delta_1-2\Delta}F\left(-\frac{n-1}{\lambda}, -\frac{n+1}{\lambda}, -\frac{n}{\lambda}; x\right) \tag{70d}$$

$$\lambda = c_v + k = n + k \tag{71a}$$

$$\Delta = \frac{n^2-1}{2n(n+k)} \tag{71b}$$

$$\Delta_1 = \frac{n}{n+k} \tag{71c}$$

and F is the hypergeometric function

$$F(a, b, c; x) = 1 + \frac{ab}{1!c}x + \frac{a(a+1)b(b+1)}{2!c(c+1)}x^2 + \cdots \tag{71d}$$

At last, crossing symmetry (replacing x by $1-x$) fixes h to be

$$h = \frac{1}{n^2}\frac{\Gamma\left(\frac{n-1}{n+k}\right)\Gamma\left(\frac{n+1}{n+k}\right)\left\{\Gamma\left(\frac{k}{n+k}\right)\right\}^2}{\Gamma\left(\frac{k+1}{n+k}\right)\Gamma\left(\frac{k-1}{n+k}\right)\left\{\Gamma\left(\frac{n}{n+k}\right)\right\}^2} \tag{71e}$$

These results are worth comparing to correlators of the WZW's g-field obtained by Knizhnik and Zamolodchikov,[16] using $g(x)$ as a primary field obeying

$$\left(J_{-1}^a\tau^a + \frac{1}{2\pi}(n+k)L_{-1}\right)g = 0 \tag{72}$$

with a current $J(z)$ (and also $\bar{J}(\bar{z})$). The results are as follows. For the two-point function

$$\langle g_{ij}(z,\bar{z})g_{kl}^{-1}(w,\bar{w})\rangle = \delta_{ik}\delta_{jl}(z-w)^{-2\Delta}(\bar{z}-\bar{w})^{-2\Delta} \tag{73}$$

a result following only from conformal invariance, which, comparing with (66)-(71), implies

$$\langle g_{ij}(z,\bar{z})g_{kl}^{-1}(w,\bar{w})\rangle = \langle\psi_{1i}^{+}(z)\psi_{1k}(w)\rangle\langle\psi_{2j}(\bar{z})\psi_{2l}^{+}(\bar{w})\rangle \tag{74}$$

A sufficient condition for (74) to be valid is

$$g_{ij}(z,\bar{z}) = \psi_{1i}^{+}(z)\psi_{2j}(\bar{z}) \tag{75}$$

We now check this relation for four-point functions. It is enough to borrow $k\,z$ formulas:

$$\begin{aligned}
&\langle g_{ii'}(z_1,\bar{z}_1)g_{jj'}^{-1}(z_2,\bar{z}_2)g_{kk'}^{-1}(z_3,\bar{z}_3)g_{ll'}(z_4,\bar{z}_4)\rangle\\
&\quad = \mu^{-8\Delta}(z_{14}z_{23}\bar{z}_{14}\bar{z}_{23})^{-2\Delta}\{[\mathscr{F}_1^{(0)}(x)\mathscr{F}_1^{(0)}(\bar{x}) + h\mathscr{F}_1^{(1)}(\bar{x})\mathscr{F}_1^{(1)}(\bar{x})]\delta_{ij}\delta_{kl}\delta_{i'j}\delta_{k'l'}\\
&\qquad + [\mathscr{F}_1^{(0)}(x)\mathscr{F}_2^{(0)}(\bar{x}) + h\mathscr{F}_1^{(1)}(x)\mathscr{F}_2^{(1)}(\bar{x})]\delta_{ij}\delta_{kl}\delta_{i'k'}\delta_{j'l'}\\
&\qquad + [\mathscr{F}_2^{(0)}(x)\mathscr{F}_1^{(0)}(\bar{x}) + h\mathscr{F}_2^{(1)}(x)\mathscr{F}_1^{(1)}(\bar{x})]\delta_{ik}\delta_{jl}\delta_{i'j'}\delta_{k'l'}\\
&\qquad + [\mathscr{F}_2^{(0)}(x)\mathscr{F}_2^{(0)}(\bar{x}) + h\mathscr{F}_2^{(1)}(x)\mathscr{F}_2^{(1)}(\bar{x})]\delta_{ik}\delta_{jl}\delta_{i'k'}\delta_{j'l'}\}
\end{aligned} \tag{76}$$

We compare (76) with previous results and verify that it corresponds to the replacement $g_{ij} \sim \mu^{-1}\psi_{1i}^{+}\psi_{2j}$, with crossing obeyed by the composite field, a requirement that fixes the value of h as given by (71e).

3. THE THIRRING MODEL AND STRINGS

3.1. Abelian Symmetry

As a preliminary we consider a one-dimensional compactification. Suppose that one coordinate field operator $X(z,\bar{z})$ is compactified on a circle of radius R. The mode expansion is

$$X(z,\bar{z}) = X(z) + \bar{X}(\bar{z}) \tag{77a}$$

$$X(z) = x^{(0)} + \frac{i}{2}p\ln z + \frac{i}{2}\sum_{n\neq 0}\frac{1}{n}\alpha_{-n}z^{-n} \tag{77b}$$

$$\bar{X}(\bar{z}) = \bar{x}^{(0)} - \frac{i}{2}p\ln\bar{z} + \frac{i}{2}\sum_{n\neq 0}\frac{1}{n}\tilde{\alpha}_{-n}\bar{z}^{-n} \tag{77c}$$

and we define, as before,

$$\tilde{X}(z,\bar{z}) = X(z) - \bar{X}(\bar{z}) \tag{77d}$$

with currents

$$J(z) = \frac{i}{\sqrt{\pi}}\frac{\partial X}{\partial z} \tag{77e}$$

$$\bar{J}(\bar{z}) = \frac{-i}{\sqrt{\pi}} \frac{\partial \bar{X}}{\partial \bar{z}} \tag{77f}$$

The energy momentum tensor is

$$T(z) = 2\pi :J(z)^2: \tag{78a}$$

$$\bar{T}(\bar{z}) = 2\pi :\bar{J}(\bar{z})^2: \tag{78b}$$

The associated fermionic field theory is defined by the field operator

$$\psi_{\alpha,\beta}(z, \bar{z}) = :e^{i(\alpha X_i(z)+\beta \bar{X}_i(\bar{z}))}: = :e^{i\alpha(X+\tilde{X})/2+i\beta(X-\tilde{X})/2}: \tag{79}$$

We have the following operator product expansions (OPEs):

$$T(z)\psi_{\alpha,\beta}(w, \bar{w}) = \frac{\pi\alpha}{z-w} :J(z)\psi_{\alpha,\beta}(w, \bar{w}): - \frac{i\alpha^2}{4(z-\omega)^2} \psi_{\alpha,\beta}(w, \bar{w}) \tag{80a}$$

$$\bar{T}(\bar{z})\psi_{\alpha,\beta}(w, \bar{w}) = \frac{\pi\beta}{\bar{z}-\bar{w}} :\bar{J}(\bar{z})\psi_{\alpha,\beta}(w, \bar{w}): - \frac{i\beta^2}{4(\bar{z}-\bar{w})^2} \psi_{\alpha,\beta}(w, \bar{w}) \tag{80b}$$

Since $T(z) - \bar{T}(\bar{z})$ generates Lorentz transformations, we readily compute the Lorentz spin of the field $\psi_{\alpha,\beta}$:

$$s = \frac{\lambda}{2} = \frac{\alpha^2 - \beta^2}{8} \tag{81}$$

The constant β corresponds to $g/\sqrt{\pi}$ in the Thirring model with coupling constant g.

We now use the identifications (23) and (28) in (79) to obtain the transformations of the field $\psi_{\alpha,\beta}$:

$$\psi_{\alpha,\beta} \to \psi_{\alpha,\beta}\, e^{i\pi R(\alpha+\beta)} \tag{82}$$

under (23), and

$$\psi_{\alpha,\beta} \to \psi_{\alpha,\beta}\, e^{i\pi(\alpha-\beta)/2R} \tag{83}$$

These transformations correspond in string language to modular transformations. Modular invariance of the theory requires that well-defined operators be invariant under the foregoing transformations. Therefore, in general, we are required to study products of those operators. If we have a bound state of F ψ's, we require, at the same time that the following equations be satisfied:

$$FR(\alpha + \beta) = 2n \tag{84a}$$

$$F\frac{\alpha - \beta}{2R} = 2m \tag{84b}$$

where n and m are integers and the spin

$$s = \frac{\alpha^2 - \beta^2}{8} = \frac{mn}{F^2} \tag{85}$$

is a rational number, unless we take an infinite number of ψ's to build bound states.

Two simple examples are (1) the free field case $\beta = 0$, $s = \alpha^2/8$, where $s = 1$ and the compactification radius $R = \sqrt{2}$ ensure invariance of ψ itself, and (2) $s = \frac{1}{2}$, $R = 1$ require bound states of two ψ's.

These results are readily generalized to a symmetry $[\mathrm{U}(1)]^d$.[7] In this case we come back to Minkowski space, recalling expressions (9) and (21)–(28), which imply the quantization of momentum

$$P_0^a = M_\mu \tilde{E}^{a\mu} \tag{86}$$

$$\tilde{P}_0^a = 2L^\mu E_\mu^a \tag{87}$$

where M_μ and L^μ are integers.

The corresponding fermionic model is described by the action

$$S = \frac{1}{\pi} \int d\sigma\, d\tau\, [i\psi_1^{i+} \partial^- \psi_1^i + i\psi_2^{i+} \partial^+ \psi_2^i + H_{ij} \psi_2^{i+} \psi_2^i \psi_1^{j+} \psi_1^j] \tag{88}$$

where

$$H_{ij} = F_{ai} K_{aj}$$

and the spinor is

$$\psi = \begin{pmatrix} \psi_1 \\ \psi_2 \end{pmatrix}$$

The formal field equations are

$$\partial^+ \psi_2 = -2\pi i F_{ai} J_+^a \psi_2^i \tag{89a}$$

$$\partial^- \psi_1 = -2\pi i K_{ai} J_-^a \psi_1^i \tag{89b}$$

and the commutation relations are given by

$$[J_+^a(x), \psi_2^i(y)] = -\tfrac{1}{2} A^{ai} \psi_2^i(y) \delta(x_+ - y_+) \tag{90a}$$

$$[J_-^a(x), \psi_2^i(y)] = -\tfrac{1}{2} B^{ai} \psi_2^i(y) \delta(x_- - y_-) \tag{90b}$$

$$[J_+^a(x), \psi_1^i(y)] = -\tfrac{1}{2} D^{ai} \psi_1^i(y) \delta(x_+ - y_+) \tag{90c}$$

$$[J_-^a(x), \psi_1^i(y)] = -\tfrac{1}{2} C^{ai} \psi_1^i(y) \delta(x_- - y_-) \tag{90d}$$

(for free fields, or critical Thirring coupling, $B = D = 0$).

We assume the energy momentum tensor has Sugawara form. Therefore, by the foregoing equations we may compute the Lorentz spin as in (80) and (81):

$$S = \frac{\lambda}{2} = \frac{1}{8}\left(\sum_a (A^{ai})^2 - \sum_a (B^{ai})^2\right)$$

$$= \frac{1}{8}\left(\sum_a (C^{ai})^2 - \sum_a (D^{ai})^2\right) \tag{91}$$

A possible solution may be expressed in terms of a matrix Z^{ai} together with its inverse $(Z^{-1})^{ai} = \tilde{Z}^{ai}$ and an antisymmetric matrix $Y^{ab} = -Y^{ba}$ (other solutions exist):[7]

$$A^{ai} = \sqrt{\lambda}(Z^{ai} + \tilde{Z}^{ai} - Y^{ab}Z^{bi}) \tag{92a}$$

$$B^{ai} = \sqrt{\lambda}(Z^{ai} - \tilde{Z}^{ai} + Y^{ab}Z^{bi}) \tag{92b}$$

$$C^{ai} = \sqrt{\lambda}(Z^{ai} + \tilde{Z}^{ai} + Y^{ab}Z^{bi}) \tag{92c}$$

$$D^{ai} = \sqrt{\lambda}(Z^{ai} - \tilde{Z}^{ai} - Y^{ab}Z^{bi}) \tag{92d}$$

The bosonized realization of the $[\mathrm{U}(1)]^d$ spinor fields is

$$\psi_1^i = :e^{iC^{ai}X_-^a(x_-) - iD^{ai}X_+^a(x_+)}: \tag{93a}$$

$$\psi_2^i = :e^{-iA^{ai}X_+^a(x_+) + iB^{ai}X_-^a(x_-)}: \tag{93b}$$

Thus, under shifts

$$\Delta X_0^a = 2\pi E_\mu^a \tag{94a}$$

$$\Delta \tilde{X}_0^a = \pi \tilde{E}_\mu^a \tag{94b}$$

we have the transformations

$$\psi_1^i \to \psi_1^i\, e^{2\pi i\sqrt{\lambda}(\tilde{Z}^{ai} + Y^{ab}Z^{bi} - B^{ab}Z^{bi})E_\mu^a} \tag{95a}$$

$$\psi_2^i \to \psi_2^i\, e^{-2\pi i\sqrt{\lambda}(\tilde{Z}^{ai} - Y^{ab}Z^{bi} + B^{ab}Z^{bi})E_\mu^a} \tag{95b}$$

$$\psi_{1,2}^i \to \psi_{1,2}^i\, e^{-i\pi Z^{ai}\tilde{E}_\mu^a\sqrt{\lambda}} \tag{95c}$$

As in the previous case, bound states must be considered.

3.2. Nonabelian Symmetry

In the nonabelian Thirring model, a complete operator solution is not known, but there are some helpful expressions that may be computed and used to bound state calculations and to fix the spin of the field.

For the product spinor-antispinor we have[10]

$$\psi_1^i(x)\psi_1^{j\dagger}(y) = \mathbf{C}(x_+ - y_+)^{-A}(x_- - y_-)^{-B} \times \exp\left\{\frac{-i}{2c_0}\left[(a+\bar{a})\int_{x_+}^{y_+} j_+(w_+) + (a-\bar{a}) \times \int_{x_-}^{y_-} j_-(w_-)\, dw_-\right]\right\}_{M^{ij}(x,y)} \tag{96}$$

where M satisfies

$$\partial^+ M = 0$$

$$\partial^- M = \frac{-1}{n+k}\left\{\frac{1}{2}\lambda^a{:}j^a M{:} + \frac{1}{4\pi}\frac{1}{i(x_- - y_-)+\varepsilon}\left[\lambda^b M \lambda^b - \frac{2(n^2-1)}{n} M\right]\right\} \tag{97}$$

with the condition

$$M(x,x) = \mathbf{1} \tag{98}$$

Thus

$$M(x,y) \sim \exp\left\{iC'\int_x^y \lambda^b y^b\, dw\right\} \sim \exp\{iC\lambda^b(X^b_{(x)} - X^b_{(y)})\} \tag{99}$$

Again, we have nonabelian fermionization and abelian bosonization formulas, which are the same as the usual (75).

Comparing abelian and nonabelian cases defined on the same compactification torus, we have the identifications

$$H_{ij} \leftrightarrow (\tau^a)_{ii}(\tau^a)_{jj} \tag{100a}$$

or

$$F_{ai} \leftrightarrow \tau^a_{ii}; \qquad K_{ul} \leftrightarrow \tau^a_{ll} \tag{100b}$$

Also

$$A^{ai} \leftrightarrow (\tau^a)^{ii} \tag{100c}$$

$$C^{ai} \leftrightarrow (\tau^a)^{ii} \tag{100d}$$

for an even self-dual lattice,[17] ψ is modular invariant, and there are no further constraints in the nonabelian piece. Only abelian pieces are arbitrary.

4. VERTICES AND STRING THEORY

Vertices in compactified string theory have been discussed in detail by Gepner and Witten.[18] As it turns out, a vertice is a product of the Minkowski space vertex and the compactified piece. We discuss the simplest case of the tachyon. The product

$$V(z, \bar{z}) = \phi^{R}_{\text{comp.}} : e^{ip^{\mu} X_{\mu}(z)} :$$

where $X_{\mu}(z)$, $\mu = 0, \ldots, D-1$ are the uncompactified variables, P_{μ} is the momentum, and $\phi^{R}_{\text{comp.}}$ is a representation of the group acting on the compactified manifold. The latter is also an element of the Verma module corresponding to the Kac-Moody algebra. Thus it may be represented by the WZW field g_{ij}, or, since it has the form $\exp\{iK^{a}X^{a}(z, \bar{z})\}$, it may be well described by (99), namely a bound state of the previously defined fermion or simply by an expression such as (93), which is the fermion. The only requirement is that of modular invariance, as we discussed previously. Correlators are a product of the Minkowski piece and the compactified part. Consider as an example a bound state

$$j^{ab}(\xi) = N[\psi^{a}(\xi)\psi^{b\dagger}(\xi)] \tag{101}$$

We have an explicit formula for

$$\langle \psi^{a}(\xi + \varepsilon)\psi^{b+}(\xi)\psi^{c+}(\xi')\psi^{d}(\xi' + \varepsilon')\rangle \tag{102}$$

given by (66). We compute (102) for ε, $\varepsilon' \to 0$, using

$$F\left(-\frac{1}{\chi}, \frac{1}{\chi}, 1 + \frac{n}{\chi}; x\right) \simeq 1 - \frac{x}{\chi(n+\chi)} \tag{103}$$

where $\chi = n + k$.

In the limit we have

$$\begin{aligned}
&\langle \psi^{a}(\xi + \varepsilon)\psi^{b+}(\xi)\psi^{c+}(\xi')\psi^{d}(\xi' + \varepsilon')\rangle \\
&\quad = \delta^{ab}\delta^{cd}(\varepsilon\varepsilon')^{-2\Delta} + h\left[\frac{\varepsilon\varepsilon'}{(\xi - \xi')^{2}}\right]^{\Delta_{1} - 2\Delta}(\delta^{ab}\delta^{cd} - n\delta^{ac}\delta^{bd}) \\
&\qquad - \left(\delta^{ab}\delta^{cd}\frac{k - n}{nk(n+k)} - k\delta^{ac}\delta^{bd}\right)\frac{(\varepsilon\varepsilon')^{1-2\Delta}}{(\xi - \xi')^{2}} + \cdots
\end{aligned} \tag{104}$$

The first contribution is trivial and should be subtracted. For $k \neq 1$, the second contribution is the only one remaining after renormalization is performed. We have

$$\begin{aligned}
&\langle N(\psi^{a}\psi^{b+})(\xi) N(\psi^{c+}\psi^{d})(\xi')\rangle \\
&\quad = \frac{h_{\mu}^{-8\Delta}}{(\xi - \xi')^{2/n(n+k)}}(\delta^{ab}\delta^{cd} - n\delta^{ac}\delta^{bd})
\end{aligned} \tag{105}$$

Therefore we have for $j^{ab} = N(\psi^a\psi^{b\dagger})$ an anomalous dimension

$$\gamma_j = \frac{1}{n(n+k)} \tag{106}$$

For $k = 1$, we have $h = 0$, and the dimension j of γ is canonical: $\gamma_j = 1$. In this case

$$\langle j^{ab}(\xi) j^{dc}(\xi')\rangle = \frac{\mu^{-8\Delta}}{(\xi - \xi')^2}\left(\delta^{ab}\frac{1-n}{n(n+1)} - \delta^{ac}\delta^{bd}\right) \tag{107}$$

Therefore the problem is nontrivial for $k \neq 1$. The case $k = 1$ has the values of free field theory for the dimensions.

5. CONCLUSIONS

We analyzed the issue of equivalence between bosons and fermions at the level of Green functions, concluding that the nonabelian Thirring model at critical coupling presents as a defining field a representation of the conformal algebra, whose bound state is the bosonic WZW field, the level of both representatives being the same (k).

Therefore, using this result, we may study vertex operators of compactified bosonic string theories, which turns out to be the elementary field operator in the fermionic language. Thus, a fermion operator $\psi \sim e^{i\alpha x}$ of spin $\alpha^2/4$ corresponds to a vertex operator of momentum α. Bound states of ψ obeying modular invariance can be computed and, for $k \neq 1$, anomalous dimensions arise naturally, as discussed in the last section.

At last, in nonabelian theory, the number of free parameters is very much reduced, contrary to the abelian case, where compactification radii are completely uncorrelated. The nonabelian symmetry group, being connected, correlates all radii, and the only freedom left is in the abelian piece. This property can have some nontrivial role in further developments.

REFERENCES

1. M. Green, J. H. Schwarz, and E. Witten, *Superstring Theory*, Cambridge University Press, 1987.
2. J. H. Schwarz, *Phys. Rep. 69*, 223, 1982, Trieste lectures, 1985.
3. A. M. Polyakov and P. Wiegman, *Phys. Lett. 131B*, 121, 1983.
4. E. Witten, *Commun. Math. Phys. 92*, 455, 1984.
5. E. Brézin, S. Hikami, and J. Zinn-Justin, *Nucl. Phys. B 165*, 528, 1980.
6. D. Friedman, E. Martinec, and S. Shenker, *Nucl. Phys. B 271*, 93, 1986.
7. J. Bagger, D. Nemeschansky, N. Seiberg, and S. Yankielowicz, *Nucl. Phys. B 289*, 53, 1987.
8. E. Abdalla and K. D. Rothe, *Phys. Rev. D 36*, 3190, 1987.

9. P. de Vecchia, B. Durhuus, and J. L. Petersen, *Phys. Lett. 194B*, 245, 1984; E. Abdalla and M. C. B. Abdalla, *Nucl. Phys. B 255*, 392, 1985; P. di Vecchia and P. Rossi, *Phys. Lett. 140B*, 344, 1984.
10. R. Dasken and Y. Frishman, *Phys. Rev. D 11*, 2781, 1975.
11. D. Goddard and D. Olive, *Int. J. Mod. Phys. A1*, 303, 1986.
12. F. Schaposnik, La Plata preprint; C. Destri and H. J. de Vega, CERN preprint.
13. M. Gomes, V. Kurak, and A. J. da Silva, São Carlos preprint.
14. E. Abdalla and M. C. B. Abdalla, *Rev. Bras. Fis.* (to be published).
15. A. A. Belavin, A. M. Polyakov, and A. B. Zamolodchikov, *Nucl. Phys. B 241*, 333, 1984.
16. V. G. Knizhnik and A. B. Zamolodchikov, *Nucl. Phys. B 247*, 83, 1984.
17. D. Gross, J. Harvey, E. Martinec, and R. Rohm, *Nucl. Phys. B 256*, 253, 1985.
18. D. Gepner and E. Witten, *Nucl. Phys. B 278*, 493, 1986.

Index